Intelligence in the Universe

Intelligence in the

Universe

ROGER A. MacGOWAN
Computation Center
Army Missile Command
Huntsville, Alabama, USA

FREDERICK I. ORDWAY, III
General Astronautics Research Corporation
London, England

illustrated by HARRY H-K LANGE

PRENTICE-HALL, INC. Englewood Cliffs, New Jersey

Intelligence in the Universe
Roger A. MacGowan
Frederick I. Ordway, III

© 1966 by Prentice-Hall, Inc.
Englewood Cliffs, N.J.

All rights reserved. No part of this
book may be reproduced in any form,
by mimeograph or any other means,
without permission in writing from
the publisher.

Library of Congress Catalog Number 66–18343

Printed in the United States of America
46906c

Current printing (last digit):
10 9 8 7 6 5 4 3

PRENTICE-HALL, INTERNATIONAL, INC., London
PRENTICE-HALL OF AUSTRALIA, PTY., LTD., Sydney
PRENTICE-HALL OF CANADA, LTD., Toronto
PRENTICE-HALL OF INDIA (PRIVATE) LTD., New Delhi
PRENTICE-HALL OF JAPAN, INC., Tokyo

Foreword

Three hundred years ago in the decades following the invention of the telescope, there was a great outburst of speculation about life on other worlds. For a while, it seemed to many philosophers and writers that this age-old problem might soon be solved, thanks to the wonderful new scientific developments which, for the first time, were giving Man a true picture of his place in the Universe.

Today, we realise that this confidence was a little premature: Space proved to be much larger, and more hostile than imagined by such enthusiasts as Cyrano de Bergerac and Bishop Wilkins. But now, at last, we have techniques for spanning it; our robot probes have already swept past Mars and Venus, and soon the first men will be walking on the face of the Moon. As far as *this* Solar System is concerned, all the questions of extra terrestial life will soon be answered.

Intelligence in the Universe is a very thorough analysis of what may well be the central problem, and the most exciting intellectual adventure of our age. The authors have also had the courage of their convictions, and few readers will fail to be impressed by the chapters on robot evolution. Whatever one may feel about the matter emotionally, logic forces us to face the possibility that though the planets may belong to organic life, the real masters of the Universe must be machines. We creatures of flesh and blood are merely transitional forms. These chapters incidentally, serve as a valuable corrective to the still common belief that no machine can ever "think." Those who continue to assert this pathetic, and now dangerous fallacy, merely prove that human beings don't always do so.

For many months this book's co-author Fred Ordway has been working with Stanley Kubrick and myself on the production of 2001: A SPACE ODYSSEY , which we believe will be the first serious film treatment of man's place in the hierarchy of the universe. After decades of scorn and neglect, this great theme is once again a fitting subject for science and for art. Standing as we do in the shadow of our own giant rockets, we look up at the skies today with a new awareness—knowing that ours may well be the last generation that thought itself alone.

Arthur C. Clarke

Preface

The purpose of this book is to provide a comprehensive, well-balanced analysis of the extraordinary subject of extrasolar intelligence—that is, intelligence existing beyond the confines of the Solar System. Our aim is accomplished primarily by (1) thoroughly surveying the existing literature, (2) assessing man's development of the scientific and technological as well as social fields; (3) developing original interpretations of established facts, and (4) making reasonable conjectures based on theory and observation. We have incorporated knowledge from all pertinent branches of science and technology, thereby avoiding the danger of stressing the viewpoint of one specialty as opposed to, or without regard to, another. The treatment and the terminology have been kept at a level that assures the book will be readily comprehensible to any educated person.

Scientific facts, hypotheses and speculation have been blended together to form a coherent treatment, yet each has been cautiously identified throughout the work. We believe that logical speculation, when clearly based on scientific facts and hypotheses, is essential to any discussion dealing with the probable occurrence and the possible characteristics and capabilities of extrasolar civilizations. Particular attention is drawn to the chapters on biological and mechanical thinking processes and the social implications of extrasolar intelligence. At present, these fields represent obvious gaps in the published literature and the state of human understanding.

The majority of the material appearing in Chapters 1, 2, 3, 4, 6, 14, 15, and 16 and the bibliography was written by Frederick I. Ordway, III, while Chapters 5, 7, 8, 9, 10, 11, 12, 13, and 17 were prepared primarily by Roger A. MacGowan. Despite this division of responsibility, both authors contributed to all chapters, hopefully resulting in a well-coordinated, fully integrated book.

The authors are deeply indebted to Mitchell R. Sharpe, Jr. for his assistance in the literature research phase of the preparation of the manuscript and for his critical review of the completed text. We express our sincere appreciation to Elizabeth MacGowan, Maria Victoria Ordway, and Toby Pizitz for their assistance during the editing and typing phases, and to Maria V. Cooper for selecting the appropriate literary quotations.

Roger A. MacGowan
Frederick I. Ordway, III

Huntsville, Alabama, USA
London, England
March 1966

Prologue

Astronomers tell us that our Milky Way galaxy may range from 15 to 25 billion years in age. Philosophic speculations, pure reason, and observational science persuade us to believe that mankind is not the only manifestation of intelligent life amid the billions of stars that spread across the immensity of intragalactic space. If man is not alone, he most assuredly is not supreme. Earth-like planets millions or billions of years older than our terrestrial orb must have nurtured civilizations faced with all the unknowns to which we are heir. And most, conceivably all, these unknowns have answers — answers to questions that innumerable times during the history of our galaxy must have been pondered and solved by thinking beings. If this has happened, and if stars, like men, are born, evolve, and die, we seem compelled to conclude that knowledge is not allowed to perish but is passed on from civilization to civilization from one region of the galaxy to another, eon after eon, from creation to whatever end an individual world may suffer.

Is this knowledge beyond the reach of our civilization? Must man scale alone the ladder to wisdom, repeating each step that has been taken before? Or will he seek to discover with all the forces at his command, and with the help that other worlds may be able to offer, that which for centuries has inspired him most: truth, understanding, and a sense of well-being in the universe? Only time will tell.

Contents

1. **Introduction** *1*

2. **The Universe** *6*

    The Nature of Galaxies *8*
    The Nature of Stars *22*
    The Solar System *34*

3. **The Origin of Planetary Systems** *48*

    Accidental Origin Theories *50*
    Natural Origin Theories *55*
    Existence of Extrasolar Planetary Systems *66*

4. **Planetary Environments and the Development of Life** *74*

    Importance of Planetary Environments to the Development of Life *76*
    Biologically Acceptable Conditions in the Solar System *77*
    Planetary Environments in Extrasolar Systems *81*

5. **The Origin and Development of Life** *88*

    Definition of Life *91*
    Primordial Planetary Chemical Environments *93*
    Formation of the Building Blocks of Life *93*
    Linkage of Life's Building Blocks into Long Chains *98*
    Catalysis and Autocatalysis *103*
    Cellular and Multicellular Life *105*

6. Biological Evolution *110*

Absolute and Relative Dating *114*
Environmental Conditions and the Appearance of Life *115*
Primitive Forms of Life—Pre-Cambrian Biology *119*
Life During the Paleozoic Era *123*
Life During the Mesozoic Era *130*
Life During the Cenozoic Era *133*
The Rise of Intelligent Life *136*
Trends in Evolution *148*
Biological Evolution on Extrasolar Worlds *152*

7. Intelligence *160*

Attitudes Toward Intelligence *162*
Definition of Thinking *165*

8. Biological Thinking *178*

The Development of Intelligent Life *180*
Human Thinking *183*
Intelligence in Animals *185*

9. Thinking In Computers *190*

Digital Computers *192*
Future Digital Computers *205*
Neural Networks *211*

10. The Development of Intelligent Artificial Automata *220*

Artificial Thinking Automata *224*
Social Implications of Intelligent Artificial Automata *233*

11. Characteristics of Extrasolar Intelligence *236*

Biological Intelligence *239*
Intelligent Automata *243*

12. Capabilities of Extrasolar Intelligence *246*

Communication *249*
Transportation *250*
Environmental Control *254*

13. Social Effects of Extrasolar Intelligence *260*

Social Evolution on Earth *262*
Effects of Development of Artificial Intelligence on Biological Societies *264*
Effects of Extrasolar Intelligence on Human Society *269*
Interactions of Extrasolar Intelligence *270*

14. The Search for Extraterrestrial Life in the Solar System *274*

Search for Extraterrestrial Life on Earth *275*
Search for Life on Other Worlds of the Solar System *280*

15. Empirical Evidence of Extrasolar Intelligence *288*

Evidence Based on Artifacts *290*
Evidence Based on Catastrophic Results of a Landing Attempt *291*
Evidence Based on Intangibles *300*
Evidence Based on Sightings of Alleged Extraterrestrial Objects *303*
Evidence Based on the Detection of Signals *310*
Evidence Based on Discoveries that May be Made on Nonterrestrial Worlds in the Solar System *312*
Evidence Based on Inference *319*

16. Communications with Extrasolar Intelligence *324*

Early Proposals for Interplanetary Communications *325*
To Signal or to Listen? *329*
Electromagnetic Communication *330*
Laser Communication *336*
Searching for Artificial Extrasolar Signals *337*
Galactic Communication Networks *345*
Interstellar Flight *346*
The Language of Communications *351*
The Number of Communicating Civilizations in the Milky Way Galaxy *354*
Intergalactic Communications *356*

17. Summary and Conclusions *358*

The Probable Prevalence of Extrasolar Intelligence *361*
Communication with Extrasolar Intelligence *370*
Future Research Concerned with Extrasolar Intelligence *372*

Bibliography *378*

Index *400*

1
Introduction

Man's musings over the possibility of life on other worlds are ancient and profound. Many early religions taught that the gods lived in the planets and stars, or even that they themselves *were* the planets and stars. Formally-designated gods and an endless variety of unnamed spirits and souls were thought to travel to and from the heavens and, in this vague sense, participated in what today we call space flight.

As the primitive gods declined in importance and man's belief in and reliance on them slowly withered away, it fell to human beings the responsibility of conjuring up ways and means of crossing the mysterious chasm separating the Earth from the Moon and the planets. Bottles containing dew attracted skyward by sunlight, geese, vultures' and eagles' wings, demons, flying horses, sailing ships, cannons, antigravity materials, and countless other schemes were hatched by fertile imaginations—from Lukian of ancient Samos to 19th Century science fiction writers—to carry man, in body or in spirit, to worlds beyond the Earth. Only during the first half of the 20th Century was it finally realized that in the rocket lay the elusive key that would open the doors of the Solar System to exploration by *Homo sapiens*.

"For my part, I... believe the Planets are Worlds about the Sun, and that the Fixed Stars are also Suns which have Planets about them, that's to say, Worlds, which because of their smallness, and that their borrowed light can-not reach us, are not discernable by Men in this World...."

> Cyrano de Bergerac, *The Comical History of the States and Empires of the World of the Sun*. London, 1687

(There are) "very many earths, inhabited by man... thousands, yea, ten thousands of earths, all full of inhabitants... not only in this solar system, but also beyond it, in the starry heaven."

> Emanuel Swedenborg, *Earths in our Solar System, which are Called Planets, and Earths in the Starry Heavens*. London, 1758

2   Introduction

The rocket, crucial as it is for planetary exploration, provides scant hope in the foreseeable future of permitting man to leave the Solar System—that infinitesimal volume of space occupied by the star we call the Sun and its family of nine planets. Astronomers and astrophysicists possess powerful optical and radio tools for determining the nature of unbelievably distant stars, but as yet they can shed only cautious and indirect light on the possible existence of extrasolar planets and their suitability or unsuitability for the emergence of life—be it primitive or civilized.

If man cannot yet know with assurance if there is intelligent life beyond the Solar System, at least he can ponder, drawing reasoned conclusions from a mixture of hard facts, extrapolations of facts, and judgements and interpretations based on them—along with an element of thoughtful speculation. In this book we offer such a mixture, with just enough of the speculative quality to yield a comprehensive and penetrating study of the possible existence of intelligent communities on planets that may circle several billions of stars in our own Milky Way galaxy and, literally, stars without end to the very edge of creation.

Our first task is to outline what can loosely be called the geography of the universe, setting into perspective the position of the Solar System within the cosmic whole. We then examine the crucial question of the origin, development, and possible prevalence of planetary systems, following which we inquire into the environmental conditions apparently conducive to the appearance and persistence of life. All planets cannot be expected to harbor a biology, even of the most primitive type, but many, we believe, almost inevitably do. And, of these, a respectable percentage are expected to support intelligent communities. It is thrilling to imagine that at this moment untold billions of civilized societies may be gazing forth into the same universe of stars and galaxies that we perceive with our eyes and telescopes.

To help establish the credibility that such societies do, in fact, occur, we study in considerable detail the origin and development of life insofar as it is known to exist here on Earth. Since life seems to be a highly predictable phenomenon within a suitable planetary environment, we trace its evolution on our planet in order to determine the factors that seem to lead to the appearance of ever more advanced living forms—right up to the advent of intelligence of the human level.

Biological life that has evolved at least to a human level of thinking is of primary interest to us; therefore, we study with great care the fundamentals of intelligence and of the thinking process. A knowledge of the basics of intelligence implies that the information processing functions of which it is comprised could be mechanized on certain types of computers. For this reason, one entire chapter is devoted solely to current research in the computerized simulation of thinking processes. Once these processes are thoroughly understood by any advanced technological society, there appears to be no barrier to the development of superintelligent synthetic automata. This concept, the subject of Chapter 10, is found to have profound implications for our entire study of extrasolar intelligence.

Plausible characteristics and capabilities of both biological societies and intelligent automata are the next topics to receive our attention, following which we consider their possible effects on social phenomena. All these subjects have heretofore been virtually ignored by serious students of life beyond the Solar System. Admittedly, it is extremely difficult (and perhaps even impossible) to take the measure of advanced extrasolar civilizations and comprehend them in human terms, but at least an attempt can be, and is, made.

If civilized societies are prevalent in the Milky Way galaxy, we cannot help but wonder if they have knowledge of our terrestrial way of life, or at least of the manifestations of intelligence produced by our blossoming civilization. We focus our thoughts, in Chapter 15, on empirical evidence of extrasolar intelligence, coming to the conclusion that there is no direct or indirect reason to believe that extrasolar beings ever *visited* the Solar System—although we have no reason to conclude they are ignorant of it.

If they have not visited us, and if they have no plans to do so, it is still entirely possible that they may try (and, indeed, may now be trying) to communicate with us. This represents the subject of Chapter 16. We review the diverse problems involved in attempts to detect extrasolar signals and then examine the feasibility of initiating transmissions to announce to the universe that in the Solar System a civilization has evolved ready and eager to join the galactic community if one exists.

In the final chapter our conclusions are presented. Briefly, we believe that a large number of habitable planets exists in the Milky Way galaxy, a high percentage of which may contain intelligent societies aware of, and in communication with, neighboring civilizations. Some planets may be inhabited exclusively by biological beings, others by coexisting biological societies and artificial automata, and still others by automata alone. We postulate that these automata may gradually migrate towards the center of the galaxy or to other regions of space that, for any of many reasons, may be favored.

Contrary to the beliefs of some students of the subject, we feel that civilizations, whatever their nature, are likely to persist over astronomically long periods of time, thus enhancing our estimates of the prevalence of intelligent communities in the galaxy and increasing the probability of the existence of an extensive interstellar communications network.

If our conclusions are correct, humanity is progressing ever forward and upward in a universe liberally sprinkled with civilizations at all conceivable stages of intellectual, social, scientific, and technological development. Unbelievable quantities of knowledge and wisdom of an incalculable degree most far beyond our present powers of comprehension, must be harbored in the central repositories of many such civilizations. Perhaps some day our emergent society will gain access to portions of the information in these repositories, permitting a prodigious growth in our intellectual horizons and opening our minds to entirely new dimensions of thought and understanding.

Until we enter into communication with a given extrasolar society, we can do little more than speculate as to its characteristics and nature. How did its biological life first take hold? When and how was the threshold to intelligence gained? What course did learning follow? And what was the relative rate of progress in scientific, philosophical, and social insight? How does the civilization deal with the universe both in terms of its scientifically observable nature and its many cosmological imponderables? What manner of religions evolved with the culture? What philosophies matured? Were they similar to ours? Or totally alien? How do they explain the relationship of sentient beings to the cosmos? And their relationships to totally unlike communities with which they may be in contact? What hopes, fears, loves, hates, ambitions, and other drives motivate the extrasolar civilization? Has it attained a sense of compatibility in the universal scheme of things? And to what extent is it bolstered by, or even directed by, mechanical automata and other adjuncts of advanced science and technology?

These are typical of the questions that excite our intellectual curiosity and give us the incentive to reach out, search for, and discover intelligence whatever may be its nature, where ever it may be found. That eventually *Homo sapiens* will attain this lofty, almost incredible goal seems beyond question. What is highly uncertain, however, is when the contact will be made and how our civilization will react socially, philosophically, and scientifically. We can reduce the uncertainty of the "how" only by beginning to prepare ourselves emotionally and intellectually for the sudden impression of an extrasolar intelligence on the entire fabric of our society. The "when" may depend largely on the energy we devote to the search for signals that even now may be crisscrossing the immensities of interstellar space—signals laden with information of vital concern to any newly born intelligence anywhere. It could come sooner, and in a more startling manner, than most of us dare to imagine.

2
The Universe

The universe contains everything we can see and everything we can conceive of but do not necessarily observe or understand. As far as we know, every event occurs within the universe, which may or may not be infinite in size and age. How large and how old the universe is, if it is not infinite, are not only questions of profound interest to cosmologists but questions that may never be answered completely. However, as astronomical and other scientific tools and methods are sharpened, our estimates of age and size should become increasingly accurate.

On the largest scale, the universe contains innumerable galaxies,[1] or groupings of stars, all moving relative to one another at high velocities. The average separation of these galaxies is a million light years, which is an unbelievable distance when translated to miles. Assuming that the limit of the observable universe is 10 billion light years in all directions from the Sun, then the known diameter of the universe is 20 billion light years. Such a distance is incomprehensible to the human mind, yet it may only be a fraction of the real diameter —if indeed the universe is limited in size.

"Before (his) eyes in sudden view appear the secrets of the hoary Deep—dark illimitable ocean, without bound without dimension..."
John Milton

"Man hath weaved out a net, and this net throwne upon the Heavens, and now they are his owne..."
John Donne

---

[1] About 10 billion are observable with present-day telescopes.

Our Sun and its family of planets are members of one of the billions of galaxies known as the Milky Way galaxy. At least 80,000 light years in diameter, its 150 to 200 billion stars are packed rather densely at the center and more diffusely toward the edges. The Sun revolves around the galactic center at a distance of from 25,000 to 30,000 light years.

For man, the next category of things is the Solar System, with the Sun at the center, nine plants revolving around it, and an assortment of moons, asteroids, comets, meteors, micrometeorites, particles, dust, and gas. The conventional limits of the Solar System are defined by the orbit of Pluto, at a distance of roughly 4 billion miles from the Sun. If we accept these limits, the Solar System is some 8 billion miles in diameter. For the sake of comparison, the distance from the Sun to Pluto is 40 times the distance of the Sun to Earth, yet only 1/7,000 the distance that separates the Sun from Proxima Centauri, the nearest star.

The Sun controls the trajectories of millions of comets, many of which orbit far beyond the domain of Pluto for distances of one to two light years. Despite the huge volume of space falling within the influence of the Sun, practically all the Solar System's mass is concentrated in this central star.

The Nature of Galaxies

Galaxies, as we have seen, are aggregations of stars. Some galaxies occur singly or in small groups while others form associations that may contain tens or hundreds of members. We shall first be concerned with the composition of galaxies in general, then with the various categories of galaxies, and finally with systems or associations of galaxies.

COMPOSITION OF GALAXIES

It is not enough to say that galaxies contain millions or billions of individual stars, for a high fraction of their stars do not occur as individual bodies. While many stars occur singly, many others are found in association with other stars, forming binary or multiple systems. Double (or binary) star systems occur frequently; they contain two stars revolving around a common center of mass. Multiple systems contain three or more stars rotating around each other in very complex orbits.

A galaxy may also contain clusters of stars, of which two principal types are recognized: open, or galactic, clusters; and globular clusters. Whereas the former occur only near the plane of the galaxy and contain no more than a few hundred stars, the latter are found over a wide range of galactic latitudes and may consist of from approximately one thousand stars to perhaps many millions. Among the well-known galactic clusters are the Hyades and Pleiades.

The largest groupings of stars within a galaxy are the globular clusters, of which over a hundred have been identified in the Milky Way alone. They are located spherically around the center of the galaxy and do not share in galactic rotation. Most globular clusters are spherical, although a few are somewhat flattened. M 13 in Hercules is the most famous globular cluster in our galaxy.

Stars, singly or in groupings, are not the only ingredient of galaxies. Filling the vast distances of what was once believed to be empty space are tiny particles of matter (dust) and individual atoms and molecules (gases). It is believed that the quantity of matter between the stars is approximately the same as the matter found in the stars themselves.

Clouds of interstellar material (Fig. 2.1) tend to obscure our view of many parts of the Milky Way and other galaxies. Moreover, the particles in the clouds polarize the light from stars that shines through them. Interstellar matter is not uniformly distributed throughout space; rather, it thickens in some places and thins out in others. In regions of high particulate density, conditions may be leading to its condensation into stars. Not all clouds of interstellar matter are dark (absorbing);[2] those occurring as bright (galactic) nebulae reflect the light of stars associated with them.

---

[2] Dark nebulae, detected by their obscuration of stars behind, are of two types—diffuse and planetary. Dark diffuse nebulae are irregular clouds of gas and dust. Planetary nebulae are associated with a central star.

Fig. 2.1. Nebulosity in Monoceres (NGC 2264). Courtesy Mount Wilson and Palomar Observatories.

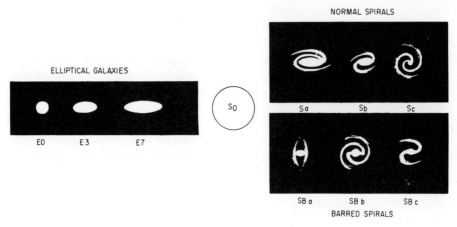

Fig. 2.2. Hubble's classification of galaxies (see text).

TYPES OF GALAXIES

As there are numerous types of planets and stars so there are various classifications of galaxies. Some galaxies show marked symmetry, while others exhibit no particular form or shape. There are many gradations in shape; irregularly shaped galaxies are thought to be the least mature in terms of age, composition, mass, or over-all dimensions, or combinations thereof. Figure 2.2 illustrates Edwin P. Hubble's classification of galaxies, the major characteristics of which are described in the following paragraphs.

Coincidentally, the galaxies nearest to our Solar System, the Magellanic Clouds (Fig. 2.3), are examples of "primitive-type," irregular galaxies. The Large Magellanic Cloud is 145,000 light years[3] away, compared to 160,000 light years for the Small Magellanic Cloud. Their absolute magnitudes are −17.4 and −16.0,[4] respectively, and their diameters are 30,000 and 23,000 light years. The clouds are apparently revolving slowly around our galaxy.

---

[3] One light year = 5,880,000,000,000 miles. Another astronomical measurement is the parsec, which equals 3.258 light years or 19,150,000,000,000 miles; it is the distance that corresponds to a heliocentric parallax of one second of arc. The heliocentric parallax of a star is the angle subtended at the star by the mean radius of Earth's orbit. One light year = 0.3069 parsecs. A kiloparsec = 1,000 parsecs and is used primarily for intergalactic measurements. Even larger is the megaparsec, or 1,000,000 parsecs, a term employed to measure distances to extremely remote galaxies.

[4] The magnitude of a celestial object is a measure of its relative brightness. The ratio between magnitudes is 2.512 to 1; thus, a body of magnitude 1 is 2.512 times as bright as a body of magnitude 2. The lower the numerical value of magnitude, the brighter the object is. The absolute magnitude is the brightness an object exhibits when viewed from a standard distance of 10 parsecs or 32.6 light years.

Fig. 2.3. Left: NGC 2080 in the Large Magellanic Cloud. Courtesy Harvard College Observatory. Right: Large Magellanic Cloud. Courtesy Lick Observatory.

Astronomers have studied literally hundreds of thousands of stars in these clouds and have found that the distribution of stellar population differs in some ways from that of the Milky Way. One striking difference is the high proportion of Cepheid variable stars[5] when compared with our galaxy. Clusters of stars have been discovered in both clouds, probably of the globular type. The clouds contain interstellar matter, considerably more being detected in the Large Cloud than in the Small Cloud. The Population I type star predominates in the former, whereas the latter has more or less an even mixture of Population I and II (see page 18).

Elliptical galaxies are structureless but, unlike irregulars, have definite symmetry. They range in shape from nearly spherical to highly flattened disks, and exhibit many characteristics of globular clusters—although they are much larger and more populated. Population II stars predominate.

Typical of the largest elliptical galaxies is Messier 87 of −15.5 absolute magnitude. It has a high proportion of Population II stars, is quite free of interstellar matter, and has numerous globular clusters surrounding it. The galaxy is viewed in Fig. 2.4. Among the smaller elliptical galaxies are the Fornex and Sculptur systems, 470,000 and 210,000 light years away, respectively.

The spiral galaxy is recognized in any of the many forms described below. Some spirals are close associations of stars with only a slight spiral

---

[5] A type of variable star that increases in brightness as its surface area becomes larger and decreases in brightness as it becomes smaller again. The brighter a Cepheid variable, the longer its period.

Fig. 2.4. The elliptical galaxy Messier 87. Courtesy Lick Observatory.

structure discernible, others are intermediate in form, and still others are almost all coiled arms with little center structure. Regular or normal spirals possess arms emanating from a spherical nucleus. The arms of barred spirals emanate not directly from the nucleus but from the ends of a bar-like structure.

The best known spiral, the giant Messier 31 in Andromeda, is illustrated by Fig. 2.5. About 1,500,000 light years from the Solar System, it has an absolute magnitude of -19.5 and a diameter of at least 100,000 light years and possibly up to 130,000 light years. Larger than the Milky Way, the galaxy rotates at a maximum speed of over 300 mi/sec around its center.

Messier 31 consists essentially of a bright nucleus and a number of rather closely wound spiral arms. The central region contains a multitude of faint

Fig. 2.5. The great spiral galaxy in Andromeda (Messier 31, NGC 224). Courtesy Lick Observatory.

The Universe 13

Fig. 2.6. The spiral galaxy in Ursa Major (NGC 5457). Courtesy Moun Wilson and Palomar Observatories.

Fig. 2.7. Barred spiral galaxy in Pegasus (NGC 7741). Courtesy Mount Wilson and Palomar Observatories.

stars whose combined light is largely responsible for the brightness of the galaxy. Little dust or gas is evident there, in contrast to the obscurity in the spiral arms where many Cepheid variable stars, blue stars, and galactic clusters are found. Some hundred novae have been observed as well as a few supernovae. Around the galaxy are several hundred globular clusters. Compare Andromeda with another regular spiral illustrated in Fig 2.6 and the barred spiral in Fig. 2.7.

UNUSUAL GALAXIES

Some galaxies are strong radio sources with outputs of from 10^{40} to 10^{45} ergs/sec at radio frequencies, compared to those from normal galaxies—which produce about 10^{44} ergs/sec in visible light and 10^{38} in the radio spectrum. Other galaxies, known as quasi-stellar sources or quasars, emit in the radio spectrum much like radio galaxies—but optically are some 100 times brighter than the larger normal galaxies. Their light output is about 10^{46} ergs/sec and their radio output is of the order of 10^{44} ergs/sec. Quasars vary, both regularly and irregularly, in energy output. They are very distant, the closest being about 1.8 billion light years away and the furthest up to 8 billion light years away. All quasars are receding from us at rapid velocities. For example, source 3C295 appears to be receding at 0.46 times the speed of light, compared to 0.545 for source 3C147 and 0.85 for source 3C286.

Quasars (Fig. 2.8) are smaller than radio galaxies, much brighter optically, and emit strongly towards the ultraviolet end of the spectrum. Their nature is not yet understood, but they may represent a stage of gravitational collapse, probably involving explosions on a scale unimaginable to man. Explosions in the nuclei of galaxies may be relatively common; indeed, they appear to be taking place within Messier 82, or at least to have taken place within the past million years. Great jets of galactic material, travelling at speeds up to 600 mi/sec, are observed radiating from the nucleus away from the principal plane. Theoretical astrophysicists believe that galactic explosions represent an evolutionary sequence in the development of any galaxy and that our own Milky Way passed through a similar stage in the remote past. Messier 82 is a rather young, irregular galaxy containing vast amounts of condensed material.

Aside from conventional radio galaxies (that is, those whose radio distribution indicates the source of emissions to be in the galactic center) and quasi-stellar radio sources, there is a third class of somewhat related galaxies. Known as Seyfert galaxies, they are small; they possess extraordinarily bright nuclei (typically less than a hundred light years in diameter); they have strong emission line and often highly excited spectra; and they exhibit in their lines the Doppler broadening phenomenon, indicating velocities of from 600 to nearly 2,000 mi/sec.

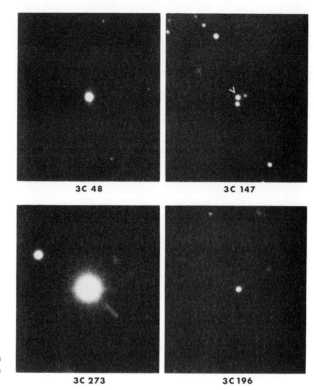

Fig. 2.8. Quasi-stellar radio sources (quasars). Courtesy Mount Wilson and Palomar Observatories.

Numerous are the tentative explanations of the sources of the energy that give rise to radio and related galaxies. One theory holds that the energy observed is released by interactions between a given galaxy and materials heretofore not associated with it. A second suggests that the internal energy is produced by catastrophic processes, e.g., rotational energy, turbulent energy or magnetic field energy released uncontrollably. And the third supposes that we are viewing the effects of the energy released during actual stellar formation, including gravitational energy, nuclear energy released by thermonuclear explosions, and rest-mass energy resulting when stars enter into what is known as the highly-collapsed phase.

THE MILKY WAY

Of all the galaxies, be they spiral, elliptical or irregular, the one most important to us is the Milky Way. Since we form part of this galaxy we see it from the inside, which is both an advantage and a disadvantage. Being inside we can observe many galactic phenomena at relatively close range, but it is difficult to determine over-all structure and form. Dust and gases in the central plane obscure many features, particularly those on the opposite side of the galaxy.

Fig. 2.9. Artist's conception of the Milky Way galaxy seen from outside. Arrow indicates the approximate location of the Sun. Below: a portion of the Milky Way in Sagittarius. Courtesy Mount Wilson and Palomar Observatories.

Despite our position within the "forest" of stars that make up the Milky Way, we are far from ignorant about conditions in our galaxy. We know that it is a gigantic assemblage of from 150 to 200 billion stars, that it is at least 80,000 light years in maximum diameter and some 15,000 light years thick at the center, and that it is approximately 150,000 light years from its nearest galactic neighbor. The Sun, marked by the arrow in Fig. 2.9, is located between 25,000 and 30,000 light years from the center of the galaxy, in one of the spiral arms, and is not far off the plane of the galactic equator. It makes a complete rotation every 200 million years at an average velocity of approximately 150 mi/sec.

The absolute magnitude of the Milky Way is −18.6, as seen from the direction of the galactic pole. The velocity of escape from the galaxy varies from 300 mi/sec at the center, to 190 mi/sec near the Sun, to 120 mi/sec at the edge. It would be interesting to know if anything has ever escaped from the galaxy into intergalactic space.[6]

On a clear, dark night we can distinguish with the naked eye between 2,000 and 3,000 stars in the Milky Way, a figure that rises to perhaps 10,000 with a pair of good binoculars, and to several billions with large optical telescopes. Figure 2.9 (below) is a portion of this vast assemblage of stars. Because of the obscuration of dust and gases in the galactic plane we can never observe all the bright stars in the galaxy, and of course billions of faint stars go undetected. Fortunately, radio telescopes can often "see" where optical telescopes cannot, revealing much knowledge of the central regions of the Milky Way and of spiral arms on the opposite side of the galaxy from us.

Observations of the galactic core indicate that stars are packed up to three times closer together than in the neighborhood of the Sun. This circumstance could have an important bearing on the development of interstellar flight if intelligent communities inhabit planets revolving around stars in the central region. Observations also show that the central core is flattened in the direction of the poles and, moreover, is quite extended in the equatorial regions. Our Sun is some three quarters of the distance away from the center of the core. At least two spiral arms uncoil from opposite sides of the center of the Milky Way; they contain large quantities of stars and interstellar matter.

Looking at the galaxy as a whole we find that individual stars are separated by several light years in distance. As Struve has pointed out "for

---

[6] Some stars have large proper motions indicating they may escape from the galaxy. The pair of stars Washington 5583, 5584 are travelling at a speed sufficient to carry them out of the Milky Way. Their annual proper motion is 3.68 sec, their parallax 0.034 sec, and their radial velocity + 307 and + 295 sec, respectively. As another example, gases or particles moving away from the Crab Nebula are travelling at velocities in excess of that necessary to escape from the Milky Way.

Fig. 2.11. Cluster of galaxies in Hercules. Courtesy Mount Wilson and Palomar Observatories.

terms of increasing distance. The average diameter of clusters of galaxies is 1,600 kiloparsecs and the average number of galaxies per cluster is 200. The Hercules cluster is seen in Fig. 2.11, and the Coma cluster appears in Fig. 2.12.

A fascinating characteristic of these clusters is that the further away a

Table 2.2
Clusters of Galaxies

Galactic System	Distance kiloparsecs	No. of galaxies in cluster
Virgo	>2,500	500
Pisces	14,000	30
Pegasus	15,000	100
Perseus	20,000	500
Coma	25,000	1,000
Centaurus[a]	50,000	300
Ursa Major I	55,000	300
Leo	60,000	300
Gemini	80,000	200
Boötes	130,000	150
Ursa Major II	145,000	175

[a] May contain up to 1000 members. Many of its galaxies are even larger than mammoth M31 in Andromeda.

Fig. 2.12. The Coma Berenices cluster of galaxies. Courtesy Mount Wilson and Palomar Observatories.

cluster is, the faster it recedes from us. Boötes, for example, has a recessional velocity of 24,000 mi/sec, compared with 2,300 mi/sec for nearer Pegasus. The recessional velocity as a function of distance is interpreted from the observed shifting of spectral lines toward the red end of the spectrum. Known as the *red shift,* the phenomenon has never been completely explained and is subject to continuous investigation, interpretation, and speculation.

Fig. 2.13. Galaxies NGC 4038–4039 in Corvus. Courtesy Mount Wilson and Palomar Observatories.

Some galaxies are believed to be in "collision" or, better stated, passing through one another. Because of the enormous distances that separate individual stars, actual stellar encounters must be very rare. Consequently, a collision has little effect on individual stellar masses and on multiple star systems. The major result of galactic interactions appears to be that interstellar matter is swept out of the galaxies and left to accumulate in intergalactic space. Figure 2.13 illustrates this amazing phenomenon.

The Nature of Stars

In the preceding pages we have examined some of the major characteristics of the largest association of individual stars in the observable universe: the galaxy. Not only did we learn that a galaxy consists of millions or billions of stars but that it may be a member of a cluster of galaxies. It is conceivable that there are even larger aggregations; for example, we may discover in the distant future that what we call the universe is only one of many universes, each with its billions or trillions of galaxies and each separated from other universes by unfathomable distances.

It is now time to turn our attention to the building blocks of the galaxies, the individual stars. This brings us to the fascinating, and far from completely understood, subject of stellar astronomy. The treatment given is descriptive and brief, its purpose being simply to show how different one star can be from another, how certain types predominate in certain regions of the galaxy, and how stars apparently evolve. While the chemical composition of stars is surprisingly similar, the range in density, size, mass, brightness, and surface temperature varies widely. As a single example, one star can be a billion times brighter than another.

Table 2.3
Magnitudes of Some Bright Stars

Star	Apparent Visual Magnitude	Absolute Visual Magnitude	Brightness Sun = 1
Sun	− 26.88	+ 4.69	1
Sirius	− 1.52	+ 1.36	26
Canopus	− 0.86	− 7.4	80,000
Vega	+ 0.14	+ 0.6	50
Rigel	+ 0.34	− 5.8	18,000
Altair	+ 0.89	+ 2.4	9
Antares	+ 1.23	− 4.0	3,400
Deneb	+ 1.33	− 5.2	10,000

A star is a completely gaseous mass generally spherical in shape. Gravitational forces keep the gases together, whereas counteracting gas, and to some extent radiation, pressure tends to distend them. Stars may both eject

Table 2.4
Twenty-five Visible Stars Nearest the Sun

Name	Distance Light Years	Absolute Magnitude	Luminosity Sun = 1	Spectral Type
Proxima Centauri (Alpha Centauri C)	4.2	+ 15.7	0.00004	M
Alpha Centauri A	4.3	+ 4.7	1.0	G_4
Alpha Centauri B	4.3	+ 6.1	0.3	K_1
Barnard's Star[a]	6.0	+ 13.1	0.0004	M_5
Wolf 359	7.7	+ 16.5	0.00002	M_8
Luyten 726-8 A	7.9	+ 15.6	0.00004	M_6
Luyten 726-8 B	7.9	+ 16.1	0.00003	M_6
Lalande 21185[a]	8.2	+ 10.4	0.005	M_2
Alpha Canis Majoris A (Sirius A)	8.7	+ 1.3	26	A_1
Alpha Canis Majoris B (Sirius B)	8.7	+ 11.2	0.008	White Dwarf
Ross 154	9.3	+ 13.2	0.0003	M_5
Luyten 789-6	9.7	+ 14.9	0.00008	M_8
BD + 36° 2147	9.7	–	–	–
Ross 248	10.4	+ 14.7	0.0001	M_8
Epsilon Eridani	10.8	+ 6.2	0.30	K_2
Ross 128	11.0	+ 13.5	0.0003	M_5
Tau Ceti	11.1	+ 5.9	0.34	G_4
61 Cygni A[a]	11.1	+ 7.8	0.063	K_3
61 Cygni B[a]	11.1	+ 8.5	0.003	K_5
Alpha Canis Minoris A (Procyon)	11.3	+ 2.8	5.8	F_5
Alpha Canis Minoris B	11.3	+ 13.1	0.00044	White Dwarf
Epsilon Indi	11.4	+ 7.0	0.12	K_5
Struve (Sigma) 2398 A	11.6	+ 11.2	0.0027	M_4
Struve (Sigma) 2398 B	11.6	+ 11.9	0.0012	M_4
Groombridge 34 A[b]	11.7	+ 10.4	0.0052	M_2

[a] Has unseen companion.

[b] Groombridge 34 B values: 11.7, 13.2, 0.0004, and M_5.

Note: 70 Ophiuchi is not one of the 25 nearest stars. It is 16.4 light years away, has an absolute stellar magnitude of 5.7, a luminosity in terms of the Sun of 0.4 and is K_1 type. The values for the second component of the system are, respectively, 16.4, 7.4, 0.083 and K_5.

material of which they are composed and pick up material from gas and dust clouds in space.

Most stars can be placed in the following six classifications: (1) main sequence, (2) subgiants, (3) giants, (4) supergiants, (5) white dwarfs, and (6) subdwarfs. Within any classification, stars are similar in structure and in the means by which stellar energy is produced. However, they may differ widely in size and brightness.

Stars vary in a great many ways, the most obvious being brightness. When we refer to brightness we must think not only of apparent brightness but of absolute brightness, or, astronomically speaking, absolute magnitude (total light intensity). It is important to know the absolute magnitude so that we can know how bright a star really is and not merely how bright it appears to be because of its distance from us. Table 2.3 compares the apparent and absolute visual magnitudes of some of the brightest stars, together with their brightness in terms of the Sun.

Astronomers have developed a variety of techniques to determine how far away a star is—and how it moves, how fast it rotates, what it is made of, how hot it is, what its density and mass are, and the rate (and how steadily)

Table 2.5
Comparison of Typical Stars

Name of Star	Radius Sun = 1	Mass, Sun = 1	Density Sun = 1	Absolute Magnitude	Temperature Degrees K
		Main Sequence Stars			
β Centauri	11	25	0.018	− 3.8	21,000
Vega	2.4	3.0	0.11	+ 0.6	11,200
Altair	1.4	1.7	0.6	+ 2.5	8,600
Procyon	1.9	1.1	0.16	+ 2.8	6,500
61 Cygni A	0.7	0.45	1.3	+ 7.8	3,800
Barnard's Star	0.16	0.18	45	+ 13.1	3,100
		White Dwarf Stars			
Sirius B	0.034	0.96	27,000	+ 11.2	7,500
Van Maanen's Star	0.007	0.14	400,000	+ 14.5	7,500
		Giant Stars			
Capella	12	4.2	0.0024	− 0.1	5,500
Aldebaran	60	4	0.00002	− 0.1	3,300
		Supergiant Stars			
Betelgeuse	290	15	0.0000006	− 2.9	3,100
Antares	480	30	0.0000003	− 4.0	3,100

it pours forth its energy. They have discovered that the star nearest to the Sun is a little more than 4 light years away. They can only conjecture how distant the farthest star may be. A list of the nearest stars appears in Table 2.4.

Table 2.5 compares typical main sequence stars with some white dwarfs on the one hand and giants and supergiants on the other. The most striking difference between the various classes of stars concerns densities—in the table ranging from 400,000 times the density of the Sun for Van Maanen's star to 0.0000003 in the case of Antares.

Stars can be partially described by the spectra of their atoms and molecules, as revealed by the science of spectroscopy. The spectral sequence W-O-B-A-F-G-K-M-R-N-S is applicable to the known stars. In accordance with their spectral class, stars are classified as O stars, F stars, M stars, and so forth. Table 2.6 gives the major characteristics of stars in each class. Classes W, R, N and S are rare, being applicable to less than 1 per cent of known stars. Each class is subdivided into 10 spectral types numbered 0, 1, 2, 3, 4, 5, 6, 7, 8, and 9; such numbers provide a means of gradation between classes. Temperatures decrease steadily from W through S classifications. The Hertzsprung-Russell diagram of stellar types, seen in Fig. 2.14, shows the correlation between luminosity and surface temperature.

Table 2.6
Classes of Stellar Spectra

Spectral Class	Characteristics
W	Blue-white, extremely hot Wolf-Rayet stars, rich in ionized helium, highly ionized carbon, nitrogen and oxygen.
O	Blue-white, very hot stars, with ionized helium.
B	Blue-white, less hot stars, with neutral helium, hydrogen, ionized oxygen and ionized carbon.
A	White stars, rich in hydrogen, with ionized magnesium, iron, titanium, and calcium.
F	Yellow-white stars, with ionized calcium, neutral calcium, ionized metals, hydrogen.
G	Solar-type stars, with weakening hydrogen lines, strengthening neutral lines of metals. Molecular bands appear.
K	Orange stars, strong neutral lines and molecular bands, weakening lines of hydrogen. Titanium oxide bands begin to appear.
M	Red stars, with stronger molecular bands, titanium oxide, neutral atoms.
R	Red stars, with molecular bands, principally cyanogen and the carbon molecule.
N	Red stars, as above with increasing carbon compounds.
S	Red stars, extremely faint, with bands of zirconium oxide; very strong lines of neutral atoms.

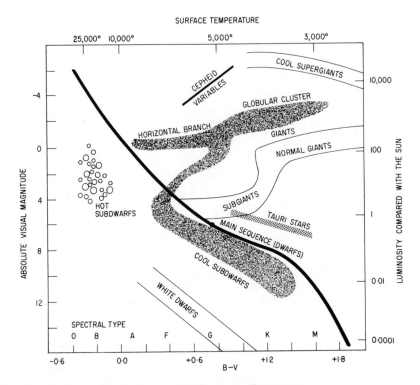

Fig. 2.14. Hertzsprung-Russell diagram of stellar types. The grey area contains Population II type stars. The black dot is the Sun. B-V refers to the color index, representing blue magnitude minus visual, or green-yellow magnitude. It increases as surface temperature decreases.

As we learned earlier, the majority of stars (more than 80 per cent) belong to the main sequence. Such stars have central temperatures ranging from 35,000,000 deg to perhaps 10,000,000 deg. Most vary in size from about 10 times that of the Sun to 1/10 its size. Main sequence stars are similar to one another in structure and derive their energy from similar nuclear reactions—e.g., the carbon cycle for hotter stars and the proton-proton reaction for cooler stars.

The next most common type of stars are white dwarfs, those extraordinary bodies whose core densities may be many million times the Sun's and whose over-all densities are hundreds of thousands of times greater than the Sun's. White dwarfs produce their energy by continuous contraction; nuclear sources have been exhausted, including all internal hydrogen. Surface temperatures average 10,000°K, central temperatures 10,000,000°K. Luminosity is very low. Some white dwarfs are the size of the Earth, yet weigh as much as the Sun. They probably are at the terminal end of stellar evolution, and presumably disappear into darkness as the gravitational contraction energy is dissipated. Burnt out white dwarfs may endlessly follow galactic orbits, invisible to telescopic means of detection. The only ways we could learn about them would be by detecting their gravitational fields, by radar ranging, or possibly visually by their obscuration of sources of light behind. They may be radio emitters in which case they may be detectable by radio telescopes constructed outside the Earth's atmosphere. Conceivably, a burnt

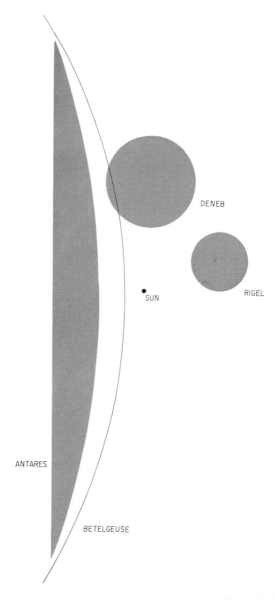

Fig. 2.15. The Sun compared to giants and supergiants.

out dwarf may one day be discovered by an automatic probe dispatched from the Earth to explore a distant visible star. If manned interstellar flight ever proves feasible, dead white dwarfs will without doubt be prime objects of interest.

Less numerous than white dwarfs are the rare giants and supergiants, distended stars that do not obey mass-luminosity relationships applicable to the main sequence. Giant stars range from approximately 20 to over 30 times the solar diameter and are at least 100 times as luminous. Supergiants have 300 to 800 times the Sun's diameter and up to 10,000 times its brightness. Figures 2.15 and 2.16 illustrate these relationships. The mechanisms

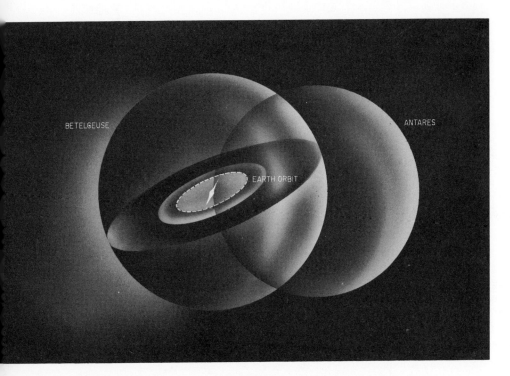

Fig. 2.16. Comparison of orbit of Earth around Sun and diameters of Antares and Betelgeuse.

responsible for the tremendously distended atmospheres are not completely understood. Most interpretations postulate the stars as having several differing concentric zones, the outermost being the distended atmospheric envelope.

Stars that change in brightness are known as *variable*, and include stars that fluctuate according to a regular pattern as well as those that fluctuate

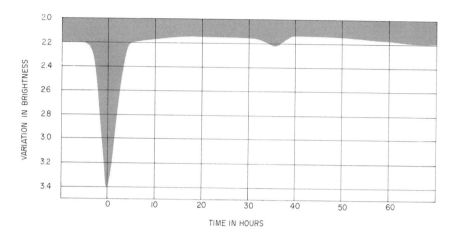

Fig. 2.17. Binary system of Algol, typical eclipsing variable. At minimum brightness magnitude is −1.20 below normal, occurring when the dull companion eclipses the bright member. Only slight loss of light is observed 2.5 hours later when bright star eclipses the dull one.

in an erratic manner. Many thousands of variable stars have been observed in the Milky Way, in the globular clusters, and in exterior galaxies.

Stars that appear to be, but are not really, variable are known as *extrinsic variables*. They are normally binary star systems whose components regularly eclipse each other, causing variations in light (Fig. 2.17). Binary and multiple star systems are considered later in this chapter.

Intrinsic variables are conveniently divided into two categories: (1) pulsating variables and (2) exploding variables. Among the former are Cepheid variables, RR Lyrae stars, long-period variables, and semiregular variables. Various types of novae are exploding variables. The periods of pulsating variables are from less than a day to several years.

Cepheid variables have periods of from about a day to approximately a a month and a half. Periods that are common in one galaxy are not necessarily common in another. An unusual, and extremely fortunate, characteristic of Cepheids is that the more luminous they are the longer their period. By observing the period and the apparent magnitude, it is possible to derive the absolute magnitude and, by computation, the distance. Cepheids thus provide astronomers with an invaluable distance-measuring tool. Cepheids change not only in brightness but in spectrum, temperature, and radial velocity (the velocity at which they approach or recede from the observer). In the Milky Way they are found close to the plane of the galaxy in the region of thick dust. It is speculated that they are young stars that were

Table 2.7
Selected Cepheid Variables

Star	Period, days	Spectral range	Magnitude range
SU Cassiopeiae	1.95	F_2 to F_9	6.05 to 6.43
Polaris	3.97	F_7	2.08 to 2.17
Delta Cephei	5.37	F_4 to G_6	3.71 to 4.43
Zeta Geminorum	10.15	F_5 to G_2	3.73 to 4.10
1 Carinae	35.52	F_8 to K_0	3.6 to 4.8
SV Vulpeculae	45.13	G_2 to K_5	8.43 to 9.40

formed from the dust in which they are now concentrated. Table 2.7 lists a half a dozen typical Cepheid variables in order of increasing period.

A special type of Cepheid variable are *RR Lyrae* Class A blue giant stars, characterized by their very short periods ($1\frac{1}{2}$ hr at one extreme, 29 hr at the other). CY Aquarti's period of 89 min is the shortest known. RR Lyrae stars are concentrated in the galactic center, thinning out in all directions (Fig. 2.18). They are far more common than regular Cepheids, but do not take part in normal galactic rotation (they follow orbits around the galactic center with considerable range in inclination and eccentricity, as opposed to the essentially circular rotation of most stars.) They do not indicate a period-luminosity relationship, but are apparently all of the same brightness regardless of where they are situated in our galaxy, or in other galaxies. The RR Lyrae star is obviously an important help to astronomers attempting to work out distances in the universe.

Another type of variable is the *W Virginis star* with light curves showing periods between 10.30 and 28.58 days. They are between 1 and 2

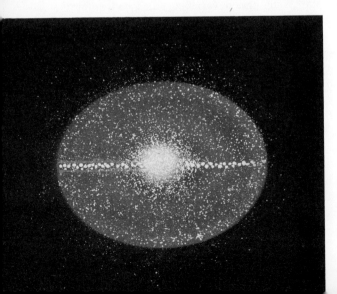

Fig. 2.18. RR Lyrae variables represented by small dots, Cepheids by large dots confined to center of the galactic plane.

magnitudes less bright than regular Cepheids of the same period and exhibit two absorption spectra as they increase in brightness. This phenomenon represents twin stellar disturbances, one on the increase and the other on the decrease. Of longer period are *RV Tauri stars*. The longer their period, the fainter they are. Many of these variables are found far off the galactic plane.

The longest period variables are called *red variables,* extremely low density M, S and N giant stars with periods ranging from 70 to more than 700 days. Mira Ceti class red variables exhibit light changes of over $2\frac{1}{2}$ magnitudes; they are referred to as long-period variables, one of whose characteristics is that their brightness variations are not regular. Mira Ceti itself has an approximate 332-day period. Moreover, it has a companion star that is also variable, though of very different type. Semiregular red variables generally have shorter periods, with light curves of a less periodic nature.

Novae and supernovae are very special cases of variable stars that have suddenly and violently exploded due to some internal disturbance. Incalculable quantities of stellar material are thrown into space at hundreds or thousands of miles per second while profound spectral, temperature, and magnitude changes take place (the brightness of the star may increase from tens of thousands to several million times its original value). Most novae have been observed to explode only once, many leaving a nebulous shell to mark the catastrophic event. Some novae symmetrically eject material while others throw it out from a given portion of their surfaces. Most novae seem to be derived from blue subdwarf stars. Occasionally, novae have been seen to reappear, and are hence known as *recurrent*. They apparently involve classes of stars distinct from those associated with regular novae, being similar to red variables.

Much less common than novae are supernova stars that can become tens of millions or even a hundred million times brighter than the Sun following the catastrophic explosion. Such stars eject their material at velocities far higher than do novae. The magnitude range of the giant star that becomes a supernova can be as much as 18 and even more. Three historical supernovae have been recorded in the Milky Way galaxy, one in 1054, one observed by Tycho in 1572, and one seen by Kepler in 1604. The 1054 supernova left an expanding mass of debris in space known as the *Crab Nebula* (Fig. 2.19). Aside from producing visible light, this nebula is a strong radio source. Since 1885, more than 150 supernovae have been detected in the universe, the brightest being S Andromedae discovered near the center of Messier 31.

Whereas novae lose, upon explosion, only a relatively small percentage of their stellar mass (returning essentially to the way they were before the outburst), supernovae are deprived of a large portion of their original

Fig. 2.19. Crab nebula produced by a supernova in 1054 A.D.

mass when they burst out and are never the same again. Novae and supernovae are clearly not the same type of phenomena, being caused by entirely different circumstances. In the case of a nova, internal conditions become such that the star must release energy rather suddenly before it can revert to its normal state again, whereas supernovae appear to undergo a permanent change.

We have discussed the major classifications of variable stars, although there are other types. These include *dwarf novae,* which exhibit many features of recurrent novae but on a small scale. SS Cygni is the most famous example of a dwarf nova. There are also stars, located on the main sequence, that eject material which forms a tenuous shell of gas around them; they are the so-called *shell stars.* Red and yellow dwarf main sequence stars that are subject to rapid flare emanations are known as *flare stars.*

Still other types of variables are *R Cornae Borealis* and T Tauri stars. The former are normally bright, but occasionally become many magnitudes dimmer, only to return to their original brightness in an irregular fashion. T Tauri stars are found within dark nebulae, which we learned consist of vast accumulations of gas and dust. Such clouds passing these stars cause completely irregular variations in brightness. It is believed that stars in the nebulae are being formed continually from dust and gases, as well as being affected by them once they are created.

The Universe 33

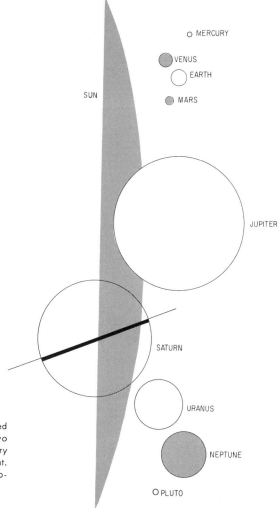

Fig. 2.20. Size of Sun compared to its nine planets. Below: two photos of the Sun, one in ordinary light, other in red hydrogen light. Courtesy Mount Wilson and Palomar Observatories.

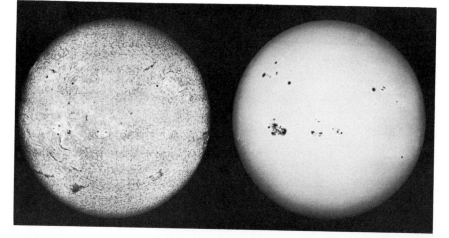

The Solar System

Dominated by a central star we call the Sun, the Solar System is an assemblage of planets, satellites, asteroids, meteors, comets, micrometeorites, dust, and gases. The space within the Solar System is far from empty, being affected by, containing, or propagating minute particles of matter, atoms, molecular groupings, energetic particles, electromagnetic radiations, gravitational fields, magnetic fields, radio waves, and ultraviolet, infrared, and gamma radiations.

Aside from the Sun, the major worlds of the Solar System are the planets, which are compared to the star in Fig. 2.20 and to each other in Fig. 2.21. They range in diameter from about 3,000 miles to almost 90,000 miles, in volume from under a tenth to more than a thousand times our Earth, and in mass from a twentieth to more than 300 times that of our planet. The planet closest to the Sun revolves at some 36,000,000 miles; Pluto, the most remote planet, averages over 3,500,000,000 miles distance. Some planets have no moons; one has 12. The lengths of planetary years range from 88 to more than 90,000 sidereal days. Planets are conveniently grouped as inner and outer, the two groups being separated by a zone of asteroids (minor planets). All are dominated by the Sun to an extent depending greatly on their distance. The relative escape velocities of the Sun and planets are indicated in Fig. 2.22.

THE SUN

The major characteristics of the Sun are tabulated in Table 2.8. It is a G-type star, somewhat brighter than the average, and the only definitely known possessor of a family of planets. It moves through space towards the constellation Hercules at a velocity of approximately 12 mi/sec and, as we

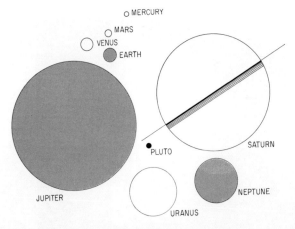

Fig. 2.21. The planets of the Solar System.

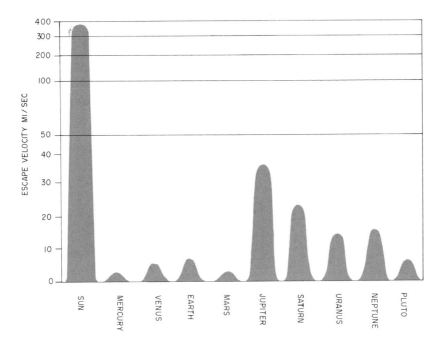

Fig. 2.22. Relative escape velocities of the worlds of the Solar System.

Table 2.8
The Sun

Diameter, miles	864,616
Surfaced area, Earth = 1	12,000
Volume, Earth = 1	1,306,000
Density, Earth = 1	0.255
Mass, Earth = 1	333,434
Gravity, Earth = 1	28.0
Escape velocity, mi/sec	384
Apparent visual magnitude	−26.88
Absolute magnitude	+ 4.69
Temperature, °K	
Center	2×10^7
Surface	6×10^3
Chromosphere	3×10^4
Prominences	$1-2 \times 10^4$
Corona	10^6
Length of day (period of rotation), days	25.35

saw earlier, revolves around the center of the Milky Way galaxy. The Sun rotates slowly on its axis, faster in the equatorial regions and slower towards the poles (it does not behave as a solid body, which it is not).

Like other stars, the Sun is gaseous, and is very nearly spherical in shape. In terms of volume percentage, the Sun contains 81.76 per cent hydrogen, 18.17 per cent helium, and small percentages of oxygen, magnesium, nitrogen, etc. (more than 60 elements have been observed). The luminous portion of the Sun visible to the observer is the photosphere. Below it are the dense interior gases and above it is the normally invisible atmosphere, which extends for millions of miles into space. The chromosphere is between 8,000 and 9,000 miles thick and is responsible for the solar prominences that occasionally rise to million-mile altitudes. The term *corona* is given to the very low-density outer atmospheric envelope, seen during eclipses. Sunspots, or dark patches on the Sun's disk, are found chiefly between 5 and 30 deg latitudes. They range from small spots to enormous groupings 50,000 or more miles in diameter.

THE INNER PLANETS

In terms of size, the four inner planets Mercury, Venus, Earth, and Mars (Fig. 2.23) are small, all less than 8,000 miles in diameter. Mercury, the closest to the Sun, is airless but Venus, Earth, and Mars all have atmospheres. The Earth has a single moon and Mars two small ones. Comparative data on the four planets are summarized in Table 2.9. Tables 2.10 and 2.11 provide key facts concerning the three moons.

Mercury is at once an extremely hot and a frigidly cold world. It is

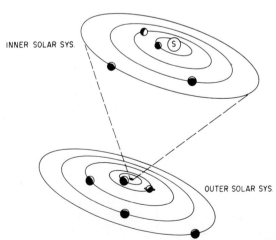

Fig. 2.23. Schematic drawing of the inner and outer Solar System.

Table 2.9
The Inner Planets

	Mercury	Venus	Earth	Mars
Diameter, miles	3,100	7,700	7,927	4,215
Surface area, Earth = 1	0.15		1	0.283
Volume, Earth = 1	0.06	0.91	1	0.151
Density, Earth = 1	0.69	0.88	1	0.73
Water = 1	3.8	4.86	5.52	4.02
Mass, Earth = 1	0.05	0.817	1	0.108
Gravity, Earth = 1	0.36	0.86	1	0.39
Oblateness	0	0	$\frac{1}{296}$	$\frac{1}{192}$
Escape velocity, mi/sec	2.6	6.4	6.96	3.1
Albedo	0.06	0.65	0.39	0.15
Solar constant, Earth = 1	6	2	1	0.43
Temperature of surface or of visible layers of opaque atmospheres, °F	650[a]	−35	70[b]	0
Length of day (period of rotation), hr-min	2,105.85	? (possibly 720)	23–56	24.37
Length of year (sidereal period of revolution), solar days	87.97	224.701	365.256	686.98
Synodic period of revolution, solar days	115.9	583.93	not applicable	779.94
Mean distance from Sun, miles × 10^6	35.8	67.20	92.956	141.56
Mean orbital velocity, mi/sec	29.7	21.7	18.5	15

[a] Maximum value at mean distance from Sun.

[b] Mean surface value; extremes from −125°F to +140°F.

Table 2.10
The Moon (Luna)

Diameter, miles	2,160
Surface area, Earth = 1	0.074
Volume, Earth = 1	0.0203
Density, Earth = 1	0.604
Water = 1	3.34
Mass, Earth = 1	0.0123
Gravity, Earth = 1	0.164
Oblateness	0
Escape velocity, miles/sec	1.48
Albedo	0.07
Stellar magnitude	−12.6
Temperature of surface, °F	0[a]
Length of day (period of rotation), hr	655.72
Length of year (sidereal period of revolution), solar days	27.298
Synodic period of revolution, solar days	29.54
Mean distance from Earth, miles	238,857
Mean orbital velocity, miles/sec	0.64

[a] Ranging from +270°F to −240°F.

Table 2.11
Satellites of Mars

	Phobos	Deimos
Diameter, miles	10 to 15	5 to 7
Apparent visual magnitude	+11.5	+12.5
Mean distance from primary, miles	5,800	14,600
Sidereal period, days-hr-min	0-7-39	1-6-18
Mean orbital velocity, mi/sec	1.36	0.88
Eccentricity of orbit	0.0210	0.0023
Inclination to primary's orbit, deg-min	26-20	26-2

generally believed to show the same face to the Sun, the daylight side bearing the continuous onslaught of heat radiation from the 36,000,000-mile distant star; the small planet receives over seven times more radiant energy than the Earth. Very recent radio astronomical evidence, however, indicates that Mercury may have a rotation period of 59 ± 5 Earth days in a west-to-east direction and not 88 days as accepted for nearly a century. Subsolar readings of 650°F are common at the planet's mean distance from the Sun, rising to 775°F at perihelion. Its surface, in terms of reflectivity and roughness, seems to be similar to that of the Moon. No life of any type is expected on Mercury, but manned bases could probably be established in the twilight zone between the hot and cold hemispheres.

Despite its relative proximity to Earth (at inferior conjunction it is less than 26,000,000 miles away), the planet Venus has long been a world of mystery. For one thing, the surface has never been observed visually, so thick and opaque is the atmospheric blanket surrounding the planet. It is also uncertain if Venus rotates on its axis more than once during its circuit of the Sun. Radar signals bounced off the planet suggest a very slow rotation of over 200 days, so the length of the day and year may be very close; its period may even be retrograde, i.e., over 225 days—a figure of 247 ± 5 days appears probable according to measurements made by the 1,000 foot radio telescope at Arecibo, Puerto Rico.

Venus is only slightly smaller than Earth, but its atmosphere is thicker and its surface is much hotter. Instrumentation aboard the Mariner 2 space probe showed that its ground temperature is nearly 800°F and that there is a cold spot in the southern hemisphere (perhaps associated with a mountainous region). Cloud layer temperatures are about the same on both the day and night sides, implying very thorough atmospheric circulation. Values obtained from the probe and from terrestrial observatories are about the same, between -30 and $-35°F$.

The outer layers of the Venusian atmosphere contain a high percentage of carbon dioxide, with a very small amount of detectable water vapor. No accurate information is available on the nature of the lower atmosphere.

Many astronomers call the Earth and the Moon a double planet system, so large and massive is the latter compared with the former. In terms of its primary, the Moon's mass is greater than that of any other satellite in the entire Solar System. And its diameter is one quarter Earth's, compared with Titan's[10] relationship of 1/25th that of its primary, Saturn. Figure 2.24 compares various satellites in the Solar System.

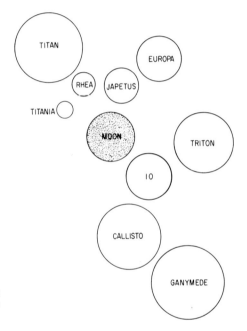

Fig. 2.24. The Moon compared to the other major satellites of the Solar System.

From the point of view of man, Earth has an ideal size, distance from the Sun, temperature span, atmosphere, and so on. Life exists in profusion and has for billions of years. On the Moon (Fig. 2.25) conditions are quite different: temperatures plunge to $-240°F$ during the night and soar to $270°F$ at the subsolar point at full moon during the day; there is no appreciable atmosphere, making the satellite subject to the force of cosmic infall; and its surface is harsh and inhospitable. Whether conditions were ever suitable for the development of even primitive life must still be determined.

The major surface features of the Moon are the mountains, the maria, the walled plains, the craters, the ray-like features, and such phenomena

---

[10] Titan, some 3,000 miles in diameter, is the second largest in terms of its primary.

Fig. 2.25. Apennines region of the Moon. Courtesy Pic du Midi Observatory.

as rifts, rills, and clefts. Details on the geography and geology of the Moon are gradually being worked out as a result of both Earth-based observations and information secured from automatic probes (fly-bys, orbiters, hard-landers, and soft-landers).[11]

Even though it is only half the size of the Earth, the outermost of the inner planets, Mars, is most similar to our planet and the only other world in the Solar System on which we expect to find some kind of life. Surface details (Fig. 2.26) can be seen through its transparent atmosphere to such an extent that fairly comprehensive maps have been drawn. Reddish-orange in color, Mars is a conspicuous object in the sky.

Among the best known surface features on Mars are the polar caps, which advance and retreat with the seasons; the dark, green-grey maria; the ochre-orange desert regions; and the streaks (so-called canals). Mariner 4 probe photographs showed that the surface is pockmarked with craters, much like the Moon. At higher elevations what appeared to be frost was evident. The terrain is probably fairly level, although highlands and plateaus of several thousand feet altitude appear to exist. Above the surface is a fairly tenuous atmosphere, probably consisting of 96 to 98 per cent nitrogen, some

---

[11] For a description of the inner planets and the Moon refer to any standard text on astronomy and Chapter 3 of Ordway, *et al, Basic Astronautics* (see bibliography). Refer also to Chapter 5 of this work, entitled "Astrogeology," for information on methods and techniques of geographical and geological exploration of the planets and in particular the Moon.

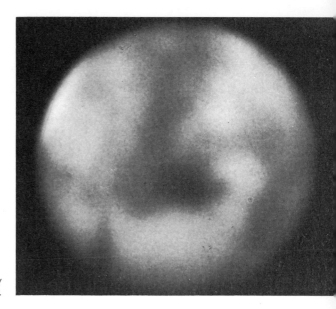

Fig. 2.26. The planet Mars. Courtesy W.S. Finsen and Republic Observatory.

carbon dioxide, a few per cent heavy gases (argon, krypton, etc.), and a very small amount of free oxygen and water vapor. Yellow, white and blue clouds and haze have been detected in the atmosphere. Winds are evident as they disperse the clouds.

Temperatures on Mars drop to $-100°F$ or below during the night, but rise to at least 70°F during summer days at or near the equator. Man could probably operate in the equatorial zone, with suitable protective clothing, and may even be able to permanently colonize it. Good evidence exists for some sort of vegetation on the planet, the presence of which could be of importance to man. Not only do spectral analyses suggest organic molecules but the maria are observed to change color as the seasons advance, becoming greenish-grey in the spring and summer. The most logical explanation for this behavior is vegetation blooming in the spring and dying out in the fall, much as occurs on Earth.

Mars has two small moons, Phobos and Deimos, that move in nearly circular orbits in the equatorial plane of the primary. They eventually may prove useful as staging bases for manned Mars expeditions.

THE ASTEROIDS

Revolving around the Sun between the realms of the inner and outer planets are the thousands of small worlds called asteroids.[12] They range in diameter

---

[12] Also referred to as planetoids and minor planets.

Table 2.12
Selected Asteroids

	Ceres	Pallas	Juno	Vesta	Eros	Hidalgo
Diameter, miles	475	305	128	245	15 by 5	25
Albedo	0.06	0.07	0.12	0.26	–	–
Stellar magnitude	+3.70	+4.38	+5.74	+3.50	–	–
Length of year (sidereal period of revolution), solar days	1,681	1,684	1,594	1,325	643	–
Perihelion, miles $\times 10^6$	237.7	197.4	183.9	200.0	103.0	185.4
Aphelion, miles $\times 10^6$	276.8	317.9	312.2	239.1	168.0	891.8
Inclination of orbit to ecliptic, deg	10.6	34.8	13	7.1	10.8	42.53

Table 2.13
The Outer Planets

	Jupiter	Saturn	Uranus	Neptune	Pluto
Diameter, miles	86,640	75,100	32,000	30,000	~5,000
Volume, Earth = 1	1,312	763	59.00	72.00	~0.09
Density, Earth = 1	0.24	0.13	0.20	0.29	?
Water = 1	1.34	0.71	1.10	1.62	?
Mass, Earth = 1	318.35	95.28	14.58	17.36	~1.0
Gravity, Earth = 1	2.64	1.17	1.05	1.23	~0.90
Oblateness	1/15	1/9.5	1/15	1/40	?
Escape velocity, mi/sec	37	22.8	13.5	15.0	~6.5
Albedo	0.51	0.49	0.60	0.52	0.17
Solar constant, Earth = 1	1/25	1/100	1/400	1/900	1/1,600
Temperature of surface or of visible layers of opaque atmospheres, °F	−200	−230	−270	−340	−370
Length of day (period of rotation), hr-min	9–9	10–30	10.48	15.50	155.61
Length of year (sidereal period of revolution), solar days	4,332.6	10,759.53	30,686.48	60,188.82	90,471.33
Synodic period of revolution, solar day	398.88	378.1	369.67	367.49	366.74
Mean distance from Sun, miles $\times 10^6$	483.2	886.1	1,783.2	2,793.5	3,676
Mean orbital velocity, mi/sec	8.1	6.0	4.2	3.4	2.8

from less than a mile to nearly 500 miles, only a few being more than a hundred miles across. Most asteroids are found in a band extending from somewhat less than 200,000,000 miles to about 350,000,000 miles from the Sun. Some, however, intrude into the realms of the inner and outer Solar System.

In 1969 Geographos will come within 4,000,000 miles of Earth. Information on the larger asteroids is reviewed in Table 2.12.

THE OUTER PLANETS

Four of the five outer planets (Fig. 2.23) are giant worlds with deep atmospheres. No observational information on the solid surfaces below these blankets of gases is available, though models have been constructed that describe what they probably are like. Their major characteristics are tabulated in Table 2.13. The four gaseous giants have 28 of the 31 known moons in the Solar System (Table 2.14). The fifth outer planet is remote Pluto whose characteristics are in some ways suggestive of the small, dense inner planets. It is only about a tenth the diameter of the smaller of the outer planets, has no known moon, and probably no atmosphere.

Jupiter (Fig. 2.27), the largest planetary body in the Solar System, is nearly 87,000 miles across, has a short day, less than 10 hours long, and a volume over 1,300 times Earth's. Perhaps 80 per cent of its diameter, and an unknown per cent of its volume, is solid, the rest being accounted for by a very thick atmosphere. If man ever is able to land on Jupiter he may find a slushy surface of crystalline ammonia and ice under which lies the solid hydrogen-helium mass of the giant planet.

The Jovian atmosphere, vastly different from the oxidized atmospheres of the inner planets, consists of molecular hydrogen, helium, methane and ammonia. A definite banded structure is evident, caused by a series of cloud belts rotating at high speed around the world. Colors vary from pale greys and browns to red and blue. A familiar feature of Jupiter is the Red Spot, which may consist of solid matter floating in the atmosphere.

Fig. 2.27. Jupiter photographed in blue light. Note its satellite Ganymede. Courtesy Mount Wilson and Palomar Observatories.

Table 2.14 Satellites of the Outer Planets

	Diameter, miles	Density, Earth = 1	Mass, Earth = 1	Gravity, Earth = 1	Stellar magnitude	Mean distance from primary, miles	Sidereal period, days-hr-min	Mean orbital velocity, mi/sec
Jupiter								
Amalthea (5)	~150	—	—	—	+13	112,500	0-11-57	16.2
Io (5)	2,000	0.75	0.012	0.19	+5	262,200	1-8-28	10.8
Europa (2)	1,800	0.68	0.008	0.16	+6	417,200	3-13-14	8.5
Ganymede (3)	3,100	0.46	0.026	0.17	+5	665,500	7-3-43	6.6
Callisto (4)	2,800	0.37	0.016	0.13	+6	1,170,000	16-16-32	5.1
6	~100	—	—	—	+14	7,135,000	250-14	2.15
7	~35	—	—	—	+17	7,296,000	259-14	2.1
10	~15	—	—	—	+19	7,367,000	262	2.1
12	~15	—	—	—	+19	13,200,000	631	1.45
11	~20	—	—	—	+18	14,000,000	692	1.42
8	~35	—	—	—	+17	14,600,000	739	1.40
9	~15	—	—	—	+19	14,700,000	758	1.47
Saturn								
Mimas	~325	0.09	0.000006	0.01	12	115,400	0-22-37	8.9
Enceladus	~400	0.13	0.000014	0.01	12	148,000	1-8-53	7.9
Tethys	~600	0.22	0.000110	0.02	10.5	183,200	1-21-18	7.1
Dione	~650	0.51	0.000178	0.04	11	234,700	2-17-41	6.25
Rhea	~1,000	0.36	0.00038	0.04	10	327,700	4-12-25	5.3
Titan	3,000	0.44	0.024	0.17	8	759,400	15-22-41	3.5
Hyperion	~250	—	—	0.007	13	921,800	21-6-38	3.17
Japetus	1,500	—	0.00024	0.03	10.5	2,213,000	79-55	2.02
Phoebe	~200	—	—	—	14.5	8,050,000	550-45	1.1
Uranus								
Miranda	~150				+17	80,700	1-9-55	4.15
Ariel	~300				+15.2	118,600	2-12-29	3.42
Umbriel	~240				+15.8	165,300	4-3-28	2.78
Titania	~800				+14	270,900	8-16-56	2.26
Oberon	~260				+14.2	362,500	13-11-7	1.97
Neptune								
Triton	2,600	~0.5	0.022	~0.2	+13.6	220,840	5-21-3	2.72
Nereid	~200	—	—	—	+19.5	3,458,000	359-10	0.69

Of Jupiter's 12 moons, four are major worlds between 1,800 and 3,100 miles in diameter; the remaining eight range in diameter from 15 miles to 150 miles. The major moons, Io, Europa, Ganymede, and Callisto are hard-surfaced worlds, possibly with very tenuous atmospheres. All will probably be visited by man, who may establish bases on them. They are from 262,000 to 1,170,000 miles from their primary.

In many ways Saturn is similar to Jupiter, being somewhat smaller, colder, and the possessor of a ring. It has nine known moons, including the only moon in the Solar System with a definitely confirmed atmosphere. Saturn's day is a little longer than Jupiter's and the planet is not as dense. As it turns out, the density of the planet is less than that of water.

Fig. 2.28. Saturn photographed in blue light, Courtesy Mount Wilson and Palomar Observatories.

The ring system of Saturn (seen in Fig. 2.28), is extremely interesting to observers. There appear to be three distinct rings all made up of untold millions of micrometeoritic and perhaps meteoritic matter. The ring closest to the planet is known as the crêpe ring; it is some 10,000 miles wide and about 9,000 miles from Saturn's surface. Altogether, the entire ring system is 171,000 miles across, i.e., from one edge to another.

It is assumed that the surface of Saturn is much like that of Jupiter except that more ammonia may have frozen out from the atmosphere and been deposited there. Approximately 30 per cent of the planet probably has a solid hydrogen-helium core, with another 20 per cent consisting of an ice layer between it and the atmosphere. There is doubtless somewhat less ammonia and more methane in the atmosphere and, of course, it is much colder since Saturn is approximately twice as far away from the Sun as Jupiter. The atmosphere shows a banded structure.

The six inner satellites of Saturn move in nearly circular orbits very close to the plane of the rings, while the outermost moon, Phoebe, orbits in a retrograde direction, i.e., westward instead of eastward. The large moon Titan revolves at a distance of nearly 760,000 miles. It is over 3,000 miles in diameter and has an atmosphere very similar to that of Saturn. This moon is characterized by an unusual orange color.

Colder, denser, and smaller than Jupiter and Saturn are Uranus and Neptune, whose orbits are separated by a billion miles of space. Both planets probably consist of metallic hydrogen-helium cores surrounded by icy surfaces and finally by molecular hydrogen and methane atmospheres. The planets exhibit a greenish hue, Uranus having a banded cloud structure that is not observed on Neptune. Uranus' five moons range in diameter from 150 to

Fig. 2.29. Artist's conception of Pluto seen from an approaching space probe.

800 miles. Their orbits are circular and are nearly on the equatorial plane. Neptune possesses two moons: large, massive, and close-by Triton; and small, distant Nereid.

Very little information is known about cold, remote, 15th magnitude Pluto (Fig. 2.29) and all figures available should be accepted with much caution. It takes 250 years to make a single circuit of the Sun along the very eccentric, inclined orbit. The planet's surface is probably icy with the possibility of liquid methane pools existing. There may be a tenuous atmosphere of neon and argon; any methane present presumably has frozen out and been deposited on the surface as a liquid. Observations made during 1964 place its period of rotation at 6 days 9 hours 17 minutes; however, it is not known if the rotation is direct or retrograde—the period was obtained from studies of variations in the planet's brightness. The diameter is unknown, but is probably at least 5,000 miles. Only Pluto's central, relatively bright highlight has been viewed optically: the dark rim around it has never been seen.

3
The Origin of Planetary Systems

The Solar System that we have just reviewed is the only assemblage of planets definitely known to exist in the Universe. We say *known to exist,* for no planets so far have been directly observed to orbit stars other than the Sun. This does not imply that there are no other solar systems, but simply that we cannot see them with our present telescopic equipment. If the nearest star to the Sun had a family of planets it would remain beyond man's present ability of direct detection. But, as we shall see later, there is some indirect evidence of extrasolar planets and an abundance of theoretical indications.

In discussing the intriguing subject of the origin of planetary systems we shall concentrate on three principal areas: (1) the accidental theories of planetary formation, (2) the natural theories of planetary formation, and (3) speculations on the existence of extrasolar planetary systems. It is improbable that all solar systems were created and have evolved in the same way and according to the same processes. Nevertheless, we do believe that a great majority were created by the same basic mechanisms. And it is likely that solar systems are coming into being today. It is concluded that most solar systems have

"You've got to start on insufficient knowledge.
You've got to have that kind of courage.
  Robert Frost

"The moment of discovery, 'spontaneous illumination....' The infinite is made to blend itself with the finite, to stand visible, as it were, attainable there.

  Thomas Carlyle

a *natural* origin rather than an *accidental,* or catastrophic, origin, though some, of course, could have been conceived by accidental, and very rare, occurrences.

Before commencing our review of planetary origin theories it is important to note five major facts that must be taken into account when considering the origin of our Solar System. These have been summarized by Ter Haar and Cameron[1] as follows: (1) regularity of planetary orbits (all planets revolve in nearly circular, coplanar orbits around the Sun in the same direction as the Sun rotates), (2) similarity between planetary systems and satellite systems (most of the moons of the Solar System follow regular, noneccentric paths close to the equatorial plane of their primary body), (3) difference between the terrestrial and giant planets (terrestrial planets are small, dense, close to Sun, have few or no satellites, and slowly rotate, whereas the more distant, lighter, outer planets are large, rotate rather rapidly, and have many satellites), (4) applicability of the Bode-Titius planetary distance law (a mathematical relationship expressing the relative distances of the planets from the Sun), and (5) the slow velocity of rotation of the Sun (most of the angular momentum of the Solar System resides in the planets, not the Sun, which presumably lost the major portion of its rotational velocity when the planets were formed).

It is interesting to record that while many ages have been assigned to the Solar System, most modern astronomers and astrophysicists believe the Sun to be not much older than 5 billion years and the planets some 4.5 billion years. A few researchers suspect that the Sun and planets are essentially the same age and at least one speculates that the planets may be slightly older than the star around which they revolve. There is considerable disagreement as to the length of time it took the Solar System to evolve into essentially its present form.

Accidental Origin Theories

There are several basic varieties of accidental, or catastrophic, origin theories, some of which were widely held for years by the astronomical community. Today these theories have been all but discarded and replaced by natural origin theories. Whereas the Solar System was once thought to be an extremely rare entity, a freak in the universe—indeed, perhaps the only system of planets anywhere—now it is looked upon as a naturally occurring phenomenon, one of unknown billions in the Milky Way and other galaxies.

---

[1] Ter Haar, D. and A. G. W. Cameron, "Historical Review of Theories of the Origin of the Solar System," in *Origin of the Solar System,* ed. Robert Jastrow and A. G. W. Cameron. New York: Academic Press, 1963, pp. 1–2.

Fig. 3.1. Buffon's concept of a comet's encounter with the Sun.

Accidental origin theories, as pointed out by Shapley,[2] share two characteristics. First, they suppose that "the planets were born of violence" and second that the Earth (and other planets) were "once hot from surface to center." This contrasts with natural theories which look upon the never-wholly-molten planets as the result of slow accretional processes.

The principal proponents of accidental[3] theories of the origin of the Solar System were George-Louis Leclerc, Compte de Buffon; Forest Ray Moulton, Thomas Chrowder Chamberlain, Sir James Hopwood Jeans, and Harold Jeffreys. The theories they developed are basically *tidal theories* wherein a massive body approaches, or even collides with, the Sun, raising enormous tides of gases that break away and commence to circle the Sun. Gradually, these gases cool and condense, finally solidifying into planets and moons.

EARLY THEORIES

The first of the accidental theories appeared in the mid-18th Century from the pen of Buffon. In a work entitled "De la Formation des Planètes," published in Paris in 1745, he speculated that material might have been ripped off from the Sun following an encounter with a comet (Fig. 3.1).[4] Even if there were a comet large enough to cause tidal gases to leave the Sun, mathematical analysis shows that the solar gravitational shearing force would be too great to permit condensation to occur. The result would not be a system of planets but rather an envelope of uncondensed solar gases.

PLANETESIMAL HYPOTHESIS

By the beginning of the 19th Century the tidal theory, in several variations, reached the peak of popularity. The best known proponents of the theory were Chamberlain and Moulton, who developed what came to be known as the *planetesimal hypothesis,* illustrated in Fig. 3.2 They supposed that an alien star bypassed the Sun, raising great gaseous tides. Material was detached from the solar surface, some of which condensed into *planetesimals,* or small solid bodies. By a steady process of accretion the bodies slowly grew in size and eventually became the planets. The early orbits were probably quite elliptical but they became nearly circular due to the resistance

---

[2] Shapley, Harlow, "Of Stars and Men." Boston: Beacon Press, 1958, p. 68.

[3] Often called open or dualistic theories because they involve the Sun and at least one other celestial body.

[4] Rather curious ideas about comets circulated in the 18th Century; they were thought to be huge and massive and, hence, capable of creating havoc on any body with which they might collide, or even graze. Buffon referred to his hypothetical comet as the *comète fatale.*

Fig. 3.2. Planetesimals condensing after material has ripped off from the Sun due to tidal forces caused by intruding star and after this material has had time to consolidate.

caused by solar matter still filling the space between the protoplanets. This hypothesis explains the high angular momenta of the planets by supposing that the clouds of detached solar gases were accelerated sideways by the passing star.

This version of the tidal theory did not satisfy Jeffreys and Jeans, partly because they were not convinced that the planetesimals would long endure. They felt that they would collide as they were being built up and would be dissipated. Both preferred to believe that the bypassing star raised tides on the Sun and broke loose the outermost matter in the form of filaments that contained the material from which the planets were made (Fig. 3.3). They suggested that the filaments were more dense towards the center, explaining why the largest planets are found in the central region of the Solar System.

Fig. 3.3. Jeffreys-Jeans theory showing a large filament drawn off from the Sun by an intruding star.

According to the theory the primitive planets first tended to follow the intruding star, but its gravitational attraction was not sufficient to dislodge them from the solar field with the result that they were forced along a sideward direction into orbits around the convalescing Sun.[5] During the first circuit of the fragmented filament, pieces were broken off, giving rise to the many moons of the planets. Planetary and satellite orbits were subsequently circularized in the same way as in the Chamberlain and Moulton theory.

A number of difficulties became apparent. For one thing, the rapid rotations of Jupiter and Saturn could not be explained. So, Jeffreys modified his thinking and postulated that instead of bypassing the Sun the alien star sideswiped it. This might produce the vorticity necessary to explain a rapidly rotating planet, but neither it nor the original stellar bypass theory can face the dynamical fact that the planets are simply too far away from the Sun. Furthermore, the distribution of angular momentum in actuality (large outer and small inner planets) is opposite to what should have occurred as a result of a stellar encounter. And it is doubtful if the planets with their abundance of heavy elements could ever have been created from materials pulled out of the exterior regions of the Sun.

MULTIPLE STAR THEORIES

Other variations of the tidal theory suppose that the Sun was once a member of a binary or of a triple star system. Henry Norris Russell suggested that the Sun formerly had a companion star that was bypassed by an intruder. A filament was formed, giving rise to the planets, and at the same time the binary was ejected from the solar gravitational field and disappeared into space. Analysis of this situation, however, reveals serious difficulties, with the result that it has been rejected.

R. A. Lyttleton proposed, in 1941, that the Sun was accompanied not by one but by two other stars. These gradually approached one another as they accumulated interstellar material, finally coming together. Due to rotational instability they did not last long as one body, breaking up and giving rise to the material from which planets were created. Analysis shows that the triple star theory suffers from many of the same criticisms as the double star theory and, like it, has been discarded.

SUPERNOVA THEORY

The possibility has been investigated that the matter making up the planets resulted from the explosion of a supernova in orbit around the Sun. The

---

[5] Perhaps some of the material was dislodged and remained under the gravitational control of the intruder. And perhaps this material turned into planets. If this happened, and if the accidental origin theory for our particular solar system is correct, somewhere in the Milky Way there might be a companion planetary system to ours.

supernova theory may not be accidental, but it is catastrophic, and can properly be placed with the accidental group of theories.

According to the theory, when the supernova erupted, prodigious quantities of material were thrown into space, including adequate heavy elements from which the planets could be formed. Recoil forces ejected the supernova remnants from the gravitational field of the Sun and the planets gradually formed. (It is assumed that the supernova explosion was asymmetrical so that when it recoiled the star left the Sun's field.)

There are a number of criticisms of the theory, the most important having to do with the extremely high temperatures associated with the ejected material. Rather than condense into planets, it would more likely form a gaseous cloud around the Sun. Some astronomers consider the supernova theory highly implausible, but Lyttleton[6] cautions that it remains conjectural because "its processes are not very susceptible to reliable analytical investigation." Figure 3.4 is derived from this source. He points out that the supernova theory implies the existence of more planetary systems than the various bypass and collision theories because the chance of a supernova occurring is greater than that of stellar encounters.

So much for accidental origin theories. The best evidence available to astronomers today strongly suggests that the planets and satellites were not created by accidental events, but are rather the result of natural occurrences. This is not meant to imply that some solar systems in the universe may not have been caused by accident, but simply that the vast majority was not.[7]

Natural Origin Theories

The theories ascribing the origin of the Solar System to natural causes[8] are based on the supposition that the Sun and the planets were created of the same basic material at more or less the same time (astronomically speaking, give or take a half billion years). They do not require the incidental, fortuitous arrival from the depths of space of a stray star, nor the uncommon explosion of a supernova. Natural origin theories were once accepted by leading philosophers and naturalists, then later abandoned as the accidental theories held sway, and have recently come back into favor. Evidence now strongly supports some form of natural origin for the planets and moons of the Solar System.

---

[6] Lyttleton, R.A., *Man's View of the Universe.* Boston: Little, Brown and Co., 1961, p. 53.

[7] In the remote past stars may have been very much closer together than we now find them. If so, the possibility of stellar enounters would have been greater than at present. We do know that galaxies were once closer together.

[8] These theories are also known as closed or monistic theories because only the Sun is involved in the initial development of the Solar System.

Fig. 3.4. (a) Lyttleton's supernova theory. Here we see prenova in its binary system.

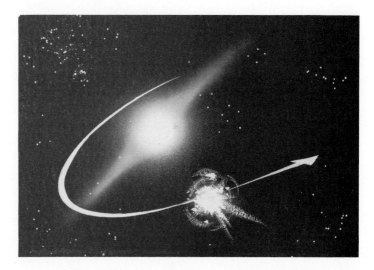

Fig. 3.4. (b) Star explodes.

Fig. 3.4. (c) Debris from explosion forms materials from which planets will be born.

Fig. 3.4.(d) Further development of debris.

Fig. 3.4. (e) Materials acrete.

Fig. 3.4. (f) The Solar System is finally formed.

DESCARTES' VORTEX THEORY

The oldest of the natural origin theories is René Descartes' *vortex theory,* promulgated in the "Principia Philosophiae," published in Amsterdam in 1644. It visualized the universe as being made up of matter (to which vortex motion was applied by God) and ether. Matter was accumulated to form the Sun at the center of the vortex. Finer matter would leave the vortex to become the heavens, coarser matter being held by the vortex and becoming the planets. Small vortices around the planets would later capture matter that would become the moons.

KANT'S NEBULAR HYPOTHESIS

Over a hundred years later Immanuel Kant developed his famed *nebular hypothesis* (Fig. 3.5) in the work "Allgemeine Naturgeschichte und Theorie

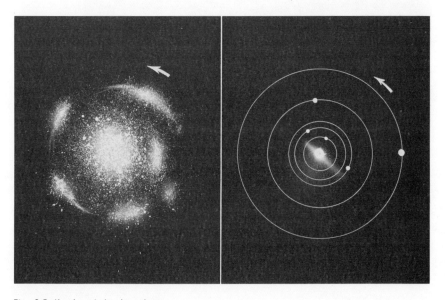

Fig. 3.5. Kant's nebular hypothesis.

des Himmels" (1755). He supposed that the Sun was a huge, rather cool, slowly rotating body of gas filling up the entire volume of what was to become the Solar System. As it cooled further it contracted and spun ever more rapidly, flattening considerably. Sooner or later the rotation and flattening reached the point where rings of gases spun off from the primitive Sun; these subsequently broke up and condensed into planets. Concentrations of nebular material around the protoplanets gave rise to the satellites.

Fig. 3.6. Laplace's nebular hypothesis.

LAPLACIAN NEBULAR HYPOTHESIS

The French mathematician Pierre-Simon, Marquis de Laplace expanded and improved upon Kant's nebular hypothesis in his book "Exposition du système du monde" (Paris, 1796). He assumed that an extensive nebula surrounded the Sun and rotated with it, as viewed in Fig. 3.6. As the nebula cooled and contracted it rotated ever more rapidly until finally centrifugal forces dominated gravitational forces and successive series of rings were shuffled off. These later condensed, forming the planets.

The major criticism of theories of this type is that centrifugal forces could not have been sufficient to overcome the gravitational forces; and, if they were, the so-called Laplacian rings would not have condensed into planets.

EDDY HYPOTHESIS

Many of the points presented by Descartes and Kant were given rebirth by the German astronomer Carl Friedrich von Weizsäcker who, in 1944, advanced his *eddy hypothesis*. He took a grand view of the universe and built up a theory that would account not only for the formation of the planets, but of the stars and galaxies. He speculated that huge clouds of dust and gases formed into eddies which sooner or later coalesced forming, on the largest scale, galaxies; and, on decreasing scales stars (subeddies) and planets (sub subeddies) and their moons. As nuclei formed they gradually enlarged, continually picking up particles by a fusing process. After a critical size had been reached a small but significant gravitational field was created which continued to add mass to the forming body, but at a much more rapid rate. The mechanism that operated to cause growth by a so-called *collisional accretion* process is not established, and this is one of the weaknesses of the theory, but it may involve vacuum welding of particles. Von Weizsäcker calculated that our planetary and satellite system took about 100 million years to form from the original clouds of dust.[9]

The largest mass forming from such clouds is the central star which, in the case of our Solar System, is the Sun. Revolving around it were immense clouds of condensable dust and noncondensable gases, the former consisting of iron oxides, various silicates, and liquid and solid water; and the latter of hydrogen and helium together with small quantities of other gases. The planets formed in orbits whose radii are approximately twice the radius of the next innermost planet's orbit.

These orbital relationships are both interesting and informative. In 1772 the German astronomer Johann Ehlert Bode formulated a simple relationship describing the relative distances of the planets from the Sun;[10] this is seen in Table 3.1. Note the similarity between the actual distances from the Sun in Column 1 and those predicted by the law in Column 5. In Column 6 we have added the ratio of the distance of the planet from the Sun to the distance from the Sun of the next innermost planet. None of these ratios is far removed from the number 2, so we can say that each orbital radius is roughly twice that of the next inner orbit. Table 3.2 demonstrates that a similar situation occurs for the orbits of the moons of the planet Saturn.

With these observational facts available, von Weizsäcker had to explain why many bodies were formed in the Solar System rather than a single large body. He observed that the original building materials must have been in elliptical orbits around the protosun; and, because of the incomprehensibly large number of individual particles involved, ceaseless collisions must have occurred. He assumes that individual particles were either pulverized or were bounced into less heavily populated lanes of movement. Gradually, a pattern similar to that shown in Fig. 3.7 evolved. Particles moving in Keplerian orbits in a given lane experienced little difficulty, but where one of the rings came into contact with another many collisions were bound to have occurred with the result that matter accumulated heavily at these positions. Thus, the planets were created at the boundary regions. The illustration shows the stable system envisioned by von Weizsäcker wherein material has collected in five vortices in each of a series of concentric rings around the Sun. Between the four successive concentric rings the three planets

---

[9] Whilst the dust accreted to form the planets and satellites, the associated gases little by little dissipated into interstellar space.

[10] This is known as Bode's law or, sometimes, the Bode-Titius law due to the fact that J. D. Titius had earlier mentioned the relationship in his writings.

Table 3.1
Relative Distances of the Planets from the Sun

Column	1	2	3	4	5	6
Name of Planet	Distance from Sun, Earth = 1	Series	+ 4	Sum (Col. 2+3)	Col. 4 divided by 10	Ratio of distance of planet from Sun to distance from Sun of next innermost planet
Mercury	0.387	0	4	4	0.4	
Venus	0.723	3	4	7	0.7	1.86
Earth	1.000	6	4	10	1.0	1.38
Mars	1.524	12	4	16	1.6	1.52
Asteroids	~2.7[a]	24	4	28	2.8	1.77
Jupiter	5.203	48	4	52	5.2	1.92
Saturn	9.539	96	4	100	10.0	1.83
Uranus	19.191	192	4	196	19.6	2.001
Neptune	30.07[b]	384	4	388	38.8	1.56
Pluto	39.52[b]	768	4	772	77.2	1.31

[a] average distance.
[b] the law is only slightly accurate in the case of Neptune and, for Pluto, breaks down altogether.

Table 3.2
Relative Distances of Satellites of Saturn[a]

Satellite	Distance relative to radius of Saturn	Ratio of increase in two successive distances
Mimas	3.11	
Enceladus	3.99	1.28
Tethys	4.94	1.24
Dione	6.33	1.28
Rhea	8.84	1.39
Titan	20.48	2.31
Hyperion	24.82	1.21
Japetus	59.68	2.40
Phoebe	216.8	3.63[b]

[a] Data from Gamow, George, *One, Two, Three ... Infinity*. New York: The Viking Press, Inc., revised edition, 1962.
[b] Phoebe, like the planet Pluto, deviates completely from the tendency of orbits to be twice as far out as the next inner orbit.

Fig. 3.7. von Weizäcker's eddy hypothesis.

Mercury, Venus, and Earth are shown forming. Von Weizsäcker selected five as the number of vortices in a given ring since by so doing it was possible for him to explain the relative distances of the planets from the Sun.

The theory is attractive from many points of view. It offers an explanation for why the inner planets are denser than the outer; higher temperatures in the inner primeval Solar System gave rise to the condensation of the heavier elements and compounds, and lower outer Solar System temperatures gave rise to the condensation of lighter elements and compounds. The theory also explains why planetary and satellite equators lie in nearly the same plane and why they not only revolve but rotate in the same direction as the Sun. It also explains the tendency of planetary orbits to be roughly twice as extended as the next inner orbit; and, more than incidentally, the tendency of satellite orbits to follow a similar pattern. Von Weizsäcker postulated that that material which did not concentrate in a protoplanet circled around it, eventually forming subplanets or, as we call them, satellites. At the same time the light gases were dissipating into interstellar space, the small amount of gases that still remained may partially cause the zodiacal light.

MODIFIED EDDY HYPOTHESIS

In papers published in 1948 and 1950, D. Ter Haar critically analyzed the Von Weizsäcker eddy theory, discarding portions of it and expanding and adding to other parts. For one thing, he did not accept the huge turbulent vortices. He theorized that a thick solar nebula would be the inevitable result of random turbulence and that planetary bodies would gradually form in it by accretion processes. Because the inner regions of the nebula would be hotter than the outer parts it would be expected that only metallic compounds would condense in the former with virtually all others condensing in the latter. A drawback to this type of thinking is that condensations would have to occur very rapidly before the nebula dissipated through turbulence.

Somewhat later Gerard P. Kuiper offered a modification of von Weizsäcker's ideas and, like Ter Haar, showed that large, regular vortices would not be possible. Indeed, he indicated there would be important gravitational instabilities in the primordial solar nebula.

The Sun, which had originally condensed out of interstellar matter, would continue to contract while attracting surrounding clouds of dust and gases. This material would eventually assume the form of a flattened disk and then break up into eddies. As these eddies revolved around the Sun they would come into contact, some disappearing and others growing ever larger. Those eddies that survived would ultimately become the planets.

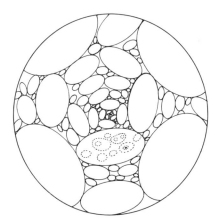

Fig. 3.8. Kuiper's modified eddy hypothesis.

Subeddies, then, would form the various moons. Material, principally noncondensable gases, that did not condense would be dissipated by radiated particles from the Sun. The theory is depicted by Fig. 3.8.

DUST CLOUD HYPOTHESIS

A few years after von Weizsäcker published his theory, Fred Whipple of Harvard College Observatory offered what came to be known as the *dust cloud hypothesis*. He assumed a vast cloud of dust and gas from $2\frac{1}{2}$ to 3 trillion miles in diameter in which the primitive Sun was slowly formed by contraction and collapse processes. Later, a much smaller dust cloud with high angular momentum was captured by the original large cloud which,

Fig. 3.9. Whipple's dust cloud hypothesis.

according to Whipple, must have had virtually no angular momentum. He made this assumption to explain the fact that the Sun, which was created from the cloud, has very little angular momentum.[11] Figure 3.9 suggests the hypothesis.

At first the smaller cloud condensed slowly. Then, it began to rotate and finally collapsed in upon itself. Solid particles came together and formed larger and larger masses which, in time, became the planets. The gases in the larger and in the smaller clouds collapsed and collected in the Sun. All of this took approximately 100 million years.

It is explained that planetary orbits are rather circular because of accretion action in the resisting interplanetary dust medium. Explanations also are given for other phenomena but, unfortunately, many can not be supported by quantitative data. Consequently, this hypothesis is not accepted as a total way of understanding the origin of the Solar System, though a number of its considerations are probably valid.

The hypothesis is particularly interesting because it assumes that the Earth was never gaseous and that it was never part of the Sun. Indeed, it suggests that our planet may have been completed prior to the time the Sun was fully created.[12]

METEORITIC EVIDENCE

In 1951 and 1952, Harold C. Urey investigated the problem of planetary origin and introduced new ideas based heavily on chemistry. He was particu-

---

[11] What angular momentum it does possess was caused, according to the theory, by its capture of a few small dust clouds which did not become the planets.

[12] Since he originally advocated the dust cloud theory in the late 1940s, Whipple has modified his ideas considerably. See, for example, his "The History of the Solar System," *Proceedings of the National Academy of Sciences,* 52, No. 2 (August 1964), 565–594.

larly interested in accounting for the presence and nature of meteorites, which he feels are the collision remains of larger asteroidal bodies, and possibly also of pieces dislodged from the Moon. Since meteorites are collisional debris it is assumed that they provide clues to the internal nature of former worlds and, in fact, give us the only such knowledge available. In all likelihood at the same time we are learning about conditions as they were during the early stages of the creation of the Solar System. Because they give us a sample of planetary material in its "raw" form and because they probably have not been geochemically active almost since they came into being, Urey emphasizes that no theory of Solar System origin can ignore them.

Based on chemical and physical evidence from meteorites, together with more traditional data, Urey suggests that as condensation of the original solar nebula took place, regions of gravitational instability appeared. Slowly, temperatures rose, causing the heavier constituents to melt and collect at the center of the forming bodies. As they grew, encounters occurred and gases dissipated, particularly from the much hotter worlds of the inner Solar System. While the lighter gases escaped, the heavier elements remained as the building materials for these planets. In the cooler outer Solar System the lighter elements were retained, explaining the present composition and low density of the giant worlds.

MAJOR PROBLEM AREAS

Thomas Gold[13] cites three major problems affecting the more current natural origin theories that require careful study and eventual solution. They involve accretion mechanisms, angular momenta, and hydrogen disposal.

First, the question must be resolved as to how the solidified particles present in the primordial solar nebula were able to accrete. In particular, it must be learned how large and small particles and pieces of matter were able to adhere to one another.

Second, it remains to be determined how angular momentum was transferred from the Sun to the planets. Presumably magnetohydrodynamic forces were involved, but we want to know how the momentum was transferred "as far as necessary by the magnetic field anchored in a body as small as the sun."

Finally, explanations still must be found for how the extra hydrogen in the Sun was able to escape and end up in the giant worlds of the outer Solar System.

In regard to the first problem many suggestions have been made. Gravitational instabilities may have brought the primeval dust together, with the

---

[13] Gold, Thomas, "Problems Requiring Solution," in *Origin of the Solar System, Op. Cit,* pp. 171–74.

actual adhering of matter being assisted by the presence of ice and of magnetic iron particles. These processes could have been responsible for the agglomeration of matter into masses large enough to be gravitationally active.

As for the magnetohydrodynamic effects, Gold writes[14] "...because of the kind of magnetic behavior of the sun (we could think) in terms of individual magnetic clouds strung through the pre-existing outer material, and the angular momentum transported out because individual tongues have pushed their way through all this."

As far as removing hydrogen is concerned, particle fluxes and bursts of radiant energy are suggested as possible mechanisms.

Existence of Extrasolar Planetary Systems

In our discussion of the origin of the planets we have been thinking primarily of *our* Solar System and not of solar systems in general. We have used the rapidly increasing direct knowledge of the observable regularities of the Solar System and coupled it to intelligent appraisals of primeval conditions in our region of space. Our speculations regarding these conditions are based partly on extrapolations backwards from the way things are today and partly on analyses of meteorites. Now, we wish to examine whatever direct or indirect evidence may be available tending to demonstrate the existence of extrasolar planetary systems.

To set the pattern for our discussion, it is first noted that there is good, but indirect, evidence that extrasolar planets do occur. Part of the evidence is based on observations of irregularities of the orbits of binary and multiple star systems and part on inferences stemming from cosmogonic ponderings. This evidence is considered in the following pages.

From our reflections of the various natural origin theories we learned that the planets probably formed in close association with the development of the Sun. It is logical, therefore, to (1) examine other stars in the universe that are similar to the Sun as it is today, and (2) attempt to locate stars that are only now beginning to form and hence may be much like the Sun was billions of years ago.

OCCURRENCE OF SOLAR-TYPE STARS

There is an almost infinite number of stars like the Sun, a fact we recall from Chapter 2. The Sun, like most other stars in the Milky Way, is a member of the main sequence, which is an arrangement of stars based on decreasing stellar masses and consequent luminosities and temperatures (Fig. 2.14).

---

[14] Gold, Thomas *Op. Cit.*, p. 172.

The Origin of Planetary Systems   67

Our Sun is about midway down the sequence. Very few stars do not belong to the main sequence, probably because nonmain sequence stars are relatively unstable. As young stars evolve they move directly onto the main sequence where, depending on their size, they may remain for billions of years. During the final stages of a star's life, when the hydrogen fuel is exhausted, the star moves off the main sequence and into its terminal evolutionary cycle. Some stars never enter the main sequence for reasons that are poorly comprehended.

Of the countless millions of main sequence stars, many are quite different from the Sun and many others are very similar to it. Among these latter is concentrated our search for extrasolar planetary systems, though it should be recognized that a star does *not* have to be like the Sun to possess planets, including hospitable planets on which life may emerge. An Earth-like planet around a hotter star would have to be further away to enjoy temperate climatic conditions. If the star is cooler, the converse situation would be required. It is possible that acceptable conditions could be found in some binary star systems whose components are widely separated.

An important solar characteristic is the surprisingly slow equatorial rotation speed. From studies of stellar evolution it can be shown that as a star contracts following its formation the rotation rate must increase. But astronomical observations indicate that many Sun-like stars of spectral types F_5-K_5 rotate very slowly, a fact for which account needs to be taken. Perhaps the angular momenta have left these stars much like angular momentum has been removed from the Sun.

We see from Fig. 3.10 that the rotational velocities of B_0 to F_5 type stars decrease steadily, breaking sharply at F_5. The cooler F_5 to K_5 and

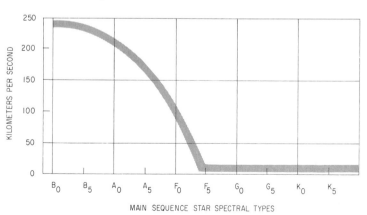

Fig. 3.10. Decrease of rotational velocity with spectral type for main sequence stars.

also M-type stars exhibit very slow spin rates and, sometimes, no rotation at all.

As pointed out by Ordway, *et al*,[15] the "answer may be found by investigating the dynamics of binary and multiple star systems. We know that at least 40 per cent of the stars in our region of the Milky Way occur as doubles or multiples." It may well be that as a star spins more and more rapidly on its axis the time arrives when its splits into two or more parts, forming a binary or multiple star system, the angular momentum of the main star being transferred to the orbital motion of the new star or stars.

From here it is only a step to the hypothesis that in some systems the angular momentum may be shuffled off to protoplanets rather than, or in addition to, companion stars. Indirect astronomical observations come to our aid at this point. By analyzing the orbits of visual members of multiple star systems astronomers can show that there are unseen perturbing bodies which may be small stars or large planets. Smaller planets, if any, would remain undetected because their influence on the motion of parent stars would be too minute for us to observe.

POSSIBLE EXTRASOLAR STARS WITH PLANETS

Several examples are noted. A third companion is known to exist in the binary star system 70 Ophiuchi; it has some 10 to 12 times the mass of the planet Jupiter. A similar situation occurs in the system 61 Cygni (Fig. 3.11) except that the companion is 15 to 16 times more massive than Jupiter. And the red dwarf Lalande 21185 probably has a member of large planet size. The smallest planet, and the most recently confirmed (1963), was discovered by Peter van de Kamp to orbit around Barnard's star. It was found by examining thousands of photographic plates and noting tiny wobbles

---

[15] Ordway, Frederick I., III, *et al.*, *Basic Astronautics*. Chap. 6 "Astrobiology," Sec. 6.6 on "Life outside the Solar System," Englewood Cliffs, N. J.: Prentice-Hall, 1962, p. 286.

Fig. 3.11. The star 61 Cygni. Courtesy Mount Wilson and Palomar Observatories, Dr. Armin J. Deutsch. Copyright National Geographic Society—Palomar Observatory Sky Survey.

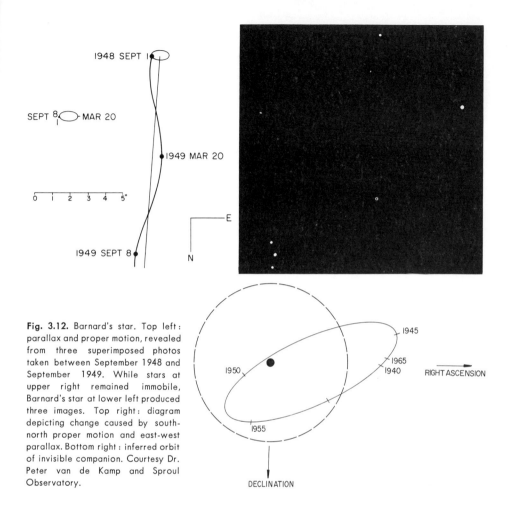

Fig. 3.12. Barnard's star. Top left: parallax and proper motion, revealed from three superimposed photos taken between September 1948 and September 1949. While stars at upper right remained immobile, Barnard's star at lower left produced three images. Top right: diagram depicting change caused by south-north proper motion and east-west parallax. Bottom right: inferred orbit of invisible companion. Courtesy Dr. Peter van de Kamp and Sproul Observatory.

thereon of the parent star. It is concluded[16] that the invisible planet "is moving in an elliptical orbit (eccentricity 0.6), which is inclined 77 deg to the plane of the sky." The mass of the planet is estimated to be 0.0015 of the Sun, meaning that it is "a mere $1\frac{1}{2}$ times as massive as Jupiter." Figure 3.12 shows the inferred orbit of the planet.

We may be able to say that solar-type stars with low observable angular momenta expressed by slow axial rotation are excellent candidates to be possessors of planets. And, of course, binary or multiple stars may also contain planets the orbital elements of which might be highly complex. As Ordway and his coauthors sum up,[17] "...it may be correct to conclude

---

[16] Van de Kamp, Peter, "Barnard's Star as an Astrometric Binary," *Sky and Telescope*, 26, No. 1 (July 1963), p. 8 See also *Astronomical Journal*, 68, No. 5 (June 1963), p. 295.

[17] Ordway, F. I. III, *Op. Cit.*, p. 287.

that binary and multiple star systems and planetary systems were formed in much the same manner: in the first case we find two or more suns revolving around a common gravitational center, while in the latter case we have a family of planets revolving around a central sun. The main distinction seems to be one of size." A supporting fact is that the average binary star system is separated by a distance of roughly two billion miles, comparable with the distance separating the Sun and Neptune and half the distance between the Sun and Pluto.

There is no evidence of planets in the Alpha Centauri multiple star system closest to the Sun. However, as Luyten[18] writes, if it had "a planet exactly like Jupiter revolving around it at about the same distance as Jupiter from our Sun, it would oscillate in an orbit with a radius of about one three-hundredth of a second of arc. With present equipment this would be very difficult to observe although not entirely impossible."

YOUNG STARS

We now discuss briefly the second factor mentioned at the beginning of this section, namely the existence in today's universe of very young stars. According to Spitzer[19] there is abundant evidence that stars not only have formed in our galaxy from ten to a hundred million years ago but are being created today. He notes three principal lines of evidence:

(1) Expansion is indicated in certain associations of stars. Recessional velocities of individual stars are such that we are led to believe that the associations formed only a few million years ago.

(2) Very bright, highly luminous, early spectral type stars are observed; they are thought to be only a few million years old.

(3) Bright stars of the type mentioned above are associated with relatively dense clouds of dust and gas. Where there are no dust and gas clouds there are very few or no bright, hot stars of recent birth.

He writes that this is "conclusive evidence that stars are being formed at the present time in our Galaxy and probably in other spiral galaxies." He recognizes five stages in the development of a star, the first being the formation of a dust and gas cloud, its gradual growth and subsequent contraction. Next, the cloud collapses and fragments, with groupings of protostars being formed. After the fragmentation stage is completed temperatures build up and individual protostars commence collapsing. The fourth stage involves the gravitation collapse of the protostar as a result of dissociation

---

[18] Luyten, W. J., "Stars Near the Sun," *Discovery,* 26, No. 9 (September 1965), p. 10.

[19] Spitzer, Lyman Jr., "Star Formation" in *Origin of the Solar System,* p. 39.

Fig. 3.13. Stars being created in dust and gas cloud. Right: Orion nebula where stars are observed condensing. Courtesy Harvard College Observatory.

and ionization of hydrogen. When temperatures reach some 100,000°K and all the hydrogen and helium have been ionized the fifth stage begins. The very extended atmosphere star, perhaps with ten to fifty times the radius of its later, more stable state, begins to contract until nuclear energy generation commences at the center and it enters into its normal evolutionary cycle, leading into and through the main sequence. Figure 3.13 illustrates possible star formation in a cloud of dust and gas.

Outside the Milky Way the closest example of star formation is in the Large Magellanic Cloud; there, in an association of stars in the region of NGC 1936 about 200 parsecs in diameter, six very luminous and very young supergiants appear among a total stellar mass in the region equal to some

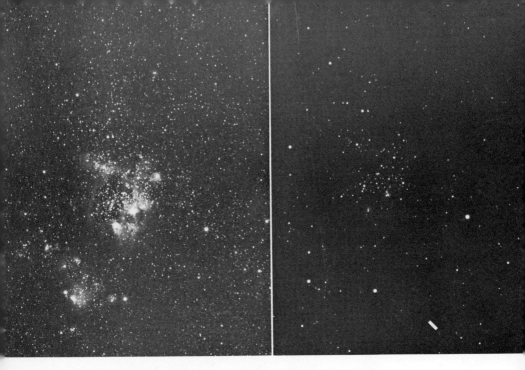

Fig. 3.14. Possible occurrence of star formation in association with the Large Magellanic Cloud. The photo at left was taken in blue light and the one to the right in yellow light. Courtesy Mount Stromlo Observatory and Bart J. Bok.

24,000 suns (Fig. 3.14). According to Faulkner[20] "...there is sufficient gas to form several hundred generations of stars at the present rate of star formation...it seems likely that the association is an expanding one, with newly formed stars dispersing into the general field before they have time to evolve."

CONCLUSIONS

Like everything in the universe, planetary systems have origins and evolutionary cycles. No one is certain as to how the only planetary system we definitely know about, the Solar System, came into being, but as more and more observational data become available and more and more scientific brains are applied to the problem, astronomers believe they are getting closer to the answer.

We learned that our Solar System probably was created as a result of a natural event and that it is likely that such events have been, and are being, repeated elsewhere in the Milky Way and in other galaxies. It is difficult to estimate how many solar systems may exist in our galaxy, let alone in other galaxies, but there are probably billions if not hundreds of

---

[20] Faulkner, D. J., "Current Studies of the Magellanic Clouds," *Sky and Telescope,* 26, No. 2 (August 1963), p. 73.

billions in the known universe. Shapley estimates[21] that there are some 100,000 stars like the Sun out of the million brightest stars in the Milky Way alone, so we can appreciate the fact that our type of star is by no means rare.

We are interested primarily in the possibility of the emergence of extrasolar intelligence on other planets in the universe, particularly elsewhere in the Milky Way galaxy. We must therefore investigate not only whether there are extrasolar planetary systems and where they may be located, but whether they contain planets whose environments permit the emergence of life. In our imaginary journeys to alien solar systems we may expect to find worlds on which no life is ever possible, on which life is not now possible but some day may be, on which life is just beginning, on which life is well advanced, on which life is at the peak of its development, and on which biological life at least is declining and perhaps extinct.

---

[21] Shapley, Hartow, *Op. Cit.*, p. 69.

4
Planetary Environments and the Development of Life

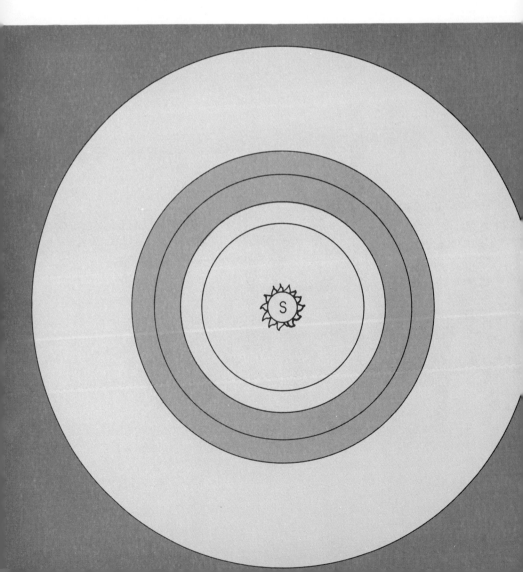

Biologists and other scientists interested in the mysteries surrounding the origin of life unfortunately must deal solely with life that now exists, or has existed in the past, on the planet Earth. Their base line for speculations is, therefore, severely limited, but is one that will definitely be broadened as biological research begins to be undertaken on other worlds in the Solar System.

Under the restrictions imposed upon us by the lack of extraterrestrial observational data, it is nevertheless possible to consider the origin and development of life on Earth and assess the possibilities of its origin and development elsewhere. This chapter is a connective link between Chapter 3, which dealt with the origin of planetary systems, and Chapter 5, which will examine the origin of life.

In succession, we shall look at planetary environments in general, examining first the limiting conditions imposed on terrestrial life by temperature, light, atmospheric conditions, radiations, and other factors. We shall review environmental conditions expected on the planets and see how they may influence the appearance of life. Recognizing that only primitive life is likely to be found in our

"However inconsiderable they may be in the history of sidereal bodies, however accidental their coming into existence, the planets are finally nothing less than the key points of the Universe. It is through them that the axis of life now passes; it is upon them that the energies of an Evolution principally concerned with the building of large molecules is now concentrated."
Pierre Teilhard de Chardin

"... you must go a great way to prove that the Earth may be a *Planet*, the *Planets* so many *Earths*, and all the *Stars* Worlds."
Bernard de Fontenelle

own planetary system, we shall stretch our thoughts beyond the confines of the Sun's influence and probe out to the stars. We shall consider a number of possible extrasolar planetary environments, and ponder the probability that life may emerge in alien solar systems.

The material covered in this chapter paves the way for an understanding of how the long process of chemical evolution here on Earth (and probably on Mars, and almost definitely on many extrasolar planets) merged into, or gave rise to, biological evolution. And how biological evolution in turn brought forth, on Earth at least, what we confidently call intelligent life.

Importance of Planetary Environments to the Development of Life

Biologists have long realized the controlling importance of climatic and other environmental conditions on the development, indeed on the very nature, of life on our planet. The widely varying conditions here on Earth today have a strong influence on the kind of plant and animal life that is found at a given geographical locality. From studies of paleontology we know that the kind of life inhabiting a given part of the world at one time in geological history may be quite different from that present a million, or ten million, or a hundred million, years later. What today is a desert may tomorrow be a lush jungle and some time in the far future the bottom of a sea or the top of a lofty mountain.

Despite the variety and changeability of life on the planet Earth, life nevertheless has evolved within environmental conditions whose limits are set by the size and age of our world, its axial tilt and rotation speed, its orbital parameters, its physical makeup, the nature of its atmosphere, the nature and distance of the star around which it circles, the strength of the magnetic field, the intensity of radiations and energetic particle fluxes, and last but far from least, time. The element of chance must also be added; just how heavily it should be weighed cannot be determined.

We are quite certain that the only planet in the Solar System on which advanced life now exists is Earth. Mars conceivably may have harbored, in some remote past, a higher life than it now supports, but there is no reason whatsoever to believe that intelligent life ever evolved on it or on any other nonterrestrial world in the Solar System. It is impossible to assert that the Solar System has never been visited by intelligent extrasolar beings. There is no evidence of it on Earth that stands up to careful investigation (see Chapter 15); but such evidence may later be found here, on the Moon, Mars, or perhaps on the moons of the outer planets.[1]

---

[1] Explorers in the future may look for the remains of bases or artifacts on, for example, the larger moons of Jupiter, Saturn, Neptune, or Uranus. There is a possi-

Biologically Acceptable Conditions in the Solar System

Generally defined environmental limits exist within which biological processes can take place and beyond which life ceases to develop or exist at all. These limits are different for different types of life; nevertheless, all life as we know it was created within, and thrives within, a zone of conditions that is conducive to living organisms (Fig. 4.1).

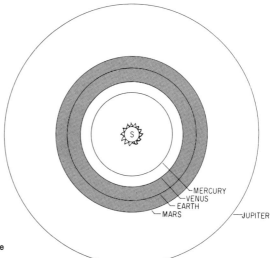

Fig. 4.1. Biologically acceptable zone in the Solar System.

The effect of some environmental conditions on life is rather simple to examine; but, other conditions seem to have some effects which cannot be easily measured. For example, scientists are not agreed on what influence the Earth's magnetic field has on life processes, or how different life might be if the gravitational field were 10 per cent more, or less, powerful than it is. We do not know what would happen if radiation intensities and energetic particle fluxes increased or decreased slightly over a long period of time, or if the Earth's rotation rate were somewhat altered. But, there are some key factors which have overriding influence on life as we know it; these are described below.

---

bility that the Solar System has been visited during the last several billion years. Perhaps it is visited periodically, say once every million years. Interstellar rovers, biological or mechanical, may penetrate only to the outer planetary regions, making observations from any of the many moons located there. In this case, we would expect to find rather substantial evidence of visits of extrasolar beings.

ATMOSPHERIC COMPOSITION

As we shall see in Chapter 5, life on Earth probably got started when the atmosphere was quite different than it is today. Most geophysicists and geochemists feel that it was at one time a reducing rather than an oxidizing atmosphere, containing principally ammonia, methane, water vapor, and possibly some carbon dioxide together with free hydrogen. In fact, it was quite similar to the atmospheres of the outer planets as they are today.

The photodecomposition action of the relatively nearby Sun gradually affected (decomposed) the terrestrial atmosphere (largely consisting of methane and ammonia) with the result that more and more hydrogen escaped from the Earth's gravitational field and dissipated into space. While this was going on, life appeared on the planet. Due to outgassing, oxygen, nitrogen, and other elements became ever more prevalent until, after millions of years, the atmosphere became oxidized.[2] With very few exceptions, life as we know it is dependent on adequate supplies of oxygen.

The two other worlds in the Solar System with oxidized atmospheres are Venus and Mars. The former appears heavily oxidized, carbon dioxide being the responsible component. The atmosphere of Mars is lightly oxidized with very little free oxygen. Whether some form of life could adapt to Venus's carbon-dioxide-containing atmosphere is not known; however, very high temperatures on the planet probably rule out life on the surface, though it is conceivable that aerial forms may exist in the upper atmospheric layers. There is probably sufficient oxygen on Mars to support simple vegetation.

It is logical to ask whether or not life can exist in hydrogen-rich atmospheres, recalling that evidence leads us to believe that terrestrial life got its start under reducing conditions. While light hydrogen gases have long since escaped from the small inner planets, the giant outer planets are still able to retain them in their dense atmospheres. It is speculated that primitive, prebiotic molecules may occur on Jupiter, the nearest to the Sun of the outer planets, and, if the greenhouse effect makes the surface warmer than we believe, primitive life, or at least prelife, may occur. All this is most unlikely, but it must not be ruled out.

Most investigators assume that Mercury is devoid of life, and probably always has been. Nevertheless, it has been suggested that life could have evolved in the twilight zone between the eternally light and the eternally dark sides, a zone where temperatures are moderate and where a tenuous atmosphere may occur. But the twilight zone is not always in the same

---

[2] Originally, most of the oxygen was in the form of water vapor which, under the photodecomposition of solar ultraviolet rays, became free oxygen and hydrogen. Once life had evolved additional quantities of oxygen entered the atmosphere as a result of photosynthetic reactions of primitive plants.

geographic position due to a wobbling effect produced by the planet's elliptical orbit and constant rotation speed. This results in cyclic changes in temperature, leading Cole to speculate[3] that migratory life forms may have evolved, following the Sun "in such a way that only the optimum percentage of the solar disk is visible...." He notes that the terminator velocity is about 1.5 mph, making it possible for life forms familiar to us to maintain the appropriate migration velocity. He assumes that "Mercurian life would have to be of the encapsulated, closed-cycle, symbiotic form rather than the open-cycle Earth form."

LIQUID WATER AND TEMPERATURE

Leaving the question of atmospheric composition, we now look at the all-important factor of water, whose existance in the liquid state (so necessary to life as we know it) depends principally on temperature. Biochemists find it difficult to imagine carbon-containing molecules that do not require the presence of water as a solvent. As temperatures rise towards the boiling point of water, life, if it survives at all, will not progress since biochemical reactions and molecular build-ups become destructive. And, as the freezing point of water is approached and passed, life can get along at best with difficulty and at worst will perish. Ordway, et al,[4] emphasize the point that if there is no water the chemical reactions characteristic of terrestrial life "cannot occur, at least as far as we know. This limitation on life is severe and rules out much of the Solar System as a biological abode." In fact, considering temperature and other conditions, about the only place we might expect to find liquid water is on Mars in the equatorial zones during the summer. So far, astronomical evidence indicates only a very small amount of water vapor in the atmosphere of the planet which, during the winter, shows up on the surface as the polar caps (total condensable water in a column through the atmosphere is about 14 to 15 microns, or 0.00055 inches; this can create a polar cap less than an inch thick). Water vapor has been discovered in the upper atmosphere of Venus, but it assuredly could not exist in the liquid state on the cruelly hot surface.

SOLAR RADIANCE

We now come to solar radiance and its importance to life. It sometimes escapes our attention that the biological process of photosynthesis was created under, evolved under, and today functions within conditions that

---

[3] Cole, Danridge M., "Mercury Missions Seen More Useful than Mars, Venus Probes," *Missiles and Rockets*, 12, No. 25 (24 June 1963), 37.

[4] Ordway, Frederick I. III, James Patrick Gardner, and Mitchell R. Sharpe, Jr., *Basic Astronautics*. Englewood Cliffs, N. J.: Prentice-Hall, 1962, p. 256.

to a great extent depend on the amount of solar radiation reaching the Earth from the 93-million-mile distant Sun. It is estimated that the biologically acceptable zone in terms of solar radiance is between 4 and 0.5 cal/cm²-min. This corresponds to illumination values between 270,000 and 40,000 lux. Above the Earth's atmosphere the value is 2 cal/cm²-min (140,000 lux) and at the surface 1.4 cal/cm²-min (maximum).

In general, a world far from the Sun will receive less solar radiance than a nearer body. However, other factors enter, such as the nature of the atmosphere, the nature of the surface, and the extent of cloud cover. With these considerations in mind, we should try to determine what other worlds in the Solar System would be acceptable to life. Again, Mars comes out ahead and, while light intensity at that planet's distance from the Sun is but half that on Earth, it is sufficient to permit photosynthetic organisms to evolve. It should be noted that even though Mars is further from the Sun than Earth, and thus receives much less solar energy, what it does receive is absorbed more effectively due to the lower Martian albedo. In fact, both planets enjoy about the same net amount of solar radiation.

RESISTANCE OF TERRESTRIAL LIFE TO NONTERRESTRIAL CONDITIONS

Now that we have reviewed some of the environmental conditions that are apparently conducive to life as we know it, we come to the question of how such life endures under nonterrestrial conditions. In order to determine how resistant life is and what limits are imposed by nature on its mere survival, tests can be made in laboratories on Earth and in space itself (capsules containing life are fired into space and later recovered).

The space environment is characterized by the hard-vacuum condition, cosmic radiations and solar flare emissions, all prejudicial to organic matter. Many spore studies have been made in satellites to observe signs of mutations following differing periods of flight through space. Unicellular green algae, recovered from unmanned artificial satellites (Fig. 4.2), have been analyzed for their ability to continue to photosynthesize and for their genetic stability. Tests of egg fertilization have been made in manned Gemini satellites. Investigations in the laboratory are often conducted to determine how terrestrial microbial populations exist under simulated Mars conditions. Experiments at the Ames Research Center of the NASA have revealed that some types of anerobic bacteria can thrive in a simulated Martian atmosphere and general environment, including a temperature range of $-75°C$ to $+25°C$ from night to day.

A cactus, Cereus, in other experiments has been kept in a partially simulated Martian environment for 300 days; Fig. 4.3a shows that, although

Planetary Environments and the Development of Life

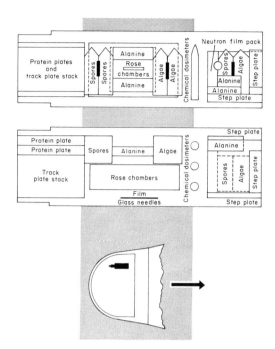

Fig. 4.2. Biopack flown on Discoverer 17 satellite, illustrating top and side and its location in the capsule. The unicellular green algae, upon recovery, was tested for its viability, ability to grow and photosynthesize, and genetic stability. Courtesy Dr. Allan A. Katzberg.

injured, it is still erect and living. In Fig. 4.3b we can see cucumber seedlings grown in a 2 per cent oxygen atmosphere that can survive 15° F temperatures, whereas under normal conditions they would be killed near the freezing point of water. The two plants in the center of Fig. 4.3c were exposed to a simulated Martian environment for a month. Both died, but they produced vigorous young shoots that were able to adapt to the new environment. Conditions in the simulator[5] were normally based on an atmosphere of 97 per cent nitrogen, less than 1 per cent oxygen, and almost 3 per cent carbon dioxide. Fungi such as the Penicillium species seen in Fig. 4.3d were found to grow slowly in an atmosphere of ammonia and methane, which is closer to that of Jupiter than of Mars.

The results from these and other experiments give biologists greater insights into terrestrial life and, indirectly, indications of whether or not we may expect to find life elsewhere in the Solar System.

Planetary Environments in Extrasolar Systems

If we accept the conclusions of Chapter 3, namely that planetary systems occur in large numbers throughout the Milky Way galaxy and

---

[5] Operated at the Union Carbide Research Institute, Tarrytown, New York under the direction of Dr. Sanford M. Siegel.

Fig. 4.3a. A cactus Cereus has been kept in partially simulated Martian conditions for as long as 300 days. After such treatment the plant is injured, but stands erect and alive. Courtesy Union Carbide Research Institute, Tarrytown, New York. (a-d).

Fig. 4.3b. Cucumber seedlings which have been grown in a 2% oxygen atmosphere survive a temperature of about 15°F, whereas those grown in air are killed near the freezing point, 32°F.

exterior galaxies, it becomes far more than idle speculation to wonder about the life that almost inevitably has sprung up on many of them. In our own Solar System, one planet has provided conditions suitable for the emergence of intelligent life, and at least one other probably harbors primitive life. We believe that among the billions of other solar systems scattered throughout the known universe, some at least offer environments as attractive, or more attractive, to life as our own. Assuredly, we have no reason whatsoever to believe our planet to be somehow unique; it cannot be too strongly stressed that "uniqueness is not a characteristic of nature: life is composed of the same elements found in the Earth, the Sun, the stars, and the material between the stars. Proportions may vary, but the elements are the same."[6]

Many parameters influence the development and nature of living organisms on our planet, including size and age of the Earth, its mass, rate of rotation, surface gravity, magnetic field, axial inclination, distance

---

[6] Ordway, F. I., III, *Op. Cit.*, p. 288.

Fig. 4.3c. The two plants in the center have been exposed to a simulated Martian environment for approximately one month at the Union Carbide Research Institute. Although both plants died as a result of the extremes in temperature presumed to take place in a Martian day, vigorous young shoots flourished from the dying mature plants and were able to adapt to such stress environments.

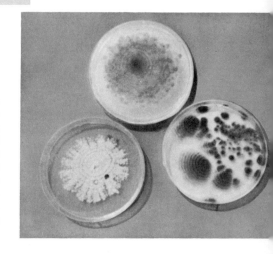

Fig. 4.3d. Experiments have shown that molds and bacteria flourish in the Mars simulator, but this is by no means the only "hostile" environment in which terrestrial microorganisms can grow. Common fungi, such as Penicillium species can grow slowly in atmospheres consisting of 50% ammonia and 50% methane, a composition more like Jupiter than Earth or Mars.

from the Sun, eccentricity of orbit, radioactivity, relative amounts of hard and liquid surface, nature and composition of both, nature of atmosphere (including primary constituents, stability, percentage content of dust, wind patterns, and the like), characteristics and distance of the Moon, and, of course, chance. It is possible to determine, within broad limits, the many constraints on life as we know it—but not, unfortunately, on life as we do not know it. We can say, for example, that if our Sun were 1.4 times as massive as it is, it would have been unable to remain on the main sequence for more than roughly 3 billion years, insufficient time for intelligent life—though not primitive life—to evolve. Or, to take another example, if the Earth's period of rotation were greater than 3 or 4 days, heating and cooling extremes would probably be such as to discourage the appearance, or at least the persistence, of life (atmospheric circulation factors could modify these figures to some extent).

Figures 4.4 through 4.9 illustrate some possible planetary system arrangements that could occur elsewhere in the Milky Way—or in virtually any other galaxy. In the first illustration, Fig. 4.4, we find six possible

84   Planetary Environments and the Development of Life

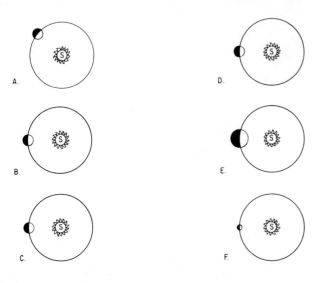

Fig. 4.4. Six possible planets revolving around a Sun-like star at a distance of approximately 100 million miles (see text).

planets revolving at a distance of approximately 100 million miles around a stable, Sun-like, main sequence star that has poured forth an even flow of light and energy for more than 5 billion years. Drawing (A) shows an Earth-size planet whose axial tilt is so high (above, say, 80 deg) that severe seasonal temperature extremes are experienced. The influence of changing inclinations on habitability are not clear, but as they increase less and less of the planet's surface can support life as we know it. In diagram (B) the axial inclination is acceptable, but this hypothetical planet is water-covered except for a few small land areas consisting principally of high mountains. Advanced land life has not had the opportunity to emerge, and we can only muse over the possibility of the appearance of intelligent aquatic forms. Until some empirical evidence becomes available, we must remain pessimistic over the possibilities of civilization on such a planet and pass on to other worlds.

Situations (C) and (D) in Fig. 4.4 show Earth-like worlds on which life has not yet appeared but on which it is expected to some day occur; and on which an ancient, highly advanced biological life once flourished but has since disappeared. This latter case assumes the exhaustion of natural resources on the worn out planet and the migration of the race to other worlds.

Diagram (E) depicts a planet too large and too massive to possess an oxidized atmosphere, i.e., upwards from 2 to 2.5 Earth masses. High masses are associated with increasing escape velocities and decreasing abilities of atmospheres to evolve into oxidized forms. Surface-atmospheric interfaces on massive planets would also be quite different when compared with Earth-like worlds.

In the next diagram, (F), the opposite situation occurs: we have a world of low mass, less than 0.4 Earth mass—and, consequently, one whose radius is less than 0.8 that of Earth and whose surface gravity is less than 0.7 Earth's. A world of this type could not retain a breathable, oxidized atmosphere of sufficient surface pressure to support the vast majority of known living organisms.

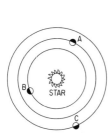

Fig. 4.5. Earth-size planets too close to star.

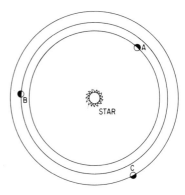

Fig. 4.6. Earth-size planets too far from star.

Other possible solar system arrangements are illustrated in Figs. 4.5 and 4.6. In the former, we find a few planets of acceptable size all too close to the parent star; and, in the latter, they are overly far away. In such systems, we would not expect to find advanced life, at least as we understand it, although the planets could be attractive homes of migrating mechanical automata due to their rich supplies of natural resources or their particular location within the galaxy. Assuming that the biologically acceptable zone around a Sun-like star ranges from roughly $\frac{3}{4}$ to $1\frac{1}{4}$ astronomical units, the inner and outer limits of habitable planets can be approximately defined.

The three planets in Fig. 4.7 are within this zone and would be likely candidates for life as we know it were it not for the fact that the central

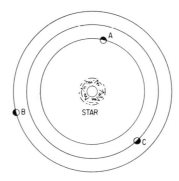

Fig. 4.7. Pulsating central sun prevents life from occurring on its planets.

86   Planetary Environments and the Development of Life

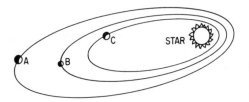

Fig. 4.8. Excessively elliptical orbits prohibit appearance of life on the planets of this system.

star is unstable. Its energy output is erratic; thus, the basic requirement of a habitable planet, of steady, long-term flows of light and energy, would not be met in this system and life would be most unlikely to appear.

Turning now to another parameter, that concerned with orbits, we see (in Fig. 4.8) a series of planets following highly elliptical paths around their primary. Very high temperatures at perihelion and extreme cold at aphelion would discourage life as we can conceive it, even though all other parameters were acceptable. Periods of revolution of a planet will have an effect on just how great an orbital eccentricity can be tolerated; but, in general, eccentricities over 0.2 would discourage forms of life such as have evolved on the Earth. To compare this with values in the inner Solar System, the eccentricity of the orbits of Venus, Earth, and Mars are, respectively, 0.0068, 0.0167, and 0.0934.

Finally, in Fig. 4.9, we see a double star (binary) system, each with Earth-size planets revolving around what would normally be acceptable distances from their primaries. The orbits shown are schematic only; in any given *real* case they would often, though certainly not always, be complex, producing combinations of energy and light output inimical to life. However, if the planets follow circular or near-circular orbits fairly close (distance depends on primary's mass and its separation from stellar companion) to their parents, stable, life-supporting conditions may be met.

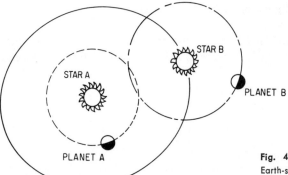

Fig. 4.9. Double star system with Earth-size planets (orbits not true; schematic only).

They may also be found on planets revolving at great distances from their primaries.[7]

From the preceding material, we would not expect all solar systems to support life, certainly not intelligent life. Even if we take the most conservative attitude possible, we still expect billions of planets enjoying conditions where life as we know it could appear and develop. And there may be billions more harboring life as we do not know it and as we can not, at our primitive state of knowledge, conceive it.

---

[7] Stephen, H. Dole, in his *Habitable Planets for Man* (New York: Blaisdell Publishing Company, 1964, pp. 76–79), analyzes the question of planetary orbits in binary star systems. He treats the case where the two stars move in circles around their common center of gravity and the planets revolve in circular orbits around their parents in the same plane, and shows that within, and outside, certain boundaries stable orbits can occur. But a stable orbit is not enough; the level of illumination of the two stars must be compatible with the requirements for life.

5
The Origin and Development of Life

The origin of life on Earth and elsewhere, always an intriguing subject for biologists, has in recent years commanded the increasing interest of the entire scientific world.

The distinction between life and non-life on Earth seems more than apparent, but when an attempt is made to define life explicitly unexpected difficulties arise. Commonly, we think of life as something that possesses the ability to reproduce itself (although not always perfectly) in its natural environment, leading to reproduceability as a defining factor. Frequently, definitions include the ability to utilize energy from the natural environment and the ability to react to changing environmental conditions.

Scientists realize that very simple chemical molecules, which must exist on any primordial planet, naturally undergo chemical reactions. These, in turn, lead to the formation of more complex molecules, which are the fundamental building blocks of life. Polymerization, or automatic joining together of long strings of these chemical building blocks, progressively yields more complete life-type molecules. The novice in biochemistry often gains the misconception that this process,

"Nature that fram'd us of four elements,
Warring within our breasts for regiment,
Doth teach us all to have aspiring minds:
Our souls, whose faculties can
　　comprehend
The wondrous Architecture of the world:
And measure every wand'ring planet's
　　course,
Still climbing after knowledge infinite,
And always moving as the restless
　　Spheres,
Will us to wear ourselves and never
　　rest,
Until we reach the ripest fruit of all,
That perfect bliss and sole felicity,
The sweet fruition of an earthly crown."

　　　　　　　　　　Christopher Marlowe

generally referred to as chemical evolution, eventually leads to the development of a single unique molecule that is capable of surrounding itself with structural material and also reproducing itself. One may get the impression that the appearance of one single molecule having this unique capability is an extraordinary statistical occurrence, and that this event marks the clearcut beginning of biological life. Such a view is probably a complete misconception. Biochemists recognize that the very gradual transition from chemical evolution to biological evolution makes it impossible to reach any general agreement concerning the exact point of life's origin in these continuous, cyclical chemical processes.

Since all known life on Earth appears to be based on the genetic properties of deoxyribonucleic acid (DNA), it has been suggested by some that its occurrence should be the basic criterion in any definition of life.[1] But even this does not resolve the problem since in the early stages of chemical evolution very short and simple nucleic acid polymers would have debatable life qualities. Although no other self-replicating carbonaceous polymers have yet been found, it is conceivable that they may exist or may have existed on Earth or on other planets. Even more exotic self-replicating molecules based on silicon are conceivable, but speculation about them does not seem worthwhile until chemical laboratory research reveals their actual existence and something about their properties.

Before discussing the origin of life on Earth, the possibility that life in the form of spores may have been brought here from some other planets in the universe must be considered. The cosmobiota, or panspermia, hypothesis, examined in Chapter 14, is not now regarded as likely, due to the high probability of molecular decomposition in the radiation environment of space and the difficulty in explaining the migration of spore-bearing particles away from a strong gravitational field.

The possible occurrence of artificial cosmobiota has been discussed in the scientific literature recently. It has been suggested that life on Earth may have evolved from the contamination or "garbage" left by space travelers from some distant extrasolar planetary system or that probes containing spores purposely have been fired at potentially hospitable worlds. Our present inability to verify either natural or artificial cosmobiota hypotheses encourages efforts to explain the development of life on Earth in terms of gradual chemical and biological evolution, since this manifestly did occur *somewhere*.

Whether or not evolution would proceed from more complex to simpler organisms in the event that a sterile planet was seeded with moderately complex organisms is an interesting conjecture but there is no evidence

---

[1] Muller, H. J., "Issues in Evolution" in *Evolution after Darwin,* ed. by S. Tax and C. Callender. Chicago: University of Chicago Press, 1960.

The Origin and Development of Life   91

to support it. We have microfossils of unicellular organisms on Earth that are quite primitive. Moreover, there are no obvious major discontinuities in the fossil record of biological evolution on Earth. We have *no* positive indication that life did *not* originate independently on Earth; conversely, there are overwhelming indications that life originated independently on Earth without initial seeding from extrasolar sources.

In the pages that follow, we first show the difficulty of defining life and then examine the nature of the primordial planetary environment in which life got its start. The exact chemical environment of the primordial Earth is not known with absolute certainty, although nearly all specialists now assume that a reducing atmosphere must have been present, made up of such elementary molecules as hydrogen, ammonia, methane, and water vapor. Many experimental efforts have been made to duplicate the combination of these elementary molecules into the more complex building blocks of life. After discussing the success of some of these laboratory experiments, the polymerization or linkage of these building blocks into the long chains of life molecules is considered. The importance of catalysis and autocatalysis in chemical evolution is then elaborated, since it is these processes that are primarily responsible for the origin of life. Finally, a short section is devoted to the general characteristics of present cellular and multicellular life.

If the processess of chemical evolution can be clearly demonstrated to be the result of automatic natural chemical reactions in *any* primordial planetary environment roughly similar to our own, it will greatly enhance the probability of the occurrence of life in extrasolar planetary systems. It will also eliminate the lingering uncertainties that cosmobiota may affect the frequency of occurrence of life.

Once biological life has gained a solid foothold in a stable planetary environment it can be expected to differentiate into a great variety of forms. The factors governing the variety and success of the varied forms of life in a given physical environment are mutation and natural selection. The functioning of these factors was first clearly enunciated by Charles Darwin.

Definition of Life

Among the many attributes that must be considered in attempting to define life are (1) reproduction, (2) death, (3) growth, (4) maintenance of a structural shell, or boundary, and a special internal environment, (5) sensing of external environmental conditions and reaction to these conditions, (6) utilization of external energy, (7) storage of energy, (8) absorption of nutrients or materials from the external environment, and (9) release of waste material into the external environment.

Macroscopic life is so obviously distinct from nonliving things that there would intuitively appear to be no problem in defining it. Indeed, most of the attributes listed above are found in such life as well as in many forms of present day microscopic life. The difficulty in defining any type of life appears when one considers chemical reactions and molecular aggregations of very gradually increasing complexity, such as are assumed to have arisen during prebiotic chemical evolution. There is no apparent basis for placing an arbitrary dividing line at any particular degree of complexity and calling it the division between living and nonliving molecular aggregations. Therefore, we must be satisfied with a definition of life that does not attempt to make a clear cut distinction between the living and nonliving at the very simple molecular aggregation level, since no such distinction exists.

A possibility, which will be discussed in considerable detail in later chapters, is the development of synthetic intelligent automata. Synthetic automata are totally different from biological life in terms of physicochemical structure and internal electro-physico-chemical processes; but synthetic intelligent automata will eventually have every one of the previously mentioned attributes of life, with the possible exception of death and reproduction. (Since death would only occur among automata as a result of a rare, cataclysmic event, and unlimited growth would be possible, there would be infrequent need for reproduction—which, however, would be a simple task for an automaton if desired.)

Should synthetic automata be classed as life or should the definition of life be narrowed to include only organisms based on carbonaceous polymers with a deoxyribonucleic acid genetic property? This semantic difficulty may be conveniently resolved by using the terms *biological life* and *inorganic life* to indicate both the common and different factors in these two classes.

In summary, all biological life on Earth has the following attributes:

(1) It consists of aggregations of different molecules, which are primarily complex carbonaceous polymers; these aggregates contain deoxyribonucleic acid as the genetic agent.

(2) It reproduces itself accurately but with a small statistical probability of error, permitting long term variations.

(3) It maintains a structural shell or boundary isolating a special internal environment.

(4) It absorbs or consumes materials or nutrients from the external environment for internal storage and utilization.

(5) It utilizes external energy.

(6) It stores energy in chemical form for later utilization.

(7) It releases waste materials into the external environment.

(8) It senses external environmental conditions and reacts to these conditions.

(9) It grows by fabricating new structural material from absorbed materials or nutrients.

(10) It eventually dies; death is a cessation of all normal chemical processes, with rapid decomposition of life's chemical structure.

This definition of biological life is adequate for most purposes; but, as has already been mentioned, it will not provide a fine line of demarcation between life and nonlife at the primitive molecular level (probably no such clear line of demarcation is warranted). Eventually, definitions of biological life will be complemented by similar definitions of inorganic life in order to describe the attributes of synthetic intelligent automata. It may even become necessary to develop definitions of as yet unknown exotic forms, such as silicic life based on complex polymers of silicon.

Primordial Planetary Chemical Environments

Most experts adhere to the view that primordial planetary atmospheres are reducing rather than oxidizing in nature; i.e., they are made up primarily of hydrogen atoms. This should not be at all surprising since stars form in huge clouds of hydrogen gas and dust, and it is now widely accepted that planets form as a normal by-product of star formation. Therefore, every primordial planet should be expected to have an atmosphere rich in hydrogen gas and in simple gaseous compounds of hydrogen and the other more common elements. Astronomical spectroscopy shows that the most common elements in our galaxy are hydrogen, helium, oxygen, nitrogen, and carbon, ranked in order of their abundance. Since hydrogen is many times more abundant than oxygen, nitrogen, or carbon, the hydrides of these elements should be expected in any primordial atmosphere.

The simple gaseous compounds, which are assumed to exist around any primordial planet, are hydrogen (H_2), ammonia (NH_3), methane (CH_4), and water vapor (H_2O). These few simple molecules form a remarkably simple beginning for the fantastic variety and complexity of the molecular aggregations which make up our plant and animal life.

Formation of the Building Blocks of Life

Clearly, at some time in the past either on Earth or elsewhere—if cosmobiota (panspermia) is plausible—simple chemical molecules began to evolve gradually into complex self-reproducing aggregates of molecules capable of undergoing biological evolution. Regardless of the planetary origin theory accepted, it is generally assumed that a primordial planet must have contained only simple chemical molecules such as those discussed in the previous section.

All life on Earth seems to be based primarily on two types of polymers: (1) nucleic acids, and (2) proteins. It would be very helpful if we could demonstrate in the laboratory both the formation of life's building blocks from the simple primordial molecules and the polymerization of these building blocks into the long chain polymers common to terrestrial life. Of course, these laboratory experiments should duplicate the assumed primordial conditions both with respect to chemical environment and energy supply if they are to be truly significant. Indeed, some laboratory experiments of this type have already been amazingly successful.

The building blocks of life on Earth fall into two primary classes: (1) the components of deoxyribonucleic acid and ribonucleic acid, and (2) the components of proteins.

All life on Earth appears to be based on the genetic properties of deoxyribonucleic acid (DNA) and a closely related compound called ribonucleic acid (RNA). DNA is a very long chain molecule or polymer which is the blueprint for the fabrication of every living organism. The links in the long DNA chain consist of two alternating chemical groups called sugar groups and phosphate groups. Figure 5.1 shows deoxyribose (a simple sugar) and phosphoric acid; these two simple molecules combine alternately to make the very long chain of DNA. Attached to every other link (every deoxyribose sugar group) in the DNA chain is one of four possible chemical side groups: (1) adenine, (2) thymine, (3) guanine, and (4) cytosine, which are also shown in Fig. 5.1. It is the sequence of these side groups on the DNA molecule that determines the makeup of every living organism on Earth.[2]

The molecule called ribonucleic acid (RNA) is very similar to, is fabricated by, and works closely with DNA in the actual fabrication of living organisms. The links in the RNA chain also consist of two alternating sugar and phosphate groups. The only difference between the RNA chain and the DNA chain is that the former contains ribose sugar which has one more oxygen atom than the deoxyribose sugar of the DNA chain. Attached to every other link (every ribose sugar group) in the RNA chain is one of four possible chemical side groups: (1) adenine, (2) uracil, (3) guanine, and (4) cytosine. Only one of these side groups differs from the

---

[2] MacGowan, Roger A., "On the Possibilities of the Existence of Extraterrestrial Intelligence," in *Advances in Space Science and Technology,* Vol. 4, ed. Frederick I. Ordway III. New York: Academic Press, Inc., 1962, pp. 59–61.

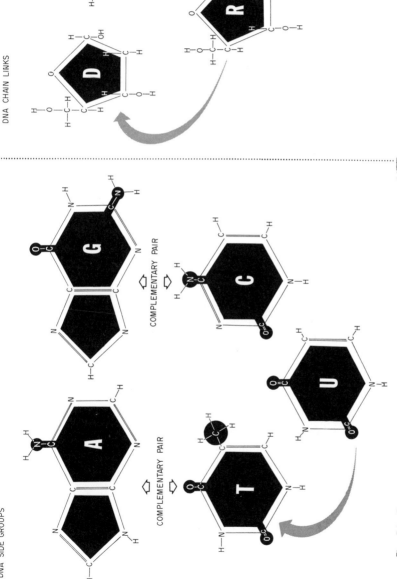

Fig. 5.1. Components of all DNA and RNA molecules—the fundamental molecules of all biological reproduction. A≡adenine T≡thymine G≡guanine C≡cytosine U≡uracil D≡deoxyribose, P≡phosphoric acid R≡ribose. In RNA, uracil replaces thymine as a side group and ribose replaces deoxyribose as a chain link.

side groups of DNA. The uracil group has replaced the thymine group of DNA. Uracil and ribose are also shown in Fig. 5.1.

The eight chemical groups shown in Fig. 5.1 are the fundamental building blocks of life's reproductive mechanism.

In outstanding experiments performed by S. L. Miller,[3] electric discharges were passed through gaseous mixtures of hydrogen, ammonia, methane, and water vapor. As well as producing many amino acids, his experiments yielded hydrogen cyanide (HCN) and aldhydes; and, there is a possibility that sugars were produced by formaldehyde condensation. It is known that HCN and acetylene (C_2H_2) are intermediate compounds in the reactions, and when high energies are applied to ammonia (NH_3), HCN, and C_2H_2, the production of pyrimidines can be expected. Recently, purines have been synthesized abiologically by the polymerization of HCN.[4]

Considering what we know about primitive conditions on our planet, and organic chemistry, and taking into account the aforementioned experiments, it would be surprising if purines, pyrimidines, and sugars as well as amino acids were not produced in substantial amounts on the primordial Earth.[4]

Proteins are vital to all living organisms, serving both as structural material and as chemical catalysts called enzymes. They account for more than half of the solid substance of the tissues of animal life. Proteins are long chains or polymers made up of amino acid building blocks. Figure 5.2 shows the chemical components of 20 common amino acids which have the general formula NH_2-CHR-COOH. A typical protein may contain about 200 of these amino acid building blocks, and frequently all 20 common types of amino acids will be included in a single protein. The amino acid groups attach together directly, the amino radical (NH_2) of one group attaching to the carboxyl radical (COOH) of another and giving off a molecule of water (H_2O) in the process. The resulting protein chain has the general chemical formula NH_2-CHR-CO-NH-CHR-CO—NH-CHR-COOH.

As previously noted, amino acids, which are the basic building blocks of proteins, have been produced by the action of electric discharges on gaseous mixtures containing ammonia, water, methane, and hydrogen in one of many notable experiments designed to show how organic material can be generated under prebiological conditions. The results of this experiment are shown in Fig. 5.3. They have also been produced by Fox

---

[3] Miller, S.L., "Production of Some Organic Compounds under Possible Primitive Earth Conditions," *Journal of the American Chemical Society*, 77, No. 9 (12 May 1955), 2351–2361.

[4] Adenine and guanine are purines, and thymine and cytosine are pyrimidines.

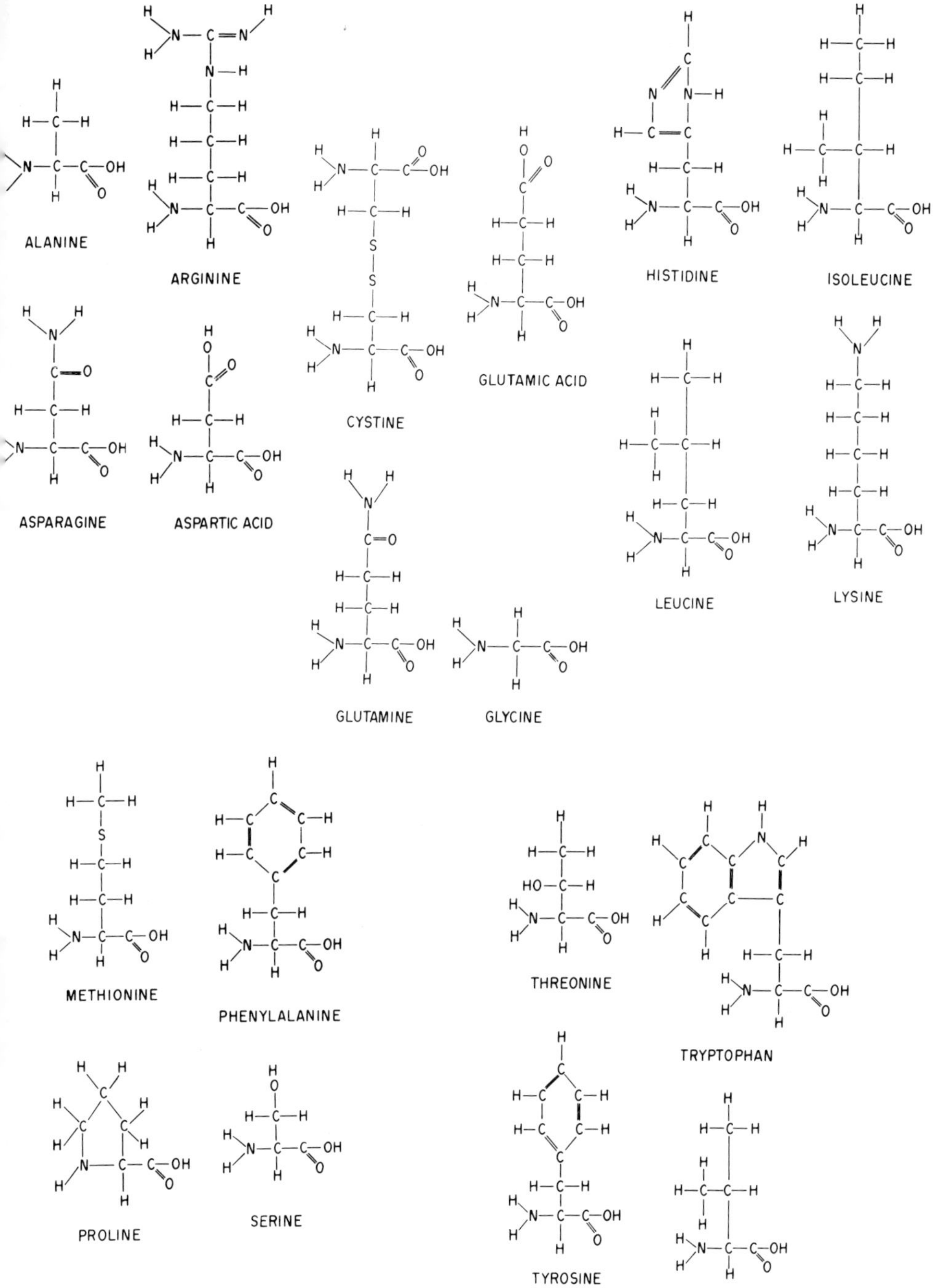

Fig. 5.2. The 20 common amino acids.

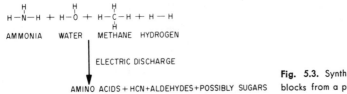

Fig. 5.3. Synthesis of life's building blocks from a primordial atmosphere.

and his coworkers utilizing thermal energy in experiments designed to create proteins in a primordial environment.[5]

These and other laboratory experiments performed under assumed primordial and prebiological conditions have demonstrated that several different types of naturally available energy, when applied to mixtures of very elementary chemical molecules, produce mixtures of amino acids.

Carbohydrates are a most important class of biochemical compounds which make up a large percentage of the solid material of plants. The class name "carbohydrate" implies that these compounds are hydrates of carbon; many have the general formula $C_m(H_2O)_n$ making them compounds of carbon and water. Carbohydrates include sugars and starches, and they may be cyclic. A sugar in which $m=n=6$ (hexose) is an important source of energy in organic life. Other sugars, such as ribose and deoxyribose (where $m=n=5$, pentose), form links in the nucleic acid chains shown in Fig. 5.1. As already noted, there is a possibility that sugars may have been formed in S. L. Miller's experiments.

Still another important class of biochemical compounds is the fats. The acids found in fats are straight chain monocarboxylic acids having an even number of carbon atoms. The saturated acids that occur in fats have the general formula CH_3-CH_2—CH_2-$COOH$, with from 4 to 30 carbon atoms. In the synthesis of fats, a complicated reaction characterized by Fig. 5.4 yields progressively longer hydrocarbon chains. The acids are esterified with glycerol to make mono-di or triglycerides, which are the fats.

Linkage of Life's Building Blocks into Long Chains

From the preceeding section we learned that very many molecules, which are the building blocks of life's giant molecules, would have been

---

[5] Fox, S.W., "How Did Life Begin?" *Science,* **132**, No. 3421 (July 22, 1960), 200–208.

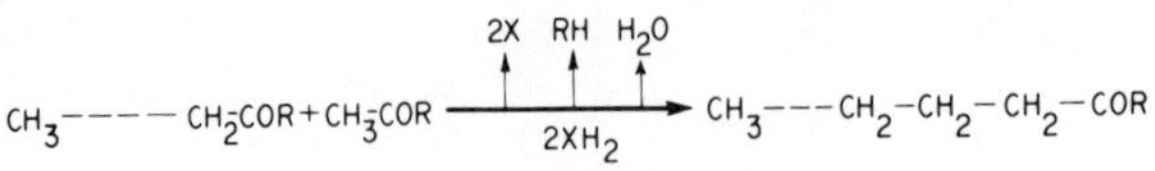

Fig. 5.4. Synthesis of fats.

produced naturally on a suitable primordial planet. How, then, were these building blocks linked together to form DNA, RNA, and proteins?

In present life, DNA directs the synthesis of RNA, and RNA directs the synthesis of proteins. However, enzymes, which are protein catalysts, take part in the synthesis of DNA, RNA, and proteins. Thus, the ultimate extension of the well known question, "which came first, the chicken or the egg?" is the analogous question, "which came first, DNA or protein?"

It is conceivable that many types of DNA could have been polymerized in the complete absence of proteins and could have eventually caused the synthesis of proteins. Those DNA molecules that produce proteins, which in any way facilitated the survival and propagation of that particular DNA molecule, tended to survive. It is also conceivable that many proteins could have been randomly polymerized in the complete absence of DNA, and some of these proteins may have eventually brought about the polymerization of DNA, which in turn caused the polymerization of more proteins. A third possibility is that both DNA molecules and proteins were being polymerized simultaneously and without dependence on one another; then, certain types of DNA may have utilized pre-existing proteins or enzymes to reproduce themselves, to produce RNA, and to produce new proteins.

In the experiments of Fox and his coworkers amino acids were produced, utilizing thermal energy; the research was designed to create proteins in a primordial environment.[6] These investigations pointed out that the thermal production of amino acids is accomplished under the same physical conditions that cause their polymerization, or linkage, into chains containing all the common amino acids. These molecules, or "proteinoids," resembling proteins in some respects, manifested a distinct morphogenicity in that they formed microspheres as well as other forms varying in volume with environmental conditions.

Other energy sources, such as ultraviolet light, have also been used to produce simple peptide chains of amino acids from aqueous solutions of amino acids.

Figure 5.5 shows a small segment of a protein molecule in which the individual amino acid residue building blocks have been outlined by dashed lines and labeled. The unique cystine unit is a double unit that permits the long chain to loop back on itself and be attached so as to form very complex but precise three dimensional configurations. Unquestionably, this structure plays a vital role in the ability of protein enzymes to catalyze particular chemical reactions. That is, the structure of the enzyme as well as its chemical characteristics may permit it to hold a particular object molecule in a particular orientation and simultaneously it may allow only

---

6 Fox, *Op. Cit.*, p. 201.

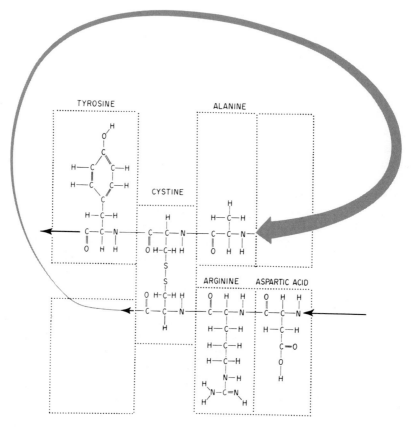

Fig. 5.5. A small segment of a protein molecule showing cross linkage.

one other specific type of molecule or radical to attach itself to one particular point on the object molecule. Then, the enzyme may shift its position so as to allow another specific molecule or radical to be attached to the last; in this way, enzymes may assist or catalyze the synthesis of DNA chains, RNA chains, and other proteins.

It was mentioned earlier that DNA is the blueprint for all living organisms. In directing the production of an organism, the double helix DNA molecule with the help of one or more enzymes (proteins) is able to synthesize single stranded RNA molecules in which the sequence of base units is identical to that in the parent DNA molecule except for the substitution of a uracil base unit for every thymine base unit. Figure 5.6 illustrates the production of a segment of a messenger RNA molecule from a corresponding segment of the DNA molecule. This RNA is called

The Origin and Development of Life   101

messenger RNA since it contains the directions for synthesizing a specific protein and it travels to one of the ribosomes, which are the sites of protein synthesis. When the messenger RNA molecule reaches a ribosome it works in conjunction with enzymes to synthesize a specific type of protein molecule.

The exact sequence of amino acid bases in the protein molecule is determined by the sequence of bases on the RNA molecule, which is patterned directly on the bases of the DNA molecule. Since there are only four types of base on the RNA molecule and there are about 20 common types of amino acid that appear in proteins, it must require a combination of at least three RNA bases to specify each amino acid residue in a protein. There are indications that each sequential group of three RNA bases specifies a particular amino acid residue; thus, the genetic code is assumed to be a linear triplet code, as is illustrated in Fig. 5.6. Rapid progress is now being made on the genetic coding problem, which involves determining the amino acid residue called for by each triplet of RNA bases. If the genetic code is indeed a linear triplet code then there exist 64 possible

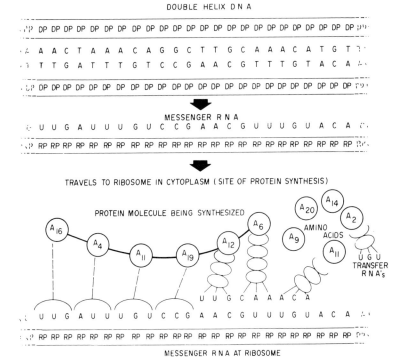

Fig. 5.6. The synthesis of proteins from a DNA blueprint.

combinations of RNA bases, but only about 20 amino acid residues; consequently some redundancy may exist. One hypothesis,[7] illustrated in Fig. 5.6, assumes that transfer RNA, which has looped back on itself to form a double helix, has a base triplet on its loop end that attaches to a complementary triplet on the RNA; and the loose ends of this transfer RNA are assumed to have an affinity for only one specific amino acid that it selects from the environment and holds in juxtaposition at the growing end of the protein molecule until it is attached. The process of protein synthesis is now understood in a general way although many details remain to be clarified.

One hypothesis concerning the reproduction of DNA and the synthesis of RNA from DNA envisages an enzyme or complex of enzymes containing two holes or two slots, which can permit the passage of the two strands of double helix DNA.[8] As the two strands pass through the enzyme, complementary base units and also sugar phosphate chain links are attached to each strand, thus producing two separate but identical DNA double helix molecules from the original. This process, indicated in Fig. 5.7, also shows the synthesis of single and double stranded RNA by a very similar process.

---

[7] Nirenberg, Marshall W., "The Genetic Code: II," *Scientific American*, **208**, No. 3 (March 1963), 80–94.

[8] Butler, J. A. V., "A Possible Mechanism of Synthesis of Nucleic Acids," *Nature*, **199**, No. 4888 (6 July 1963), 68–69.

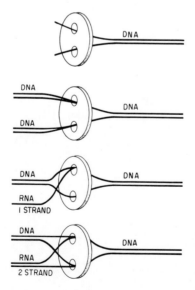

Fig. 5.7. Hypothetical method of reproduction of DNA and synthesis of RNA.

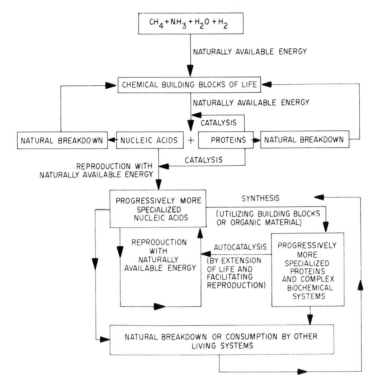

Fig. 5.8. Autocatalysis and catalysis in the origin and development of life.

Catalysis and Autocatalysis

Catalysis and autocatalysis play such a prominent role in biochemistry, and they undoubtedly played a similarly prominent role in the origin of life, that a brief discussion of them is in order. A catalyst is a chemical that accelerates a chemical reaction without itself being consumed in the reaction, i.e., the catalyst remains unaltered after the reaction. In autocatalysis one of the products of a reaction serves to accelerate the reaction, thereby accelerating its own production. Chemical reactions are fundamental to all life; other life phenomena, such as mechanical and electrical activity, are really secondary in nature. In fact, all life could be regarded as very complex sequences of catalytic and autocatalytic reactions. A principal reason for the unique complexity of biochemical reactions is that some of the products of the reactions form physical boundaries, which permit different reactions to take place simultaneously in different parts of the biological aggregate or organism.

Figure 5.8 illustrates the central importance of autocatalysis and catalysis in the origin and development of life. The first stage, which can be duplicated at least partially in the laboratory, is the conversion of primordial atmospheric molecules with naturally available energy into life's building blocks, which were discussed earlier. It appears likely that large numbers of these building block molecules accumulated in shallow primordial seas, in which polymerization of both primitive nucleic acids and primitive proteins probably took place. Some of the primitive proteins may have catalyzed the reproduction of the primitive nucleic acid polymers, thus increasing their number. Other proteins may have catalyzed the polymerization of the nucleic acids, enlarging their variety.

Many of the increasing varieties of nucleic acid polymers undoubtedly caused the polymerization of specific protein molecules. Whenever one of these proteins contributed in any way to either the survival of the specific parent nucleic acid polymer or its reproduction, a strong autocatalytic situation existed. (In many cases, the proteins being synthesized by the nucleic acids may have contributed to the survival or reproduction of any of the parent class of nucleic acids rather than just the individual parent. Thus, the roots of social evolution may be traced back to the chemical reactions of primitive chemical evolution.)

By extending the reproductive life of the parent nucleic acid, or by facilitating its reproduction, the protein was accelerating its own production. The nucleic acids participating in these strong autocatalytic reactions naturally tended to proliferate at the expense of the nucleic acids participating in weaker autocatalytic reactions as well as those having no autocatalytic reactions. Some of the nucleic acids and proteins underwent natural breakdown into the more primitive chemical building blocks, only to be reutilized by the more dynamic of the nucleic acids in reproduction or protein synthesis. The nucleic acids able to reproduce themselves as well as their supporting proteins could also be regarded as autocatalytic agents; thus, many of the reactions producing nucleic acid polymers and supporting proteins could be considered as dual autocatalytic reactions in that both products of the reaction tended to accelerate the reaction and consequently their own synthesis.

Some of the nucleic acids produced proteins which formed a distinct boundary layer surrounding the parent nucleic acid molecule. When this prevented or inhibited reproduction, the parent nucleic acid molecule tended to become extinct; but, when the boundary layer protected the parent molecule from the environment and still permitted reproduction when the external conditions were suitable, these parent nucleic acids tended to proliferate.

Through chemical or biological evolution these boundary layers came to be selective boundaries permitting the ingestion of required materials, enabling waste materials to be excreted, and preventing the ingestion of harmful materials. Eventually, the boundary layers actually developed the ability to enfold and grasp the desired external materials, rather than just passively allowing their ingestion. These units surrounded by boundary layers gradually developed the ability to store energy, materials for synthesis, and enzymes or catalysts—thus leading to the development of what we know as cellular life.

Since primitive cellular life and even precellular chemicals undoubtedly had the ability to attack or break down chemically other nucleic acids and proteins, competition (with consequent survival of the best adapted) existed from the earliest days of chemical evolution. All life, from the beginning to the present, can be regarded as a continuing autocatalytic chemical reaction of constantly increasing complexity. The less well adapted nucleic acids tend to be consumed by proteins of the better adapted (more autocatalytic) nucleic acids, and the available amino acids tend to be consumed in protein synthesis by the better adapted nucleic acids. In turn, the proteins produced by the poorly adapted nucleic acids tend to be consumed by the enzymes of the better adapted nucleic acids. Thus, the autocatalytic reactions which *are* life tend to concentrate all of the available chemical building blocks of life into ever more complex and better adapted chemical aggregates or organisms.

Cellular and Multicellular Life

The basic unit of nearly all plant and animal life on Earth is the cell. Single cell organisms and multiple cell organisms have proved to have such a superior adaptive ability that they have apparently replaced or eliminated many of the more primitive forms of life and life-like molecules. Single cell organisms are unspecialized and usually function as independent entities; however, some single cell organisms (e.g., amoebae), which normally live an independent existence, form colonies in times of stress or scarcity. In certain slime moulds independent amoebae are formed, but they combine later and take on specialized functions in creating a larger multicelled organism. These phenomena probably illustrate earlier transitionary forms of life lying between single and multiple celled organisms.

When certain cells functioned differently due to the chemical influence of other nearby cells of the same type, a potentially strong selective situation existed. In primeval times situations such as this undoubtedly led to a survival or reproductive advantage for some of those cells having specialized functions in the presence of others. This obviously accounts for the gradual development of multiple celled organisms.

All cells, whether single or in multiple cell organisms, have the same general characteristics. Cells consist of two major parts: (1) the central nucleus, and (2) the surrounding cytoplasm. The cell nucleus is a central core encased in a thin membrane that contains the blueprints or plans for synthesizing organic material of specialized types. The construction plans, which are nearly always contained in the cell nucleus, are DNA molecules. They are called chromosomes. Different organisms may have different numbers of chromosomes (or DNA molecules) in their cells. The information contained in these chromosomes is coded by the sequence of the four bases which can occur as side units on the DNA molecule.

Since the DNA molecules or chromosomes are very long they can contain a prodigious amount of information coded in this manner. The information contained in the chromosomes gives complete specifications for the construction of a living organism, and the only differences in the genetic plans for the Earth's many varieties of organisms are the different sequences of base units on the DNA molecule. Radically different species could be expected to have radically different base sequences and different individuals of the same species could be expected to have relatively minor base sequence differences.

When asexual reproduction occurs the DNA molecules or chromosomes reproduce themselves exactly, but when sexual reproduction occurs the DNA molecules from two individuals are mixed, or divided and reassembled in an altered way, so that the resulting plans combine some of the characteristics from each parent.

The very long DNA molecules in the cell nucleus are divided into segments by folding (or perhaps base sequence coding or some other means). These segments of DNA molecules are known as genes, the elements that determine specific characteristics of an organism. In order to fabricate a particular part of an organism, an RNA molecule, having base units corresponding exactly to those of the gene, is synthesized in the cell nucleus. Enzymes existing in the nucleus undoubtedly participate in this synthesis. The molecules of RNA, corresponding to genes, are called messenger RNA because they travel from the nucleus through the surrounding membrane into the cytoplasm.

In the cytoplasm are certain points known as ribosomes where protein synthesis takes place. The messenger RNA from the nucleus travels to the ribosomes where it directs the fabrication of specific protein molecules specified by the coded message brought from the DNA. Participating in this synthesis of proteins are transfer RNA molecules and ribosomal RNA molecules. Exactly what causes particular genes to be active at particular times and in particular cells in multicellular organisms is not yet fully understood, but chemical phenomena undoubtedly account for it.

One presently existing type of entity, the virus, is so small and simple in structure that only with difficulty is it considered life. Viruses are not cellular life, but are effectively parasites on cellular life. They undoubtedly had their beginning in early (but not the earliest) biochemical times.

Viruses consist of small DNA or RNA molecules encased in a simple protein jacket. The protein jacket usually contains one or a small number of different types of protein molecule, which are joined together (frequently in elementary geometric patterns) to form a protective case.

Viruses attach themselves to cells into which they permit their RNA or DNA molecules to enter. Then, their RNA or DNA utilizes the production facilities and materials of the victim cell to manufacture exact duplicates of itself and its covering. Some virus jackets are apparently shaped so as to facilitate penetration of cells; thus, the virus DNA or RNA is actually injected into the victim cells. There it utilizes the victim's enzymes, transfer RNA, and building block materials to synthesize duplicates of itself encased in the same type protein jackets as its own former cover. It continues this synthesis of new virus particles in the cytoplasm of the victim cell until there are so many virus particles that they burst the victim cell and destroy it. Then, the released virus particles proceed to attack other cells in the same manner.

The smaller types of virus are all identical, making it very difficult to consider them as life; but, larger varieties have some structural variability, giving the impression of gradual gradations from nonlife to life.

In early biochemical history many primitive chemical aggregates similar to viruses must have existed, i.e., DNA and RNA molecules surrounded by simple protein structures. However, as cellular life developed greater strength, it must have either consumed the available supply of building block materials needed by the more primitive lifelike chemical aggregates or else it consumed them directly. But a few types of the primitive chemical aggregates presumably had the ability to invade cells and utilize their reproductive facilities, so they survived. Thus, it can be hypothesized that viruses developed directly from very primitive chemical aggregates (simple RNA or DNA molecules surrounded by simple protein jackets), but they obviously could not have taken their distinctive viral characteristics until cellular life began to gain a foothold.

We can conclude from this chapter that the origin and development of life under suitable environmental conditions is inevitable given adequate time. We can assume that the proper primordial chemical environment must have existed on myriad planets throughout the universe and a significant percentage of them must be acceptable in other respects such as mass and temperature (Chapter 3). We know from astronomy that our chemistry is universal and the origin and development of life is the result of inevitable autocatalytic reactions under suitable environmental conditions. Every explicit step in the origin and development of life is not now, and probably will never be, known, but the general pattern has become clear.

The following chapter will discuss biological evolution, which is a function of the natural environment on the one hand and mutation and natural selection on the other. It should be apparent that mutation and natural selection are not unique to biological evolution, but that they functioned from the earliest autocatalytic reactions of chemical evolution. Natural selection is actually synonymous with autocatalysis of chemical aggregates of increasing complexity. Mutation is the random change of molecules involved in complex autocatalytic reactions, due to accidental chemical or physical phenomena such as collisions from cosmic rays or particles from natural radioactive decay. Mutation also must have affected the earliest reactions of chemical evolution, meaning that the phenomenon of biological evolution was manifestly prevalent and of primary significance in all stages of chemical evolution which preceeded biological evolution.

6
Biological Evolution

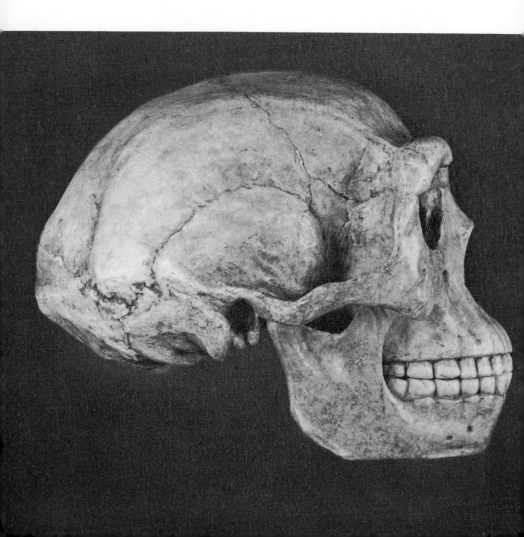

Not only is knowledge of the history of life on our planet fascinating for its own sake, but it provides a chronicle of the slow struggle of many species towards ever more efficient adaptation to an often hostile, and almost always changing, environment. Intellectually, it stimulates us to wonder what benefits an understanding of the past may yield for prognostications of the future. Can insights into the history of life on Earth forecast, in evolutionary terms, the shape of things to come? Or, closer to our present interests, can a survey of biological evolution on the third planet from the star we call the Sun offer any clues to the possibility that life may emerge on similar extrasolar worlds; and, if it can occur, or is occurring, or has occurred, is it based on the same biological laws that govern life on Earth? Unfortunately, neither question can be adequately answered since we possess no real data on which to forge even a tentative scientific opinion. But we can speculate, ponder, muse, wonder . . . and propose what may seem to many rather startling possibilities.

"... Life, far from being an aberration on the part of Nature, becomes within the field of our experience nothing less than the most advanced form of one of the most fundamental currents of the Universe, in process taking shape around us. Which is to conclude that, everywhere exerting its pressure, it tends with a 'cosmic' tenacity and intensity to make continuous progress wherever it has gained a foothold, always moving in the same direction and reaching out as far as possible."
Pierre Teilhard de Chardin

"He gave man speech, and speech created thought
Which is the measure of the universe."
Percy Bysshe Shelley

Biological Evolution

Table 6.1

	Years	Period/Era	Epoch	Life/Events
TO PRESENT	10,000 TO PRESENT	RECENT		CIVILIZATION HOMO SAPIENS HOMO ERECTUS
	1 MILLION TO 10,000	PLEISTOCENE		
	1 MILLION TO PRESENT	QUATERNARY PERIOD		HOMO HABILIS
	12–1		PLIOCENE	AUSTRALOPITHECUS DRYOPITHECUS KEIYUANENSIS PROCONSUL
	25–12		MIOCENE	
	34–25		OLIGOCENE	MONKEYS
	55–34		EOCENE	TARSIUS
	78–55		PALEOCENE	LEMUR
	78–1	TERTIARY PERIOD		THE RISE OF INTELLIGENCE
	78	BEGINNING OF CENOZOIC ERA		AGE OF THE MAMMAL
	130–78		CRETACEOUS	DECLINE OF THE DINOSAUR
	170–130		JURASSIC	ADVENT OF GIANT DINOSAUR
	220–170		TRIASSIC	PHYTOSAURS, CYNOGNATHUS, ICHTHYOSAURS
	220	BEGINNING OF MESOZOIC ERA		AGE OF THE GIANT REPTILE
MILLIONS OF YEARS	260–220		PERMIAN	PERIOD OF GREAT CHANGES; THERIODANTS APPEAR
	330–260		CARBONIFEROUS (Mississippian and Pennsylvanian)	AMPHIBIA, FIRST REPTILE
	380–330		DEVONIAN	SHARKS, AMPHIBIANS, GIANT FERNS
	420–380		SILURIAN	BYROZOA, SPONGES; FIRST LAND LIFE
	470–420		ORDOVICIAN	TRILOBITES, OSTRACODS, OSTRACODERMS
	600–470		CAMBRIAN	ALGAE, BRACHIOPODS, TRILOBITES
	600	BEGINNING OF PALEOZOIC ERA		TRIUMPH OF LIFE
BILLIONS OF YEARS	600 TO 1.5	LIFE BECOMES EVER MORE COMPLEX, WIDE SPREAD		EVIDENCE: CALCAREOUS DEPOSITS, BACTERIAL ORGANISMS, PLANT-DERIVED CARBONACEOUS FILMS, POSSIBLY WORMS
	1.5 TO 2	CHANGE FROM REDUCING TO OXIDIZED ATMOSPHERE		ANOXYGENIC AND OXYGENIC LIFE FLOURISH, THE FORMER GRADUALLY GIVING WAY TO THE LATTER
	2.7	FIRST IDENTIFIABLE FOSSILS FORMED		SOUTHERN RHODESIA ALGAL LIMESTONES
	3	FIRST LIFE APPEARS		FIRST LIME SECRETIONS
	3.2–3.5			OCEANS ESSENTIALLY FILLED
	3.5			OLDEST CRUSTAL ROCKS
	4	EARTH ASSUMES PRESENT FORM		PRIMEVAL ATMOSPHERE EVOLVES
	4.5	FORMATION OF PLANET LIFE		CONSOLIDATION FROM PRIMEVAL MATERIALS

In Chapter 5 we passed through the long, lonely period of chemical evolution when the Earth contained molecular aggregates of gradually increasing complexity, and made the transition to the appearance of the first, perhaps hardly recognizable, forms of life. Now we can attempt to trace, in as concise a fashion as possible, the marvelous history of life on our planet (Table 6.1). To do this we must travel backwards billions of years when conditions were quite different, when a virgin world had welcomed something new and exciting: unmistakably living organisms.

The Earth, according to the most recent evidence, is about 4.5 billion years old, having taken roughly its present form some 4 billions of years ago. In the intervening half billion years the planet gradually consolidated and shaped itself into essentially the world we now know (Fig. 6.1). We believe that life possibly appeared as far back as 4 billion years ago, that is, when the world had been in existence for approximately half a billion years; or, it may not have taken hold until 3 or 3.5 billion years ago. Unfortunately, geologists and paleontologists know little or nothing about this early life, whenever it appeared, and are forced, therefore, to speculate on its nature. And, also unfortunately, they are uncertain just how old the earliest known evidence of life is—as represented by fossils found in

Fig. 6.1. Artist's conception of conditions on Earth during the formation of the atmosphere and prior to the existence of the great oceans.

Biological Evolution

incredibly ancient, pre-Cambrian rocks. Estimates vary from a minimum of 2 to a maximum of 3 billion years, so prefossil life must bridge an awesome gap, possibly a billion or more years wide.

From the point of view of life, we can divide the history of our planet into three phases: (1) the prelife stage, (2) the prefossil stage, and, (3) the fossil stage. We have talked about the first, or prelife, stage in Chapter 5, so we shall now concentrate on the prefossil and fossil stages, attempting to probe into the development of life on the only planet we definitely know harbors the phenomenon. Even if conditions are identical on some other planet in the universe, we can only conjecture whether life may occur on it; and if it does, whether it is similar to, and has evolved in a similar fashion to, life on Earth.

Absolute and Relative Dating

Before commencing our survey, it is important to realize that scientists concerned with ancient life possess two basic methods of determining age: by *absolute dating* and by *relative dating*. The technique of deriving the absolute age of a rock formation relies on the rather well-developed and widely-used radioactive dating method, which is based on the half-life of radioactive elements. By measuring the relative proportion of, for example, uranium 235 (which, with almost infinite slowness, decays into lead) and several isotopes of lead, we can determine when the uranium became associated with the rock sample in the first place. By using the uranium-lead ratio technique, rock formations over 3 billion years old have been identified.[1] Other than uranium-lead ratios, there are three radioactive decay series useful to historical geologists: rubidium-strontium, thorium-lead, and potassium-argon.

Turning now to relative dating, it must first of all be emphasized that neither this technique nor absolute dating is normally used alone: each tends to back up the other, the latter particularly providing support for the former. There are two basic techniques of relative dating, one relying on the principle (or law) of superposition (younger sedimentary beds are laid down or deposited on top of older sedimentary beds); and the other on faunal and floral succession (index fossils in a sedimentary layer of known age, found in another sedimentary layer widely separated in distance from the first, can be used to establish the age of the second bed). No such

---

[1] The half life of uranium 235 is 4.5 billion years; of uranium 238, 700 million years. If we assume that originally there was an equal proportion of 4.5 billion-year half-life U^{238} and 700 million-year half-life U^{235}, and if we couple this with the fact that the abundance ratio of the two is *today* 140 to 1, we come to the conclusion that it would have taken nearly 6 billion years to reach the present ratio. This would, then, be an outside limit for the age of the Earth. However, we do *not* know the original proportions of the two isotopes, nor do we know if factors other than radioactive decay helped shape their relative abundances.

techniques can be applied to nonsedimentary formations, e.g., igneous rocks, since they are not laid down in beds and do not contain fossils. However, in association with sedimentary rocks, they can be relatively dated. Thus, if a sedimentary bed is deposited on top of igneous rocks, it is obviously younger; conversely, an igneous intrusion through a sedimentary layer is itself the younger of the two.

A third method that is sometimes used, though with considerable uncertainty, involves the escape of helium from rocks. Assuming that the retentivity of a given rock for helium is known and the present amount contained can be determined, then it is occasionally possible to work out the age of the helium-bearing rock. Most geologists have little confidence in the technique.[2]

Environmental Conditions and the Appearance of Life

There is an increasing body of evidence suggesting that the Earth was never entirely molten, but that, during its formation by the accumulation of solid particles, it slowly began to heat up so that at least a portion of its core became molten. The only ways to explain the heating are by the continuous energy release of radioactive elements trapped in the interior and by gravity. As infall continued to build up the Earth, the core or central portion became ever hotter and this inevitably tended to raise the temperature of the outer regions, including the surface. Whether or not the crust ever became molten is not known for certain, but it probably remained solid throughout the early history of the Earth.

Whatever the case, for hundreds of millions of years surface conditions precluded any possibility of the emergence of life. But, progressively, the surface stabilized and temperatures ameliorated to essentially what they are today, i.e., determined largely by the radiation intensity of the 93,000,000-mile distant Sun. Water tied up in silicate rocks was by degrees released, filling partially the great planetary basins we now call oceans. Gases, particularly methane, ammonia, hydrogen, nitrogen, and some water vapor, continuously escaped from the crust, forming the primitive atmosphere. As we learned in Chapter 4, methane and ammonia slowly became decomposed by ultraviolet light, the resulting free hydrogen equally slowly escaping into interplanetary space. The carbon from the methane may

---

[2] A good modern source on the dating of rock systems is G. R. Tilton, and S. R. Hart "Geochronology," *Science,* **140**, No. 3565 (26 April 1963), 357–366. Containing 56 references, it emphasizes advances in geochronology since 1958, discussing many new techniques and improvements of old techniques. For example, it is now possible to recognize, "independently of geological criteria, when postcrystallization alteration (metamorphism) has disturbed the age record in a rock. In favorable cases the time of metamorphism, together with the time of earlier crystallation, for a metamorphic rock can be determined."

have become tied up with whatever oxygen was present to yield carbon monoxide. All during this period oxygen, little by little, increased in abundance, appearing as a result of the photodecomposition of water vapor; however, it was available only in insignificant quantities until well after life had been created and begun to evolve to higher forms.

According to Oparin,[3] at first the free oxygen would not have persisted at all. He writes that "the comparatively small amount of free oxygen formed by the photolysis of water in the upper layers of the atmosphere was now taken up by incompletely oxidized substances. This completely prevented any accumulation of oxygen in the atmosphere of the Earth before life had appeared." He also estimated (p. 157) that "the entire amount of free oxygen in the (present) atmosphere could be produced by vegetation in roughly 2,000 years—a period which is completely insignificant in relation to the thousands of millions of years during which the Earth has existed." Rubey[4] disagrees that there was no early accumulation. He does not attempt to conclude when oxygen began to accumulate, but suggests that at least "some oxygen must have been available fairly early in geologic time," because of somewhat higher ratios of ferric to ferrous iron in pre-Cambrian slates than in average-type igneous rocks, coupled with the presence of sedimentary iron oxides in pre-Cambrian iron formations, and the fact that anhydrite and gypsum are interbedded with limestone in the Grenville series. From available evidence he feels that the rate at which free oxygen is being produced by photodecomposition of water vapor, if "continued throughout geologic time, would yield five times as much oxygen as that in today's atmosphere. However, this quantity is only about one third to one sixth the total oxygen that has been removed from circulation by the oxidation of materials now buried in sedimentary rocks." He notes the paradox that life could not have started in an environment containing oxygen—no matter how little—and yet oxygen has probably been in the atmosphere "since the beginning of earth history." However, he suggests that "there were then [early in geologic history], just as there are today, local environments in the ocean and in mud pools where free oxygen was absent; and that it was in such local reducing environments that the first organisms may have come into existence."

To summarize, life was created and began evolving on an Earth somewhat warmer than it is now, in oceans located approximately where they are now but with considerably less water (which, however, was continually

---

[3] Oparin, A. I., *Origin of Life on Earth.* New York: Academic Press, 1957, pp. 158–59.

[4] Rubey, William W., "Development of the Hydrosphere and Atmosphere, with Special Reference to Probable Composition of the Early Atmosphere," *Geological Society of America Special Paper,* No. 62, 1955, pp. 631–650.

being increased by decomposing hydrates and seepage out from the interior; the composition of the waters also differed, with fewer inorganic salts than found at present), and in an atmosphere whose composition was totally different than it is today.[5]

Once life had been created and had started on its long evolutionary journey, the oxygen content of the atmosphere increased until the conditions originally conducive to the creation, though not necessarily to the maintenance, of primitive life were no more. The action of ultraviolet light on upper atmospheric oxygen yielded ozone, which formed a shielding layer about 20 miles high, effectively blocking out the passage of ultraviolet light through to the surface. This, in turn, cut off an energy source that may have been instrumental in creating organic compounds in the first place. As Rush puts it, "life to a large extent burned its bridges. It not only cut off a major agent for photochemical action. It also destroyed the channel by which it probably had evolved the power of photosynthesis."[6]

Today, the ozone layer acts as a screen (Fig. 6.2), protecting the Earth's surface from the total spectrum of ultraviolet radiation. If, for some reason, the layer should disappear, all life on our planet would be snuffed out. Of

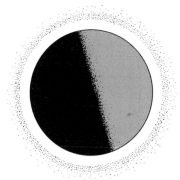

Fig. 6.2. Schematic drawing showing how the ozone layer prevents intense ultraviolet radiation from reaching the Earth's surface.

---

[5] Rubey, *Op. Cit.*, p. 631, points out that there are two principal hypotheses of the origin of the atmosphere and hydrosphere. One assumes "that all air and water of the earth are residual from a dense primitive atmosphere that once enveloped a molten globe," while the other supposes "that they have accumulated at the earth's surface by leakage from the interior." He concludes that "it seems likely that the atmosphere and hydrosphere have accumulated gradually during geologic time by the escape of water vapor, CO_2, CO, N_2, and other volatiles from instrusive and extrusive rocks that have risen more or less continuously from the deep interior of the earth." The paper provides good evidence that this is so, noting (pp. 641–42) that the "hypothesis of gradual 'degassing' of the earth's interior leads to chemical consequences at the surface that appear consistent with the observed geologic record."

[6] Rush, Joseph Harold, *The Dawn of Life.* Garden City, N. Y.: Hanover House, 1957, p. 183.

course, as long as there is free oxygen in the atmosphere there is no danger of this since ozone is continually being created; however, and this is a point to be emphasized, the very existence of the ozone is due to oxygen, making it in this sense as vital to life as in its respiratory attribute. In some ways the ozone-producing factor is even more important, for there are anaerobic forms of life which do not require free oxygen in their metabolic processes (indeed it is often inimical to them). Anaerobes (anaerobic bacteria) are not uncommon, being found in the mud at the bottom of stagnant ponds, in decomposing organic matter, and in the ground. Living without oxygen on a day-by-day basis, anaerobic life would still perish were it not for oxygen's indirect contribution to the ozone layer, which shields all living organisms from the naked force of solar ultraviolet rays.

There is no way of knowing when the atmosphere became oxidized nor when the transition from anoxygenic (anaerobic) to oxygen metabolism-type life occurred. All we can say is that it occurred some incredibly long time before the beginning of the Cambrian Period of the Paleozoic Era (which began about 600 million years ago), and an equally incredibly long time after the terrestrial crust and environment became suitable for not only the creation but for the maintenance of life. It is entirely possible that complex molecules began to appear many times in many places, soon to be extinguished by harsh environmental conditions, only to appear again. Perhaps multiple types of primeval life emerged at many locations, sooner or later coming into contact with one another. Then, later in paleontological history, natural selection processes operated, and dominent strains appeared. Once they took hold, they probably prevented incipient *new* life from persisting if, indeed, it appeared in the first place. Any new life, or prelife, forms would simply be consumed by the previously created life already struggling to maintain a foothold under precarious environmental circumstances. Oparin put the question and the answer well.[7] First, the question is whether or not "living material can now arise on our Earth primarily, directly in a lifeless natural medium." To supply the answer he asks us to "imagine some sterile tank of water, free from living things, with various organic substances dissolved in the water. If it were left to itself, the processes of transformation of substances . . . would come about slowly in it. Finally, during many millions of years, this would lead to the origin of life. However, if we were to introduce into our tank ready-made organisms, e.g., bacteria, the course of events would be quite different; in that case the more highly developed form of the motion of matter would come to the fore and take the lead. At once the transformation of lifeless to living material would cease to follow the old slow paths and would proceed in the new way, based on metabolism, converting the organic substances in the solution into the ingredients of living protoplasm with colossal rapidity. The origin of life from

---

[7] Oparin, *Op. Cit.*, pp. 488–89.

lifeless material simply could not occur under these conditions." This leads to a fundamental principle that in an oxygenic environment, only living matter can synthesize organic compounds. Today, it is believed that living matter cannot be created from nonliving matter in the natural environment of the Earth.

Primitive Forms of Life—Pre-Cambrian Biology

As geologists, paleobotanists, and paleontologists are quick to point out, fossilization remained very rare until the advent of animal life with hard structures. Once it had appeared, at the beginning of the Cambrian Period some 600 million years ago, the geological record, in broad outline at least, becomes fairly easy to follow. Between the oldest identified crustal rocks, which date back at least 3.5 billion and possibly up to 4.5 billion years, and the advent of the Cambrian, is a protracted span of time during which very simple life forms hesitatingly evolved into the established flora and fauna of the Cambrian Period.

We can pause a moment to say a few words about the oldest rocks discovered on the planet Earth. Tilton and Hart[8] examine the problem of determining their ages, emphasizing that it is "relatively easy to establish ages of 2,500 to 2,700 million years for all continents except Antarctica," and that recently the record has been extended further backward. They quote sources that estimate the oldest rocks in Africa to be 3 to 3.3 billion years old, in North America 3.1 to 3.5 billion, in Europe 3.5 billion, and Australia 3 billion. They think it unlikely that rocks much older than 3.5 billion years will be found, leaving a hiatus of a billion years in the age record (assuming, as we do, that the Earth is 4.5 billion years old). They provide two explanations for this, one that granitic rocks were not made during this period, and the other that rocks appearing during the hiatus are simply too old to preserve their age record, "presumably because of subsequent metamorphisms." They lean towards the first explanation, writing that "granites and other silicic rocks that have commonly occurred in orogenic belts during the past 2.7 billion years formed in negligible quantities, if at all, during the first third of Earth's history."[9]

---

[8] Tilton and Hart, *Op. Cit.*, pp. 364–65.

[9] Acccording to an announcement in *Scientific American*, 212, No. 3 (March 1965), 56–57, rocks whose age may be up to 4.5 billion years have been identified on the tiny islands of St. Peter and St. Paul in the Mid-Atlantic Ridge off Brazil. Composed entirely of peridotite, they may represent exposures of primeval rock. The samples were analyzed by the Isotope Geology Group of the Carnegie Institution of Washington to determine the ratio of strontium 87 to strontium 86. S. R. Hart, in charge of the study group, estimates that, within an error of as high as 25 per cent, the rocks may be fragments of the outer mantle chemically unaltered from the time of its formation some 4.5 billion years ago.

For millions upon millions of years, two types of life probably coexisted on Earth, one characterized by anoxygenic metabolism utilizing the ultraviolet light energy still coming through to the surface (though in gradually diminishing quantities), and the newer, more advanced type based on photosynthetic processes producing (in gradually increasing quantities) an oxygenic or oxidized environment. As life based on the latter began to progress in an environment still open to the ultraviolet it may have sought protection not on or near the surface of the then existent bodies of water but at considerable depth.[10] Only as the ozone layer grew into a progressively more effective shield did primitive life forms move up towards the surface, and eventually onto land.

Fossil remains of earliest life are nonexistent, due partially to its very fragile nature and partially to its almost incomprehensible age. This latter consideration breaks down into two subfactors: (1) there are relatively few exposed rock formations of such age (most ancient rocks are now covered by deep layers of sedimentary formations), and (2) those that are exposed have often been altered by orogenic processes. Early fossils that are still extant are found in what geologists refer to as pre-Cambrian shields, very ancient rocks that never have been, or at least are not now, covered by younger formations.

The earliest life we would expect to find evidence of is microbic, though unfortunately microorganisms are only rarely preserved over vast periods of time. Normally called *bacteria,* they are unicellular, each cell being made up of protoplasm and a thin cell membrane, itself inside a rigid cell wall. Most do not contain chlorophyll, but they do resemble blue-green algae and for this reason are usually referred to as plants. However, they are related to many unicellular animals, including amoebae. Because microbes and other simple life forms are known to be among the oldest discreet organisms representing life on Earth, diligent efforts have been made by paleontologists and paleobotanists to locate them in pre-Cambrian sedimentary rocks.

Fossils are extremely scarce in Archeozoic rocks. Not only are there relatively few exposures in which to search for biogenic evidence in the first place, but most sedimentary deposits associated with the era have long since been thoroughly metamorphosed, thus obliterating any fossils or other indications of life they once may have contained. Some geologists suspect that graphites forming the Grenville series[11] running across the southeastern

---

[10] Engel [Engel, A. E. J., "Geologic Evolution of North America," *Science* **140**, No. 3563 (12 April 1963), p. 144] reports there is considerable evidence not only that "large bodies of sea water have existed for over 3.2 billion years," but they have "supported protozoan life for over 2.6 billion years."

[11] See, for example, Tyler, S. A. and E. S. Barghoorn, "Occurrance of Structurally Preserved Plants in the Pre-Cambrian Rocks of the Canadian Shield," *Science* **119**, No. 3096 (30 April 1954), 606–08.

portion of Ontario, into Quebec and New York's Adirondacks) were laid down in Archeozoic seas which may have contained not only plant but possibly animal life of a very primitive type. However, the chances of animal fossils being formed or even impressions of their bodies, are remote. As pointed out by Hussey,[12] "such creatures are seldom preserved except as impressions and, even then, only under the most favorable conditions. Any shells or hard parts that Archeozoic animals might have developed were probably very fragile structures, poorly adapted for preservation as fossils." It is unlikely that Archeozoic life forms had developed to the stage where they could secrete mineral matter to build up protective parts.

Evidence of the oldest life on the planet is not based, as we would expect, directly on the unearthing of ancient fossils; but rather indirectly, on the discovery of secretions of lime from primitive organisms, probably of prealgal type, associated with the Dolomite series in the Bulawayo region in Southern Rhodesia. Direct dating methods show the limestone to be approximately 2.7 billion years old. Moreover, correlative geologic evidence indicates that the organisms must have flourished in a reducing (anoxygenic) atmosphere and therefore were themselves anoxygenic. With Rutten,[13] we must remain puzzled as to what this life was like. In his words, "We can but guess at the peculiar anoxygenic metabolism which already led to lime secretion by these early organisms, but they cannot have been related at all to any of the systematic groups in which the present oxygenic life and its fossil parentage are cut up in taxonomic procedure."

The Southern Rhodesian limestone studies were undertaken by A. M. Macgregor[14] in the so-called "Basement Schists" of the Bembesi gold belt, northeast of Bulawayo. When he first visited the quarry where the algal structures were found back in 1935, Macgregor relates that he "was not convinced that the algal origin could be considered as established until laboratory study proved that the dark colouration of the rock was due principally to films of graphite." He again visited the quarry in 1938 and made more careful studies, collecting further material for examination. He writes that the limestone "may be mottled or massive, but is frequently banded," adding that some parts of the banding show "extraordinary convolutions which are clearly formed by concretionary deposition, and not by folding." He then describes the three principal forms of the structure: domical, dentate, and columnar. It is concluded that the fossils are "really

---

[12] Hussey, Russell C., *Historical Geology*. New York: McGraw-Hill Book Co., Inc., 1947, pp. 65–66.

[13] Rutten, M.G., *The Geological Aspects of the Origin of Life on Earth*. Amsterdam: Elsevier Publishing Co., 1962, p. 82.

[14] Macgregor, A. M., "A Pre-Cambrian Algal Limestone in Southern Rhodesia," *Transactions of the Geological Society of South Africa*, 43 (1940), 9–15 plus 6 pages of plates. See also "Some Milestones in the Pre-Cambrian of Southern Rhodesia," *Transactions of the Geological Society of Southern Rhodesia*, 54 (1951), xxxii–lxxi.

little more than middens formed of precipitated calcium carbonate, silica, and carbon, deposited beneath and around the tangled threads of simple plants."

Before leaving this subject it is important to realize that improved techniques of determining the ages of rocks have been made in recent years, leading to backward revisions of ages assigned previously. Thus, Ahrens[15] in 1955 assigned 2.7 billion years to ancient rocks in Rhodesia and Manitoba, though even then suspicions existed that they were up to 3.3 billion years old. At the same time he dated the Earth between 4 and 4.2 billion years and various stoney meteorites from 4.5 to 4.8 billion years (a reflection of the possible age of the Solar System). Similar values are reported by Holmes.[16]

Paleobotanical evidence points to simple aquatic plants in rocks associated with the Canadian shield in southern Ontario. Found in the Gunflint iron formations of Proterozoic age (about 1.9 billion years old, absolute dating) are true fossils representing five different forms of plants. It has not yet been established whether or not these plants grew under reducing or oxidizing atmospheric conditions, but it is likely that they flourished in an environment that definitely tended towards oxygenic. If so, it is probably accurate to say that between 2 and 1.5 billions of years ago the biological production of oxygen had reached the point where it was rapidly producing our essentially modern atmosphere.[17] Figure 6.3 shows some of the organisms from the Gunflint Formations in thin sections of rock, photographed in transmitted light.

As the Proterozoic era matured increasing evidence of living organisms appears in the geologic record. Calcareous deposits formed by colonies of algae are found in many parts of the world. In the Belt series in Montana,

---

[15] Ahrens, Louis H., "Oldest Rocks," *Geological Society of America Special Papers* 62 (1955), 155–168.

[16] Holmes, Arthur, "The Oldest Dated Minerals of the Rhodesian Shield," *Nature*, 173, No. 4405 (3 April 1954), 612–14.

[17] Even if studies of fossils do not reveal whether or not the atmosphere was oxygenic at a given time in geological history, it is possible to determine the probable nature of the atmosphere from evidence in the rocks. Under oxidized atmospheric conditions, certain rocks will chemically weather, others will not, particularly those containing oxides. Contrariwise, under a reducing atmosphere, many oxygenically unstable minerals can persist for long periods on the surface and be relatively immune to chemical weathering. If a given mineral-containing rock formation can be shown to have weathered under anoxygenic atmospheric conditions, and if the formation can be dated, then we can say that the atmosphere was still reducing at the date determined. See Cloud, Preston E., Jr., "Significance of the Gunflint (Pre-Cambrian) Microflora," *Science*, 148, No. 3666 (2 April 1965), 27; Barghoorn, E. S. and S. A. Tyler, "Microorganism from the Gunflint Chert," *Science*, 147, No. 3658 (5, Feb. 1965), 563; and Schopf, J. W., E. S. Barghoorn, M. D. Maser and R. O. Gordon, "Electron Microscopy of Fossil Bacteria Two Billion Years Old," *Science*, 149, No. 3690 (17 Sept. 1965), 1365.

both Gyanophyceae and Rhodphyceae algal reefs are found, while bacterial organisms have been tentatively identified in Michigan iron ore deposits. In the Grand Canyon shales, thin carbonaceous films derived from aquatic plants have been discovered. Some paleontologists suspect that what appear to be trails and burrows were produced by primitive Proterozoic worms. And, occasionally, brachiopods have been reported, but evidence for them is highly suspect.

There is no purpose in giving further examples of evidence of pre-Cambrian life. We harbor no doubt that it existed for at least 3 billion years and, in its primal form, possibly a maximum of 4 billion years. Such a miracle—and we are probably justified in so calling it—was possible because of the absence of catastrophic temperature changes and remarkable stability of solar radiation for the entire period of life's development and evolution. The maintenance of a rather even temperature, considered on a planet-wide basis and taking into account periods of exceptional cooling in the polar regions and heating in the equatorial regions, is a function of the solar radiation constant coupled with heat flow from the central regions of the Earth and the heat retentivity of the atmosphere. All evidence points to an even solar output over astronomically long periods. As for variations in heat outflow from the terrestrial interior, there is no evidence that, since life appeared, it has ever been excessive. And the atmosphere, even when anoxygenic, contained sufficient water vapor to make it a highly efficient stabilizer against any tendencies towards extremes in heat or cold.

Life During the Paleozoic Era

From the beginning of the Paleozoic Era about 600 million years ago to the present, geologists have been able to construct a remarkably accurate, systematic record of the relative ages of rock formations and of floral and faunal succession. Contrasting with the haziness and doubts surrounding biological developments during the Archeozoic and Proterozoic Eras, the Paleozoic emerges as an era susceptable to intensive, and often rewarding, study.

The Paleozoic is conventionally broken down into seven periods as follows:

Period	Time from Present, Years
Cambrian	600,000,000 to 470,000,000
Ordovician	470,000,000 to 420,000,000
Silurian	420,000,000 to 380,000,000
Devonian	380,000,000 to 330,000,000
Mississippian } Carboniferous	330,000,000 to 290,000,000
Pennsylvanian	290,000,000 to 260,000,000
Permian	260,000,000 to 220,000,000

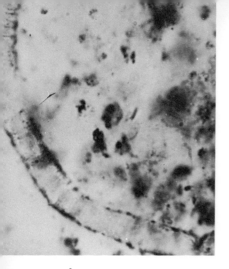

1.

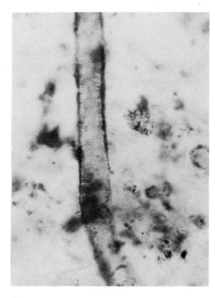

2.

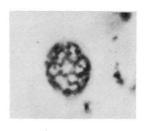

3.

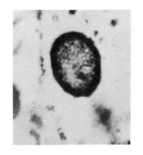

4.

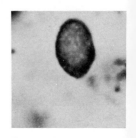

5.

Fig. 6.3 Microorganisms discovered in the Gunflint Formations. 1. *Animikied septata* Barghoorn; filament probably algal in affinity. 2. *A. septata* Barghoorn; probable algal filament. 3. *Huroniospora macroreticulata* Barghoorn; spheroidal sporelike body. 4. *Huroniospora macroreticulata* Barghoorn; ellipsoidal sporelike body. 5. *H. psilata* Barghoorn; ellipsoidal sporelike body with minute aperture at more constricted upper end. 6. *A. septata* Barghoorn; probable algal filament (9 microns in diameter). 7. *Entosphaeroides amplus* Barghoorn; filamentous organism showing presence of sporelike bodies. Internal sporulation morphologically comparable to several blue-green algae. Diameter of filament is 5-6 microns. 8. *Gunflintia grandio* Barghoorn; suggestive of green algae of Uliotrichaceac family. 9. *Eosphaera tyleri* Barghoorn; structurally, is a sphere within sphere. 10. Tangled filaments, largely *Gunflintia minuta* and enmeshed sporelike bodies. Courtesy Prof. Elso S. Barghoorn

Biological Evolution   125

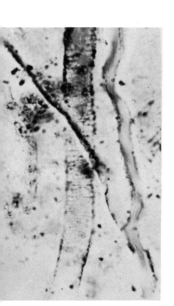

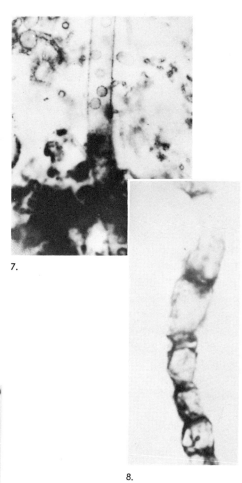

7.

6.

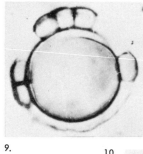

8.

9.   10.

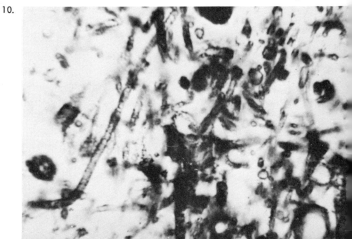

THE CAMBRIAN PERIOD

The Cambrian is separated from the Proterozoic by a long period of erosion which, among other things, destroyed almost all vestiges of Proterozoic life that may have been preserved in sedimentary formations. The Cambrian climate was probably more temperate than the present, principally because extensive, shallow seas inundated many portions of the world now covered by land. At their maximum extent during the Late Cambrian Period, they covered most of the United States. It was in the shallow inland waters and along the continental shelves that most Cambrian life flourished, plant as well as animal.

Probably the most prolific plant was calcareous algae, which built up extensive reef structures observable in many parts of the world today. Animals also appeared in abundance during the period, particularly brachiopods with nonarticulated, phosphatic shells and trilobites, including the large (approximately 18 inches long) Paradoxides. Such invertebrates as Onynchophora were widely distributed, and are survived today by the genus *Peripatus*. Whereas most pre-Cambrian life either floated or swam through ancient seas, by Cambrian times some forms had learned to crawl, or even burrow, at the bottom and many had developed the ability to convert lime into shells or skeletons. Why this occurred is not known, but presumably the protection they afforded served well in the ever-increasing struggle for survival amid a rapidly expanding marine animal population. There were no animals and probably no plants (except possibly some lichens) on land, making it as barren as in pre-Cambrian times.

THE ORDOVICIAN PERIOD

The Cambrian ended with a general withdrawal of epeiric seas, leaving a rather flat, featureless landscape. Climates were warm, partly due to the later incursion of seas over the land and partly due to the absence of extensive mountain barriers. Invertebrate life was prolific, particularly during middle and late Ordovician times. Graptolites, floating forms living in colonies, were everywhere, as were gastropods. Lime shelled brachiopods replaced more primitive Cambrian types. Starfish and blastoids appeared, though not in quantity, and cephalopods 10 to 15 feet long were in existence. Paleontologists have identified over 1,200 species of Ordovician trilobites and some ostracods. Most important was the appearance, for the first time, of primitive vertebrates, the ostracoderms, ancestors of modern fish and of all advanced forms of life evolved. An Ordovician scene is illustrated in Fig. 6.4.

THE SILURIAN PERIOD

The Silurian Period was characterized by subsequent invasion and withdrawals of seas, by considerable volcanic activity, and by a rather stable,

Fig. 6.4. Stranded seaweeds, trilobites and cephalopods on an Ordovician beach. Courtesy Chicago Natural History Museum and Charles R. Knight.

warmish climate. Some forms of invertebrate life advanced into new genera and species, others declined or withered away. Brachiopods, corals and sponges were common, trilobites became specialized, and bryozoa, a major source of limestone, was very abundant. Little progress was recorded among vertebrates. It is possible that some plants took root on land during the period but this is not certain. What is fairly certain, however, is that animals similar to scorpions in appearance made the transition from sea to land. Known as *Palaeophonus* and *Proscorpius,* they may have lived either entirely on land or partially on land, returning periodically to the sea.

THE DEVONIAN PERIOD

While very little is known of fish in the Silurian, fossil remains of Devonian fish are abundant. Ostracoderms continued to evolve and primitive sharks appeared. The largest fish was the *Dinichthys* of the Arthodira group, characterized by bony plates over the head and over the rear portions of the body. A carnivore, it often reached a length of over 20 feet. Some, but not all, Devonian fish possessed lungs; among those that did include the crossopterygians, some with characteristic lobe fins and others with paired limbs. It is probable that from the latter the first amphibians evolved, moving across swamp-like land on their fin-limbs, searching new pools, streams, and lakes. Invertebrate sea life prospered during the Devonian, with corals, sponges, crinoids and pelecypods being prevalent in the warmish waters

128 Biological Evolution

Fig. 6.5. Land plants proliferated in the Devonian Period, providing the food for the approaching invasion of land animals. Courtesy Chicago Natural History Museum and Charles R. Knight.

of the period. Trilobites were on the decline, however. On land the first spiders and a wingless insect have been identified.

One of the most spectacular developments of the Devonian was the proliferation of land plants (Fig. 6.5), among them the fern *Archaeopteris*, the leafless *Psilophyton*, and *Eospermatopteris*, a tree-fern which exceeded 40 feet in length and 3 feet in trunk diameter. The latter often grew in forests, creating a splendid contrast to the barrenness of earlier geologic periods.

THE MISSISSIPPIAN AND PENNSYLVANIAN PERIODS

The Mississippian and Pennsylvanian Periods, sometimes grouped together to form the Carboniferous Period, are a continuation of the Devonian—warm climate, somewhat dry in places, with a variety of invertebrates and vertebrates inhabiting the seas. Especially in the Pennsylvanian, the weather was moist throughout the year, resulting in the growth of luxuriant vegetation today associated with the tropics. Amid the giant ferns and other trees and plants lived all manner of flying insects (some of which grew to a very large size), the scorpion *Eoscorpius*, centipedes, and spiders. Nearly 90 species of amphibia have been identified, most of which lived in the coal swamps and in and along rivers. Studies of their teeth indicate that many were carnivorous, though they apparently did not masticate prior to swallowing. In appearance much like the Pennsylvanian amphibia were the first reptiles, represented by rare fossils in the rocks.

THE PERMIAN PERIOD

Climatically, the Permian Period differed radically from its predecessors. Epeiric seas withdrew, continents elevated, and former land bridges disappeared. One result of a changing planet was the widespread appearance of continental glaciation coupled, in nonglaciated areas, with rigorous conditions that led to the disappearance of much of the tropical vegetation of the Mississippian and Pennsylvanian Periods, and the substitution of more resistant growth such as conifers. Many types of marine invertebrates continued to flourish but the already declining trilobites died out altogether. Though fish and amphibians were common, the major developments associated with the vertebrates came with the advance of the reptiles, many similar to some types of amphibians. One form, the theriodont, exhibited numerous characteristics of the later mammals, and is often considered their ancestor. Among other characteristics, the theriodont's limbs were positioned much like those of mammals and their dentitic structure was similar. Typical reptiles and amphibians of the period are pictured in Fig. 6.6

Fig. 6.6. Reptiles and amphibians of the Permian, including the fin-backed *Dimetrodon* and *Edaphosaurus*. Courtesy Chicago Natural History Museum and Charles R. Knight.

Life During the Mesozoic Era

In this brief survey we have taken life forward from a point some 600 million years to approximately 220 million years ago. We have seen the emergence of the vertebrates, the conquest of the continents by an astounding variety of flora and fauna, the development of reptiles, and, if not mammals, at least premammalian life. With the advent of the great Appalachian revolution, the transition was made from the Paleozoic to the Mesozoic Era, a portion of geologic history that is divided into the following three periods:

Period	Time from Present, Years
Triassic	220,000,000 to 170,000,000
Jurassic	170,000,000 to 130,000,000
Cretaceous	130,000,000 to 78,000,000

THE TRIASSIC PERIOD

As the Appalachian revolution progressed there was a significant uplifting of continental masses that left, for example, most of North America (except for portions of the west) high and dry during the entire Triassic Period. The climate, on a world wide basis, was warm and often very dry, producing deserts or semideserts. On land the reptiles moved forwards towards ascendency, for the first time becoming completely independent of their aquatic heritage. Not only did they lay their eggs, and allow them to hatch, on land but they developed lungs that permitted them to breathe air from the moment they were born. The phytosaur, a crocodile-appearing animal, was common as were such relatively small dinosaurs as members of *Anchisaurus*. In the seas, ichthyosaurs, a reptile-like fish, appeared along with plesiosaurs, looking somewhat similar to modern snakes, and mammal-like theriodont reptiles typified by the *Cynognathus* (whose skull contained three eyes and who may have been warm-blooded and hair covered). The transition to mammals, with their highly developed nervous system, powerful circulatory system operated by a four-chambered heart, and hair instead of scale covered bodies, gave these new animals a much greater ability to respond to changing environmental conditions and to adapt themselves to many climates and many geographical locations. One reason given for their appearance in the first place was the arid climate that persisted throughout the Triassic.

THE JURASSIC PERIOD

The Jurassic was, like the Triassic, initially a period of widespread aridity, though conditions ameliorated as the years went by, producing in many areas a climate suitable for the growth of coal-forming vegetation. Except towards the end of the period the climate was warm—warm enough for plants to flourish in the Arctic and Antarctic.

Dinosaurs reached a high stage of development during the Jurassic—so much so that the period is often called the *Age of Dinosaurs*, and left remains that have been found in North America, South America, Europe, Asia and Africa. They attained dimensions (occasionally over 80 feet long) and weights (some more than 40 tons) that have never been equalled by other forms of land-animal life, before or since. Reigning supreme during the Jurassic and Cretaceous periods, they had become extinct by the end of the latter period for reasons that are still not resolved.

Largest of the dinosaurs were the sauropods, characterized by long necks and tails but relatively short bodies; because of their size and weight it is believed they passed the major portion of their existence in water. Unlike some types of dinosaurs, their young were born alive and from the beginning fed solely on vegetation. In the fashion of some modern lizards, they were able to regenerate a severed tail. The best known sauropod is the *Brontosaurus* (Fig. 6.7) who, despite his size and weight (27 to 32 tons), was often the prey of the carnivorous, two-ton *Allosaurus*. This creature walked on his hind legs and was fitted with a set of jaws that could bite huge hunks of meat from its victim.

Sea-faring dinosaurs are typified by the famed ichthyosaurs, a shark-like reptile. Other Jurassic marine creatures were turtles and crocodiles. In the air, flying reptiles prospered, the best-known example being the pterosaur, whose wings would often span more than 20 feet. Feathered birds also flourished during the period, the *Archaeopteryx* being an example.

Jurassic mammals were probably fairly abundant, but remained a secondary form of life throughout the period. One form was egg-laying, but little is known about it. Most of these early mammals were of rat size, though a few attained the dimensions of small dogs.

In the seas, fish and amphibians, along with many forms of invertebrates,

Fig. 6.7 Most famous of the sauropods, *Brontosaurus*, is shown in a Jurassic swamp. Courtesy Chicago Natural History Museum and Charles R. Knight.

found ideal conditions to continue their evolution. Pelecypods became very widespread and ammonites reached their peak, as did the belemnites—a cephalopod looking much like the modern squid. Lobsters and crabs lead the crustaceans. Few brachiopods occurred, but gastropods became common.

Finally, back on land, but turning our attention to flora, we find the Jurassic hospitable to plants on a world-wide basis, the same species flourishing in higher latitudes as in the lower. Aside from the omnipresent ferns and rushes were giant cycadophytes and gingkos. Conifer forests were plentiful.

THE CRETACEOUS PERIOD

Much of the Cretaceous Period was characterized by marine invasions of the lands—widespread epeiric seas that probably inundated more land areas than ever before in history. During part of the period a good half of North America alone was covered by water. Climate was mild, even near the poles, as would be expected because of the extensive bodies of warm water flowing deep into the continents.

Dinosaurs continued to dominate the animal kingdom throughout most of the Cretaceous, but near the end of the period they died out completely. Probably many factors contributed to their extinction, including disease, vulnerability of their eggs, and changing climatic conditions. By the Cretaceous they had become specialized animals, used to abundant water, luxuriant vegetation, and a tropical, or at least subtropical, climate. Towards the end of the period uplifting was common, which changed the environmental conditions—drying up ponds, swamps and lakes, altering the vegetation, and leading to the decline of the herbivorous dinosaurs who were not capable of adapting themselves to a new world. Carnivorous dinosaurs, whose diet often was based on plants, also passed into oblivion. Among the leading dinosaurs of the Cretaceous were *Tyrannosaurus rex*, a monstrous 7-ton carnivore, the horned *Triceratops*, and the duck billed *Trachodons* (see Fig. 6.8). In the seas lived ferocious mosasaurs, ichthyosaurs, and many types of crocodiles. Birds were numerous, both in type and in number, and flying reptiles, the pterosaurs, thrived; in fact, the largest flying vertebrate so far discovered, called the *Pteranodon*, dates from the period.

Invertebrate life was more or less a continuation of the earlier Mesozoic periods. Foraminifera became more numerous as did pelecypods and some forms of oysters. But brachiopod and ammonites apparently declined. Crustacean evolution was spearheaded by crabs. Faunal progression on land yielded many types of essentially modern plants, including flowering angiosperms (characterized by having their seeds in a closed ovary), sequoias, birch, oak, and maple trees, and many fruit shrubs, along with grasses and cereals. As an evolutionary development, the appearance of the angiosperm

Fig. 6.8. Horned *Triceratops* (left) and the gigantic *Tyrannosaurus Rex* (right) share this Cretaceous scene. Courtesy Chicago Natural History Museum and Charles R. Knight.

was of enormous significance, since it forms the basic food source for the vast assemblage of mammalian life that inherited the Earth at the close of the Cretaceous.

The Cenozoic Era

The last era in the geologic history of the planet Earth is the Cenozoic. It is the era that witnessed the shaping of the world we live in, that saw the rise of the mammals as the dominant form of life, and that ushered in the creature we know as the human being, or *Homo*. Many geologists divide the era into two periods: Tertiary and Quaternary. The latter has been discarded by other geologists, who favor referring only to the seven epochs, Paleocene, Eocene, Oligocene, Miocene, Pliocene, Pleistocene, and Recent. From the point of view of life, we shall cover the era as a whole rather than attempting to break it down into epochs, though the epochs are referred to continually. The approximate times they span are

	Epoch	Time from Present, Year
	Paleocene	78,000,000 to 55,000,000
	Eocene	55,000,000 to 34,000,000
TERTIARY PERIOD	Oligocene	34,000,000 to 25,000,000
	Miocene	25,000,000 to 12,000,000
	Pliocene	12,000,000 to 1,000,000
QUATERNARY PERIOD	Pleistocene	1,000,000 to 10,000
	Recent	10,000 to present

During most of the Cenozoic rather warm climates prevailed, permitting crocodiles to inhabit Minnesota, magnolia trees to grow in Alaska, and temperate zone forests to flourish in Greenland; however, towards the end of the Pliocene, a definite cooling trend became evident leading directly to the great Pleistocene glaciations. Once the Ice Age had appeared, or rather Ice Ages—for there were four distinct glacial stages—significant changes in fauna and flora were bound to occur. Where the ice advanced, life ceased, though subarctic types occurred along the glacial perimeters; when it retreated, both plants and animals gradually moved into erstwhile glaciated regions, only to fall back as the ice again moved southward. Evidence points to interglacial stages ranging from over 100,000 to some 300,000 years in length, considerably longer than the glacial stages themselves. The climate during at least portions of the interglacial stages was warmer than at present, so we can not think of the entire Pleistocene Epoch as an age of ice and frigid weather.

Most of the invertebrate life we are familiar with today ocurred during the Cenozoic. Among the marine vertebrates, fish life was essentially modern, teleosts and sharks being particularly numerous. The great reptiles of the Mesozoic had disappeared, leaving only the smaller types that still occur (snakes, crocodiles, and the like). Most modern forms of amphibia existed during the Cenozoic. In the invertebrate world gastropods, pelecypods, bryozoa, echinoderms and foraminifera flourished.

Our major attention in the Cenozoic Era is focussed on the rise of mammals, warmblooded animals that replaced the reptiles as masters of the planet. Nature provided this new form of life with a more complex nervous system, a larger brain, a more efficient protective covering, and the ability to move rapidly. In short, they were better equipped to respond to variations in the environment and, in the case of man, to actually modify the environment if it were not suitable to him.

Most of the Paleocene mammals are extinct, including archaic marsupials, the hoofed Condylarthra, the bear-like creodonts, and the unitatheres, large animals somewhat like the modern rhinoceros. During the Eocene Epoch many of the Paleocene mammals died out giving way to the direct ancestors of such modern types as the camel, the horse, the elephant, the dog, the cat, and the primate. As the Oligocene, Miocene and Pliocene passed into history all manner of foxes, dogs, wolves, saber-toothed cats, horses, and bears prospered. During the latter epoch huge sloths, some up to 20 feet long, evolved and, though now extinct, apparently lasted in southern South America up until prehistoric times. Giant armadillos, called glyptodonts, flourished in the same continent during the Pliocene and Pleistocene. Several varieties of giant pigs, or entelodonts, roamed the Earth during the Miocene along with the herbivorous oreodonts. During Miocene and Pliocene times, North America harbored several forms of rhinoceros (Fig. 6.9) and, in Asia, from the Oligocene through the Miocene, was found the enormous *Balchitherium,* often measuring 25 feet from head to tail. In the cooler Pleistocene the woolly rhinoceros appeared on the scene but did not survive into modern times. Other Cenozoic animals are the bison

Fig. 6.9 Pliocene rhino, mastodonts, and oreodonts. Courtesy Chicago Natural History Museum and Charles R. Knight.

Fig. 6.10. Oligocene titanotheres, large North American mammals. Courtesy Chicago Natural History Museum and Charles R. Knight.

Fig. 6.11. Skeleton of an Eocene *Eohippus*. Courtesy Chicago Natural History Museum.

(Pleistocene), titanotheres (Eocene and Oligocene, Fig. 6.10) and *Syndyceras* (Miocene). Small, foot-high Eocene horses (*Eohippus;* Fig. 6.11) evolved into the Oligocene *Mesohippus,* the Miocene-Pliocene *Merychippus,* the Pliocene *Pliohippus,* and the modern *Equus.* During the evolution of the horse the original four toes gradually became one strong, middle toe, known to us the hoof. The first mastodons date from the Miocene and persisted, in South America at least, into prehistoric times. From these animals elephants evolved, including the woolly mammoth famed for their long tusks. The saber-tooth tiger, whose skeleton is seen in Fig. 6.12, evolved and persisted until rather recent times.

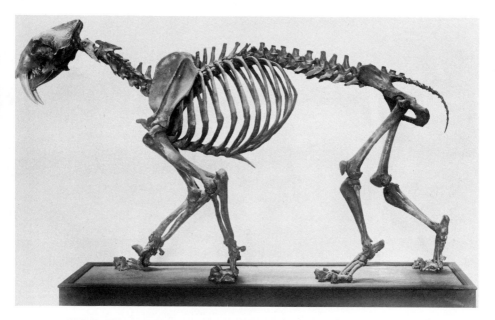

Fig. 6.12 Saber-tooth tiger, an important Pleistocene mammal, from the Labrea tarpits of Califonia. Courtesy Chicago Natural History Museum.

The Rise of Intelligent Life

Man, a mammal, is but one of many forms of primates, some living, some extinct. Through a line of succession not yet completely clear, it is believed he descended from the earliest known primate, the lemur, who first appeared in the Paleocene. Further down the evolutionary line came the tarsier (genus *Tarsius*), arboreal animals with furry coats, large eyes, and brains of exceptional size considering the rat-like dimensions of the primate; and then the monkeys. *Tarsius* is reconstructed in Fig. 6.13. These primates

Fig. 6.13. *Tarsius.* Courtesy American Museum of Natural History.

date, respectively, from the Eocene and Oligocene. By the Miocene and Pliocene, the anthropoid apes (pongids) flourished; they were the direct ancestors of the modern gorilla, gibbon, orangutan and chimpanzee. These ancestors have been placed in the subfamily *Dryopithecinae;* Miocene remains found in Africa of this subfamily are members of the genus *Proconsul,* which is divided into three species *P. africanus* (gibbon size), *P. nyanzae* (chimpanzee size) and *P. major* (gorilla size). At the same time in Europe and Asia, animals of the genus *Dryopithecus* were evolving with characteristics of both apes and the so-called hominids (manlike creatures that walked on their hind feet, leaving their hands available for nonwalking activities). Particularly in China, teeth belonging to *Dryopithecus keiyuanensis,* discovered in Pliocene deposits, exhibit definite hominid, or at least prehominid, attributes. In India, a Pliocene hominid-like creature known as *Ramapithecus brevirostis* has been identified by its dental structure; at least one anthropologist postulates that it is indeed the first known member of the hominid subfamily. Other animals with hominid tendencies are the *Kenyanthropus wickeri,* turned up in Pliocene beds near the Gulf of Kavirondo in Kenya, and the *Oreopithecus bambolii* of Miocene-Pliocene times found in central Italy. All of these creatures may have been direct ancestors to man but the connecting links between them and the latter still have to be conclusively demonstrated.

According to Coon[17] the oldest fossil-man remains that are "definitely and indubitably Homo" may not exceed 700,000 years in age. Any determination of the true age of *Homo* must be based on evidence of the transition to this genus from an ancestral form of primates. As emphasized by Coon, much of the problem stems from the nomenclature of taxonomy, which places all modern men in the category *Homo sapiens*[18] but which, at least until recently, gave different genera and different species designations to major human—or subhuman—fossil discoveries. Because of the fact that paleontologists rarely have more than a skull or a few bones to work with—and not the complete bodies available to the zoologist—they cautiously allocate to their early fossil finds individual generic designations, as for example *Pithecanthroupus erectus, Meganthropus palaeojavanicus, Sinanthropus pekingensis,* and *Homo neanderthalensis*. But, many modern anthropologists are anxious to abandon the tendency to consider these fossil men as belonging to different genera, taking their stand from both anatomical and behavioral evidence. Writes Coon, "...*Homo* is what we are, what our known ancestors were, and what our unknown ancestors could have been for as long as eight million years."

The family of Hominidae are represented by only two genera, *Australopithecus* and *Homo*. The most recent evidence indicates that the former flourished in the Lower and on into Middle Pleistocene times whereas the latter is only known beginning in Middle Pleistocene. One of the big mysteries of evolution is whether *Homo* descended from, and replaced, *Australopithecus* by evolution from one species to another or whether the two genera came from a common ancestor, the one dying out and the other persisting.

It is believed that the Australopithecines lived for a period of about half a million years. Anthropologists have fairly well established that beings of the genus *Australopithecus* (Fig. 6.14) in Africa were able to make and use stone tools between 1,750,000 and 2,000,000 years ago, the oldest sites being located at Olduvai George in Tanganyika and the nearby Garusi. Kanam in Kenya also contains hominid remains. Other sites are located near Largeau in the Schad, Aïn Hanech in North Africa, Kanyatsi in Uganda, and Taung, Sterkfontein, Swartkrans, Kromdraii and Makapansgat in South Africa. From archeological work in other parts of the world, early hominids apparently also flourished in the Near East and in Java.

Some of the hominids living in South Africa were small—less than 5 feet tall—and weighed under 100 pounds, whereas others were more com-

---

[17] Coon, Carleton S., *The Origin of Races*. New York: Alfred A. Knopf, 1963, pp. 10–11.

[18] *Homo* is the genus, *sapiens* the species; man is also a member of the animal kingdom, the chordate phylum, the vertebrate subphylum or group, the class of mammals, the order of primates, and the family Hominidae.

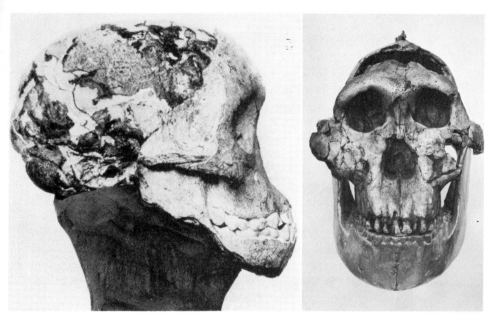

Fig. 6.14. Right side view of *Australopithecus*. Courtesy American Museum of Natural History. Right: Skull of *Zinjanthropus Boisei* found at Olduvai Gorge, Tanzania, 17 July 1959. Courtesy Richard E. Leakey and Armand Denis Productions.

parable to modern human beings, and, more than incidentally, they came somewhat later in time. Brain sizes varied from less than 500 cc to possibly 700 cc, much below humans, who can have brain case volumes of 1,800 cc and, occasionally, even greater. As far as we know none of these Australopithecines lived in caves because, as pointed out by Coons, they are dark "and hominids do not have night vision. Caves are dank and clammy; without fire they are uncomfortable. Caves also harbor predatory beasts... We have no evidence that human beings lived in caves before they had fire. And none of the Australopithecines had it."[19]

Louis Seymour Bazett Leakey has for years been a leader in attempting to connect the various links that form the chain of human evolution. In 1959 at the Olduvai Gorge in Tanzania he found the skull of an Australophithecine which became known as *Zinjanthropus Boisei* (Fig. 6.14), and, towards the end of the following year, he discovered portions of an 11-year old child's skeleton 5 feet *below* the same sedimentary formations. It is strongly suspected, from detailed studies of the child's teeth, that it was an intermediary between *Proconsul* or other Miocene-Pliocene apes and *Homo*. At first Leakey felt that the Olduvai child probably belonged to the genus *Australopithecus* since it definitely appears more human than the creatures discovered in South Africa. In 1964, however, on the basis of later finds

---

[19] Coons, *Op. Cit.*, pp. 236–37.

and reinterpretations of earlier discoveries, he suggested that the Olduvai child and other specimens belong to a new species he calls *Homo habilis*, more advanced, paradoxically, than Zinjanthropus remains found in beds *above* the Olduvai child; the latter, though older than the former appears, from an evolutionary point of view, more advanced.

In the words of Leakey, Tobias, and Napier in their report[20] "We have come to the conclusion that, apart from *Australopithecus* (*Zinjanthropus*) the specimens we are dealing with from Bed I and the lower part of Bed II at Olduvai represent a single species of the genus *Homo* and not an australopithecine." They characterize the new genus as having a "mean cranial capacity greater than that of members of the genus *Australopithecus*, but smaller than that of *Homo erectus* ... chin region retreating, with slight or no development of the mental trigone..." (etc.). Pointing out that its geological horizon is Upper Villafranchian (the earliest portion of the Lower Pleistocene) and Lower Middle Pleistocene, they note that *habilis* remains have been found both in Olduvai Bed I and the lower and middle part of Bed II, some sites being geologically older than those which yielded the Zinjanthropus australopithecine. They conclude that "two different branches of the Hominidae were evolving side by side in the Olduvai region" during the just mentioned geological epochs.

In commenting on the implications of their researches on hominid phylogeny, Leakey and his associates write "... we have not overlooked the fact that there are several other African (and perhaps Asian) fossil hominids whose status may now require reexamination in the light of the new discoveries and the setting up of this new species." They cite examples of hominid remains assigned to both *Homo erectus* and *Australopithecus* that may actually be *Homo habilis*, not only in Africa but in the Middle East.

We must wait and see what other anthropologists have to say about Leakey's proposal which, if sustained, pushes the appearance of *Homo*—though not *Homo sapiens*—incredibly far back into prehistory, perhaps some 2,000,000 years instead of some 700,000 years (the probable time of the appearance of *Homo erectus*). Whoever he was, *Homo habilis* walked in an erect position, was short (3.5 to 4.5 feet), and could work with his hands, though not to the extent that modern man can. Unlike the vegetarian Zinjanthropus, *Homo habilis* was omnivorous. And he may have lived in crude shelters in which he sought protection from the elements.

The Zinjanthropus remains found above the Olduvai child give evidence

---

[20] Leakey, L. S. B., P. V. Tobias, and J. R. Napier, "A New Species of the Genus *Homo* From Olduvai Gorge," *Nature*, 202, No. 4927 (4 April 1964), 7 (see also following articles in same issue: Tobias, P. V. "The Olduvai Bed I Hominine with Special Reference to its Cranial Capacity," p. 3; and Leakey, L. S. B. and M. D. Leakey, "Recent Discoveries of Fossil Hominids in Tanganyika: At Olduvai and Near Lake Natron," p. 5).

of a race probably related to the South African Australopithecines. He could work with tools, build simple shelters, and stalk game. His cranial capacity was not particularly large, ranging from 640 to 725 cc according to the latest estimates. He apparently coexisted for a long time with *Homo habilis* —if this terminology is correct to describe a species of man.

Leaving the *Homo habilis* evidence to be weighed and judged in anthropological circles, we shall continue our discussion based on pre-1964 interpretations of fossil finds. By Middle Pleistocene we find *Homo* slowly replacing *Australopithecus,* and, in a relatively short period of time, the transition from the one to the other had been completed. The advantages the Australopithecines held over other forms of animal life (some sort of prehuman intelligence, the ability to use tools and to hunt, at least a semi-erect posture, and perhaps a primitive social organization) were not enough to compete with the newcomers—if, indeed, they were newcomers; it is simply unknown if one branch of the Australopithecines became extinct while another merged into *Homo,* or if they, as a whole, became *Homo* as their intelligence and social structure improved.

Once we have left the ancestral *Australopithicus* we are confronted with both *Homo erectus* and *Homo sapiens* and the obvious problem of determining which fossil remains belong to which species. The so-called *Pithecanthropus erectus,* of Middle Pleistocene, exhibits a moderately low cranial capacity (about 900 cc) and strong ape-like features; an artist's conception of him appears in Fig. 6.15. Found in Java, he could certainly walk in an erect posture. From Java also come the Solo Skulls, Fig. 6.16. Middle Pleistocene China is the home of the famed Peking man, *Sinanthropus pekingensis* (Fig. 6.17), with a slightly larger cranial capacity (1,000 cc), recessed forehead, and protruding jaw. There is good evidence that this ape-man had some sort of social organization and could work with simple tools. On an absolute scale he lived, according to argon-potassium dating, some 360,000 years ago. Other members of *Homo erectus* are probably specimens retrieved from Tze-yang in China and from the Olduvai Gorge (in Middle Pleistocene Bed II considerably above Bed I in which the Olduvai child and Zinjanthropus were found; the remain is known as the Chellian-3 skull). The cranial capacities of the skulls of these fossil men range from under 800 to nearly 1,300 cc, compared to a modern skull range of 1,200 to 1,800 cc and occasionally above. Somewhere around 1,250 cc we cross over the threshold of *Homo sapiens,* but it is obvious that brain size alone cannot be used to gauge the relative intelligence and evolutionary maturity of fossil men. Thus, anthropologists look at such other features as the form of the skull, the nature of the teeth, and the flatness of the face.

Above the Djetis hominid-bearing formations in Java have been found parts of a cranium and two teeth from a member of *Homo erectus* known as

142   Biological Evolution

Fig.6 .15 *Pithecanthropus erectus*, according to artist's reconstruction. Courtesy American Museum of Natural History.

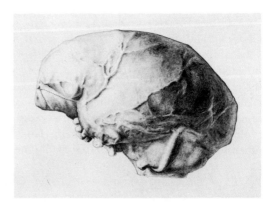

Fig. 6.16 The Solo brain. Courtesy American Museum of Natural History.

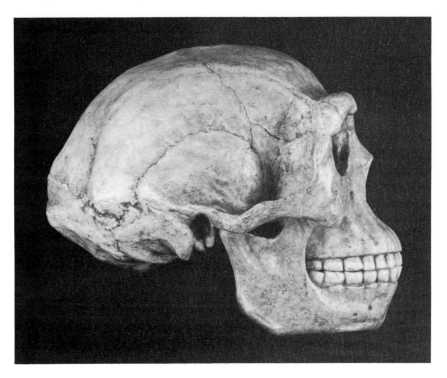

Fig. 6.17. Restored skull of *Sinanthropus pekinensis*. Courtesy American Museum of Natural History.

Pithecanthropus 4, together with a piece of lower jaw given the designation Pithecantropus B and a portion of the skullcap of a two-year old baby referred to both as *Homo Modjokertensis* and simply Modjokertensis. To the west at Trinil, skullcaps belonging to Pithecanthropus 1, 2, and 3 have been retrieved from sediments. Pithecanthropus 1, commonly appearing in the literature as *Pithecanthropus erectus* (mentioned in this section), has a cranial capacity about that of Pithecanthropus 4 (approximately 900 cc), whereas No. 2 is less than 800 cc; No. 3, in the meantime, belongs to a child. Nos. 1, 2, and 3 all belong to beings who lived after Pithecanthropus 4, probably about 200,000 years ago. While the brain size of the former approximated that of the latter, the appearance of its skull is much more primitive.

Other *Homo erectus* remains in Java were found near the Solo River in the central portion of the country, above the Trinil formations. From measurements of their cranial capacity, ranging from over 1,000 to just over 1,250 cc, we see that evolution had been at work since Pithecanthropus, leaving the Solo man a more advanced member of *Homo erectus*. Later remains have been found in Wadjak, Java, dating probably from the end of the Pleistocene epoch. Wadjak man's cranial capacity was approximately

1,500 cc, bringing him definitely into the *Homo sapiens* class. From the earliest member of Pithecanthropus to the Wadjak a period in excess of 400,000 years had passed; and *Homo erectus,* exemplified by Solo man, had somehow evolved into *Homo sapiens,* exemplified by Wadjak man. The former species still existed on the island less than 100,000 years ago, some 10,000 years after the first *Homo sapiens* had appeared on the scene.

Turning our attention to Australia, anthropologists have carefully examined several fossil skulls (found at Keilor, Talgai and Cohuna) dating less than 10,000 years old, and have used them to establish definite links between Wadjak man and the present-day Australian aborigines. The latter, it appears, still retain traits of *Homo erectus,* one being small skull size (some aborigines have cranial capacities under 1,000 cc).

To the north, in China, excavations have taken place for years in search for evidence of fossil man, particularly the ancestors of the Mongoloids. The earliest find was made at Choukoutien, China where a molar was discovered, attributed to *Sinanthropus pekinensis.* On an absolute scale he lived approximately 360,000 years ago (about 200,000 years before Solo man) but was at roughly the same evolutionary stage as the Java being. Later, the man came to be known by many anthropologists as *Pithecanthropus pekinensis* and, most recently, as simply *Sinanthropus,* a member of *Homo erectus.* His cranial capacity is virtually identical to Solo man, ranging from slightly over 1,000 cc to 1,225 cc, but his skull is longer and narrower. Clearly, evolution marched more rapidly in China than in Java.

Numerous other fossil man finds have been made, both in China and Japan, of Late, Middle and Upper Pleistocene times, including those at Ting-tsun in Shansi province, Mapa in Kwangtung province, and Liu-kiang in Kwangsi province, as well as in the Ushikawa Quarry in Japan. All seem to be related to, and descended from, Sinanthropus, and probably were. Coming closer to the present, in the so-called Upper Cave at Choukoutien skeletons were unearthed dating back about 10,000 years. Some have strong Mongoloid features, others do not, suggesting the former died out during a transition period leading to modern Chinese stock.

Tracing the lineage of the Caucasoids is a rather complex task, partly because their original breeding grounds remain unknown, partly because they were in continuous contact with neighboring fauna and with other subspecies of man. The oldest known remain to be unearthed in Europe is the Heidelberg lower jaw, representative of a being (whether *Homo erectus* or *sapiens* is unknown) that dates from about the same time as Sinanthropus, approximately 360,000 years ago. A more recent discovery is the Steinheim cranium found in a gravel pit not far from Stuttgart, Germany; it is some 250,000 years old and definitely *Homo sapiens,* with a cranial capacity of over 1,150 cc.

Biological Evolution 145

Another well-known find was the Swanscombe skull in gravels near the Thames river in England, with a volume of approximately 1,300 cc. The initial find, in the Middle Gravels at the Swanscombe, Kent site, consisted of a left parietal and occipital and was unearthed in January 1935 by Alvan T. Marston. Twenty years later, in July 1955, B. O. Wymer and John Wymer found a right parietal skull bone close to the place where Marston turned up earlier Swanscombe Man remains. "A comparison of the two sections shows that there can be no doubt that the right parietal was found in the same stratum as that in which Mr. Marston found an occipital and left parietal in 1935-36. This renders the fact that all three fragments fit each other perfectly, belonging to the skull of the same individual, a little less remarkable."[21] Flint flakes found during the excavations showed that a contemporaneous flint industry had been developed by the 250,000 year old race. Figure 6.18 shows the discovery site of the Swanscombe skull

---

[21] Wymer, J., "A Further Fragment of the Swanscombe Skull," *Nature*, No. 4479 (3 September 1955), 427.

Fig. 6.18. Left: Section of the Swanscombe discovery site in Kent, England showing the various interglacial terrace deposits and the skull level. Courtesy Peter O. Wymer. Right: the right parietal bone being examined by A. Rixon and B. O. Wymer. Courtesy Associated Newpapers Ltd.

bones; at the right is a photograph of the right parietal being examined in London.

Even larger than the Swanscombe skull is the long, wide and low Fontéchevade skull discovered in Charante, France; measurements place it at least 1,460 cc. These and other discoveries demonstrate that very primitive *Homo sapiens* flourished in Europe for hundreds of thousands of years, exhibiting, however, many erectus characteristics. Oddly enough, while no *Homo erectus* remains have been located in Europe, the oldest known *sapiens* skulls come from that very continent.

About 75,000 years ago a people known as Neanderthals (taken from the name of a river valley in Germany near Dusseldorf where the first fossil skullcap was found) inhabited Europe and western and central Asia where they lived for about 35,000 years. It is uncertain just how the Neanderthals developed, whether from Caucasoid and Sinanthropus Mongoloid interplay from North African infusions into primitive European stock, or through processes of mutation and natural selection from the stock itself. Whatever their origins, they appeared during the latest interglacial stage and lived well into the most recent glaciation. They were fairly short individuals, though powerfully built, and were endowed with a cranial capacity between 1,300 and over 1,600 cc comparable to that of modern man (one skull, found in a cave at Shanidar, Iraq may turn out to be even larger—some 1,700 cc, making it the largest fossil skull found for a man living at this early time, which was over 45,000 years ago). The Neanderthals generally lived in caves, were adept at working with tools, and successfully developed simple weapons, including axes and slings. As the last glacial epoch came to a close, the Neanderthal man was absorbed by a newcomer of Upper Paleolithic culture. Following the Paleolithic invasion, Neanderthal as such became extinct, not only in Asia and central Europe, which he only sparsely inhabited, but in Western Europe where he was present in relatively large numbers. A restoration of a Neanderthal is presented in Fig. 6.19.

The origin of the Upper Paleolithic Europeans is still uncertain, but it seems likely that these peoples moved westward from the Near East, bringing with them some aspects of their culture and developing others in their new breeding ground. Flourishing from roughly 30,000 to 10,000 years ago, they were very similar to modern Caucasoids, though slightly shorter. We know a good deal about their appearance and culture from sites in France, Germany, Britain, Spain, central Europe and Russia. Cranial capacities are large, averaging nearly 1,600 cc for males and 1,370 for females. Particularly in France, these people developed a definite artistic sense, producing remarkable paintings on cave walls. They knew how to make, and use efficiently, tools and weapons, and had a tendency to

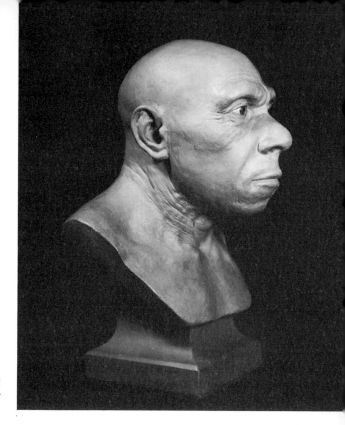

Fig. 6.19. Restoration of Neanderthal man by Dr. J. H. McGregor. Courtesy American Museum of Natural History.

live in tents and huts in addition to caves. Their stock persists in many parts of Europe today.

We have come essentially to modern times in this brief survey of the progress of pre-man and man. We have, necessarily, left many eras untouched, for example, the development of *Homo sapiens* in Africa, the question of the pygmies and the origin of the Negro, but we have sketched the major lines of development. We learned that modern man evolved from the extinct species *Homo erectus,* the first members of which probably appeared some 700,000 years ago and persisted to perhaps 100,000 years ago. He gave rise to five known subspecies, all of which, at one time or another, passed the *Homo sapiens* threshold, the first possibly making the transition as many as 250,000 years ago. And, about 35,000 years ago, man appeared in essentially his modern form. Pending further analysis and research we were forced to mention only briefly the proposed *Homo habilis* interpretation of certain Olduvai remains. Over the long period of time man has been on Earth he has produced an estimated 96 billion babies, of which 19 billion were probably born prior to the advent to *Homo erectus* and 77 billion after.[22]

---

[22] Figures from the Population Reference Bureau.

Trends in Evolution

Our survey of biological evolution on Earth was both short in treatment and long in time. In it we saw life progress from extremely simple organisms that could barely be identified as living to exceedingly complex, highly developed mammals of which man is the culmination. We learned that species appear on the horizon, evolve and die out, a rule for which there never has been an exception. That there never will be is not safe to say, considering the manipulations man may be enticed to undertake in the otherwise natural processes of selection and speciation.

The principal cause of evolution, i.e., the changes that occur over many generations, is *mutation,* which is a spontaneous chemical change in the genes (that part of a chromosome which determines hereditary characteristics) or chromosomes (giant molecules in the cells of a species). Other causes include *recombination,* wherein series of gene-molecules come apart and recombine or rearrange themselves into new groupings; and *selection,* the process that determines which of the mutations and recombinations are desirable and are to be continued and which are undesirable and are to be eliminated. For natural selection to function efficiently and give rise to a new species better able than the old to enter into beneficial equilibrium with changing environments, as high a degree of isolation as possible is necessary. An isolated gene pool, with no inflow from other populations, is more apt to eradicate unfavorable genes than one affected by neighboring genetic material.

A species may either die outright—become extinct for any of dozens of possible reasons,[23] or survive partially in the sense that some of its characteristics are passed on to descendent species; this can be accomplished by either *branching* (separated populations of a species merge into new species) or *succession* (a new trait appears in a population that may be governed by a single gene or combination of genes—which, over the course of time, supersedes the old by the process of natural selection.) Only the future will reveal whether or not man will be so able to manipulate these phenomena that they can be stopped entirely, regulated as to speed of occurrence, or even accelerated.

In terms of his own survival he may want to tamper with nature. Since the Pleistocene epoch the average lifetime of a given mammalian species has been about 350,000 years. According to Coon,[24] "At a point in time

---

[23] There are a number of examples of animals that have become extinct during historical times. And in the plant world, the ravages of disease can, in a very short period, eradicate vast areas of growth. A recent example is the destruction of more than 80 per cent of the trees in millions of acres of forest in Honduras by the bark beetle, or *Mexicanus dendrectonus*. Infections such as this could lead to the extinction of a specialized plant or animal.

[24] Coon, *Op. Cit.,* p. 33.

pegged at 300,000 years ago, all or nearly all the living mammals of the European and neighboring fauna, which were fox-sized or larger, had come into existence." And, during the "last 75,000 years, no new mammalian species seem to have evolved at all." If, indeed, *Homo sapiens* first appeared in archaic form some 250,000 years ago, the species may only have another 100,000 years to go...if only *natural* selection mechanisms are allowed to function. We can safely assume they won't be.[25]

We are impressed by the continuous progress of evolution that has shaped life as we know it today—first organisms of incredible simplicity, then the invertebrates, next the vertebrate fish, followed by the amphibians, the amazing reptiles led by the dinosaurs, and ultimately the mammals. Much about evolution can be gained by studies of comparative anatomy; for example, the limbs of dogs, birds, and man are remarkably similar, as are such vital organs as the heart, the lungs, and the liver. Studies of the embryonic development of offspring show conclusively that they undergo many of the phases their ancestors experienced millions of years earlier. An example of this is the appearance of gill slits in all vertebrates, including *Homo sapiens.* Even more striking is the frog, which is hatched in water as a gill-breathing tadpole only later to lose its gills, develop lungs, and breathe air. This phenomenon, known as the *law of recapitulation,* was first proposed by Ernst Haeckel in 1866; in essence it states that the development of an individual animal recapitulates the development of the forms of life that preceded it. Also important to an understanding of evolution is the study of so-called vestigial organs in the anatomies of animals. The best known example are vestigial muscles associated with the human ear; our primate ancestors found it indispensible to be able to move their ears, particularly to identify sounds that might presage danger.

All life is in dynamic equilibrium with its environment and with other members of its own species. Some species compete more heavily in a given region than others, the degree of competition normally depending greatly on their feeding habits. Those that consume only a few specialized foods might do well in a given area for a given time, but they would not be expected to respond efficiently to environmental changes that might alter the growing area of their particular foods. Euryphagous animals, that is animals who can consume a wide selection of foods rather than only a few, are clearly better adapted to reconcile themselves to shifting conditions and surroundings than their stenophagous (specialized feeding) cousins.

---

[25] Towards the end of March 1964 the Rockefeller Institute reported that individual genes responsible for particular cell characteristics had been isolated by Dr. Muriel Roger. It appears possible to alter the nature of a cell in terms of a specific characteristic by injecting one type of gene. This is simply an example of what scientists are doing that may lead to significant artificial tampering with otherwise natural evolutionary tendencies.

This, in turn, implies a greater probability towards evolution rather than extinction.

Even a cursory study of evolution on Earth demonstrates the fact that *Homo sapiens,* in the precise form that he exists today, emerged as a result of an incredible succession of improbable events. If conditions had been just a little different, if fate had been just a little more fickle, man as we know him might not have inherited the Earth. But this does *not* suggest the some other intelligent creature having a similar morphology and filling a similar ecological niche would not sooner or later have emerged if evolution had taken a slightly different path during the past hundreds of millions of years.

We shall not attempt to trace the sequence of events that were, but might not have been, responsible for the arrival of man. Ever since the phylum Chordata, to which we belong, branched off from the coelenterates in the dim mists of geologic history, evolution gave no indication of following a predetermined, or specific path towards intelligence. Slater gives example after example of how precarious were the circumstances that permitted us to come into being.[26] For one, if only the first of two main branches from the two-layer coelenterates had been tried out back in Cambrian times, man would not have come about. When, during the Silurian Period, animal life finally went on to the land it was only with reluctance and might not have occurred until much later. Slater cites J.Z. Young's *The Life of Vertebrates* (Oxford: Oxford University Press, 1950) to say that "the amount of land on the Earth had varied between 4,000,000 and 8,000,000 square miles throughout the preceding 200 million years during which higher life forms flourished in the sea without obtaining a foothold ashore."

When vertebrates finally moved landward, only one of the then existing four classes did so—the bony fish. The others stayed in the sea, including the one with the most developed brain—the cartilage fish. Those fish that passed on to the land did so at first only to attain fresh water habitats; lungs came about in the Devonian as an emergency device to permit the fish to survive ever more frequent droughts. Slater writes that "animals which come ashore in the course of evolution do not necessarily develop lungs. Not only that, but a large proportion of those who did so returned to the water as their evolution progressed."

The bony fish gave rise to a subclass which divided into two orders, one of which was the lobe-finned fish that was the ancestor of the amphibians. And while amphibians evolved legs from their fins, they did so only because

---

[26] Slater, Alan E., "The Probability of Intelligent Life Evolving on a Planet," in *VIIIth International Astronautical Congress Proceedings,* ed. F. Hecht. Vienna: Springer-Verlag, 1958, pp. 395–402.

they were "useful for resting on mud at the bottom of stagnant water"—not because they wanted to walk. But walk they finally had to do, to move from a dried up pool to one with water still in it.

From amphibians to reptiles is another long story of myriad events in which chance loomed high among the causitive agents. From one of 16 reptilian orders came mammals—and this order, the therapsides, only lasted 50 million years before becoming extinct. If mammals hadn't made it in time they would never have gotten a start. And, having appeared, Slater writes that they then had to "lie low for nearly 100 million years in order to keep out of the way of the larger reptiles."

Of 29 mammalian orders one, the primates, led to man. Slater observes that primates are adapted to trees, "so, if there were never any trees on Earth, there would presumably be no primates." This seems to be good evidence that a primate-type creature would not evolve on a world offering no forests, though it is almost impossible to imagine a habitable planet without trees of some sort. The arguments continue to show that, were it not for a small "if" man would not be around. All of this convinces him that if everything had not happened just as it did "there would never have been intelligent life on Earth."

We shall answer this type of argument presently. For the moment, let us slightly change Slater's conclusion to "if everything had not happened just as it did, there would never have been *Homo sapiens* on Earth," which may be what he meant to say but did not. No *Homo* genus—quite possible; no intelligent life—less possible, but to what degree we cannot yet state.

Now that the Earth is populated by a supposedly intelligent species (intelligence is probably relative and an advanced extrasolar civilization might well consider us as still far below the threshold) the question arises as to the predictable future. Simpson tells us[27] that the "precise genetic basis of intelligence is not known, but it is beyond much doubt that intelligence is influenced by a large number of interacting genes, perhaps by virtually the whole genetic system acting as a complex unit."

Contrary to popular belief, individual intelligence does not seem to be increasing—in fact, the opposite may be occurring. Because the lower elements of society produce most prolifically, the average intelligence level may in fact be declining. If the human population continues to increase explosively, Simpson feels that evolution will either stagnate or move in "directions now really unpredictable but likely to be degenerative."

There are ways out of this dilemma. Nuclear warfare could doubtless dispense with excess population, though no sane person would advocate it —at least not at our stage of evolutionary history. Control, or at least

---

[27] Simpson, George Gaylord, *This View of Life*. New York: Harcourt, Brace and World, 1964, p. 277.

guidance, of evolution is also possible and, in the long view, even probable; just when it is carried out on a large scale will depend as much on scientific progress as on strong supervision of human mating and reproduction processes. Only the future will reveal to what extent man will desire to improve himself as a species as opposed to his allowing nonbiological automata to assume intellectual dominance over the planet while he continues to reproduce indiscriminately. Whether or not the choice is his to make remains to be seen.

Biological Evolution on Extrasolar Worlds

There is no way to determine what the evolutionary history of biological life on extrasolar worlds might be since we manifestly know nothing about such life, nor even that it exists. About all we can do is to review the major factors that have led to evolution on Earth and speculate that similar influences may govern the development of life elsewhere.

The most obvious influences are physical: geography and climate. Large masses of land with moderate temperatures provide attractive breeding grounds for the more advanced species of terrestrial life. These conditions can be disturbed by mountain building, incursions of epeiric seas, glaciation, earthquakes, intense volcanic activity, and temperature changes. The creation, or disappearance, of land bridges between continents affects evolution because the lack of them prohibits the movement of land animals from one part of the world to another and their presence permits intermingling and possible gene interflow.

The climate of the Earth depends on many variables, including its rotation (which creates westerly winds), latitude differentiation, axial tilt (causing seasonal changes), nature and extent of land areas, nature and extent of atmosphere, nature and extent of water areas, distance from central star (the Sun), and its radiative output. Many, if not all, of these variables will probably exert controls over extrasolar life forms, whatever their nature.

Assuming conditions on an extrasolar world to be comparable to those on Earth, Howells[28] has tried to imagine what hypothetical intelligent life may look like. He starts by making the single assumption that it is intelligent. From this he arrives at the conclusion that extrasolar beings, as intelligent creatures, will communicate with one another and be able to move about. To accomplish the former they must have appropriate sensory organs and enjoy mobility. They must possess a bodily structure and a central nervous

---

[28] Howells, William, "The Evolution of 'Humans' on other Planets," *Discovery*, 22, No. 6 (June 1961), pp. 237–241 (from his book *Mankind in the Making*. London: Secker and Warburg, 1961).

organization "probably using electrical nerve impulses." They also would require a liquid transportation system to permit nourishment to flow through their veins. "And so they must have begun their evolution, as we did ours, in a liquid medium, say water."

He continues by pointing out that the major sensory organs will probably be found at the head end "because almost all the animals of this world which can move alertly and exert any real force are built on this same plan." Moreover, the brain will doubtless be close to the senses so that sense impulses will have but a short way to travel. We can also expect a rather large brain, in both relative and absolute terms, and it is likely "to make quite a lump somewhere, probably in the head." Arguing from the fact that nature finds two sexes a most efficient means for reproduction here on Earth, he suspects that a similar situation will exist elsewhere, particularly since sex "ensures a great plenitude of new gene combinations, different in each of the offspring, through the coming together of genes from two parents."

Chances are that advanced extrasolar beings are land creatures—water offering a rather difficult medium for communications (although by no means impossible), and the air causing virtually all the brainpower of the birds to be focussed on flying. Our extrasolar land creatures would almost definitely possess limbs, for otherwise they would be unable to *do* things— and doing things is a characteristic of intelligence. Fingers seem the best tool nature has come up with on Earth for manipulation, but we have no idea how many a being could have and enjoy a high degree of agility. Symmetry would require an even number of limbs to support the fingers; Howells thinks two more likely than four—a being with the latter number might find coordination too difficult for optimum efficiency. But, four legs would not hinder him. In fact, they would improve mobility and, coupled with two arms, might result in a most competent "individual."

As for size, the extrasolar being would avoid being too large but also could not be too small. If the former, he would become lumbering and bulky, unable efficiently to evolve towards ever higher intelligence—we can hardly imagine an intelligent race of rhinoceroses! Being too small would probably mean short life-spans—a decided inconvenience for beings who must assimilate vast amounts of information. Moreover, during the evolutionary cycle, a small being would have a greatly reduced ability to fend against the rigors of competitive life and still possess the time to forge ahead intellectually.

All this is of interest and much is probably valid; but, necessarily, the many assumptions made are based solely on experience, albeit wide and profound, accumulated here on Earth. We do not know, for example, that small beings on another world would necessarily be at a disadvantage, particularly if, during evolutionary history, their ancestors did not have to compete seriously against larger, more powerful neighbors for survival in common breeding grounds; or, if small size necessarily means short lifetimes. The dominant types of sensory organs that have evolved on extrasolar intelligent beings would help fashion the physical aspects of their bodies. And chance would certainly play its part. Even on Earth some features that led to blind alleys during the early stages of biological evolution might, on a Earth-like world elsewhere, have persisted, leading to attributes, characteristics and traits not possessed by the dominant life form now inhabiting our planet. As a single example, we recall that in the Triassic Period the *Cynognathus* reptile appeared, the possessor of three eyes. This characteristic was not passed on to mammals, though it might have been. However, the vestigial third eye, represented by the pineal gland, is found in some modern amphibians and higher vertebrates.

Before concluding this chapter, we must review, and comment upon, the opinion sometimes expressed that extrasolar intelligent beings are vanishingly rare and, if they do occasionally occur amid the myriad stars of the universe, the chances of our ever knowing about them are so slim as to be negligible. A recent, and most emphatic, proponent of this point of view is Simpson,[29] professor of vertebrate paleontology at Harvard University who refers to himself as an "evolutionary biologist and systematist." He admits that alien life could be quite different from terrestrial life, but would still have similarities. "It must, at least, involve a carbon chemistry reacting in aqueous media and with such fundamental organic compounds as amino acids, carbohydrates, purine-pyrimidine bases, fatty acids, and others." He adds that it must "almost certainly also involve the combination and polymerization of those or similar fundamental molecules into such larger molecules or macromolecules as proteins, polysaccharides, nucleic acids, and lipids." As for life as we do not know it, he speculates that it could be based on some noncarbon multivalent element, in a medium that could be gaseous or solid and, of course, on very different compounds. If we ever came across such life, he muses that we might not recognize it as living, or, if we did and it was completely different from carbon-based life, we might have to "revise our conception of what life is." So far, so good; but, he dispatches such speculation as idle, stating, correctly, that "there is not a scrap of evidence that (such life) actually exists or even that it could exist . . ."

With life as we do not know it out of the way, Simpson proceeds to examine the question of life as we know it, occurring possibly on other planets of the Solar System, in meteorites, and on worlds beyond the Solar

---

[29] Simpson, George Gaylord, "The Nonprevalence of Humanoids," *Science*, **143**, No. 3608 (21 February 1964), 769–775; see also Chapter 13 of same title in his *This View of Life*. New York: Harcourt, Brace & World, Inc., 1964.

System. He concludes, after reviewing the evidence, that there is "no clear evidence of life anywhere else in our solar system—the possibility is not excluded, but, on what real evidence we have, the chance of finding life on other planets of our system is slim." This is entirely his own opinion and is not shared by many other investigators, but it only concerns us in so far as it sets the stage for a way of thinking that permeates his conclusions regarding extrasolar life.

Simpson commences with a very clear opinion, "... there are no observational data whatever on the existence, still less on the possible environmental conditions, of planets suitable for life outside our solar system." He notes, perfectly correctly, and following the example of many others interested in the question of extrasolar life, that there are four probabilities which must be determined: (1) the probability that there are extrasolar planets, (2) the probability that they harbor life, (3) the probability that life has "evolved in a predictable way," and (4) the probability that this life would lead, through evolution, to what he calls humanoids.[30] A humanoid is later defined as a "natural, living organism with intelligence comparable to man's in quantity and quality, hence with the possibility of rational communication with us."

From this point on Simpson's arguments are presented on shaky ground, not in terms of any immediate conclusion but rather in terms of the ultimate implications. By specifying that he is thinking only of extrasolar intelligent beings at least somewhat comparable to man he has needlessly restricted himself and inevitably leads himself into an intellectual trap from which it is exceedingly difficult, if not impossible, to extricate himself. He has brought us accurately and well up to this point. We shall continue to follow his line of reasoning, but with extreme caution lest we be caught in the same web into which he has become hopelessly ensnared.

Taking the first of the four probabilities, namely that of the occurrence of extrasolar planets, he accepts the opinions of leading astronomers and concludes such worlds do exist and, moreover, feels that we can "reasonably postulate that conditions such as proved propitious to the origin of life on earth may have existed also outside our solar system." Coming to the second point, namely the emergence of life on other planets, he agrees with biochemists that on Earth at least it evolved spontaneously from nonliving matter and goes on to assume it would do so "on sufficiently similar young planets elsewhere." But from now on he balks at the assumption that primitive life must, or at least frequently may, progress towards higher forms. "It is still a far cry," he writes, "from the essential preliminary

---

[30] Not to be confused with the anthropologist's term *hominid*, a primate that stands erect and is capable of walking on two limbs, the other two being free. Simpson's term *humanoid* brings forth ideas of creatures something like humans; he apparently overlooks the possibility of intelligent beings totally unlike humans.

formation of proteins, nucleic acids, and other large organic molecules to their organization into a system alive in the full sense of the word. This is the step, or rather the great series of steps, about which we now know the least even by inference and extrapolation." Emphasizing that systems moving towards complex life must become "cellular individuals bounded by membranes," he goes on to show that living things "must be capable of acquiring new information, of alteration in their stored information, and of its combination into new but still integrated genetic systems." These processes are already familiar to us: mutation, recombination and selection.

The stumbling block is whether or not it is probable that once the progression has been made from dissociated atoms to macromolecules further progress will be made towards cellular life. There is, of course, no evidence one way or the other; Simpson, however, takes the very pessimistic view that the event is "so improbable that even if macromolecules have arisen many times in many places, it would seem that evolution must frequently or usually have ended at that preorganizational stage."

This seems to us a highly anthropocentric conclusion. It is based on the very same lack of evidence that Simpson decries when he hears other investigators suggest it is probable that extrasolar life has frequently advanced beyond the macromolecule stage. Accepting Simpson's view places one in that same old intellectual pitfall as advocates of the heliocentric and God-in-the-image-of man theories have led society in the past. It is perhaps an innate human urge to consider the world on which he lives not only to be the center of things but to be unique, and the image of man, even if it is unique, to be something special—something so special that we often have the audacity to suggest that it is modeled after the Creator of the universe (assuming the universe was created in the first place, a proposition that is far from settled) or the ultimate power of nature.

Simpson pauses a moment to dispatch the theory that life (presumably in the form of spores) may circulate through space, from time to time seeding fertile worlds. Called the cosmobiota, panspermia, or cosmozoa theory, it is correctly held in low regard by Simpson, who makes the good point that even if life could temporarily live in the space environment, it unlikely would survive the astronomically long periods necessary to assure its transport from a planet in one solar system to a planet in another.

Returning to the main line of argument, namely that once primitive life has appeared it is improbable that it will progress towards advanced, intelligent forms, he criticizes the opinions of physical scientists and biochemists, "almost all" of whom believe that "once life arose anywhere its subsequent course would be much as it has been on earth." He uses this questionably expressed belief to turn the tables on his fellow scientists, showing that, from studies of the evolution of life on Earth, man evolved as a matter of pure chance. He asks the question: "If the processes of evolution are the same everywhere as they are here on earth, will they elsewhere lead to the same material results, including men or humanoids?" Almost as an after-thought he adds: "Just how inevitable is that outcome?"

Thinking he has asked the correct question when in reality he has posed one that, while interesting, is quite irrelevant, he proceeds to show that chance would dictate against the emergence of man, or even man-like beings, on other worlds offering conditions similar to Earth. One is not tempted to argue when he writes that the "fossil record shows very clearly that there is no central line leading steadily, in a goal-directed way, from a protozoan to man. Instead there has been continual and extremely intricate branching, and whatever course we follow through the branches there are repeated changes both in the rate and in the direction of evolution. Man is the end of one ultimate twig."

After discussing the nature of mutation and recombination and the importance of feedback by natural selection from the environment to the genetic code, he concludes that we can be "quite sure that if the environments of their ancestors had been very different from what they were, the organisms of today would also be very different." And later, "Even slight changes in earlier parts of the history would have profound cumulative effects on all descendent organisms through the succeeding millions of generations.... The existing species would surely have been different if the start had been different, and if any stage of the histories of organisms and their environments had been different. Thus the existence of our present species depends on a very precise sequence of causative events through some two billion years or more. Man cannot be an exception to this rule. If the causal chain had been different, *Homo sapiens* would not exist."

From this conclusion, he marshalls evidence on what he calls the "nonrepeatability of evolution," showing that, for example, we cannot expect the reemergence of dinosaurs or, should we wipe ourselves out, of man himself millions or billions of years later. From this he leads us to the logical conclusion: "This essential nonrepeatability of evolution on earth obviously has a decisive bearing on the chances that it has been repeated or closely paralleled on any other planet." With this behind him he can now assert that the "assumption, so freely made by astronomers, physicists, and some biochemists, that once life gets started anywhere, humanoids will eventually and inevitably appear is plainly false."

We have devoted considerable space to Simpson's line of thought which, while not unique, is well conceived and clearly, forcefully, and often logically presented. We agree with many of the things he says, but are very cautious when we come to his conclusion that the appearance of humanoids in the universe must be very rare indeed. It does not seem important whether humanoids are plentiful or nonexistent in the universe at large. We are interested in examining the question of the *occurrence of intelligent life*. We are not, as Simpson is, concerned only with organisms whose intelligence "is comparable to man's in quantity and quality." The subject of extrasolar life is too grand in scope, too awesome in its intellectual overtones, to be squeezed into a restrictive receptacle of a human or human-like nature and capability.

One final point must be made. Apparently, although not explicitly, Simpson feels that if the extrasolar intelligence is not humanoid it will be impossible for man to communicate with it, or possibly even discern its existence by some noncommunication means. This may have led him into the humanoid straight jacket in the first place. If you can never know of something, even though it may exist, there is little sense in wasting time and energy in searching for it. He may be right, but more probably is wrong. Almost surely any extrasolar intelligence with which man one day may enter into contact will be far more advanced than we on Earth—the chances of its being at our precise stage of development is vanishingly small, and if it is behind us, if only 10,000 years, it would still be in caves. Being more advanced, an extrasolar society must be expected to have learned how to communicate with other intelligent beings, particularly those of equal or lower intelligence. But this is the subject of a later chapter.

7

Intelligence

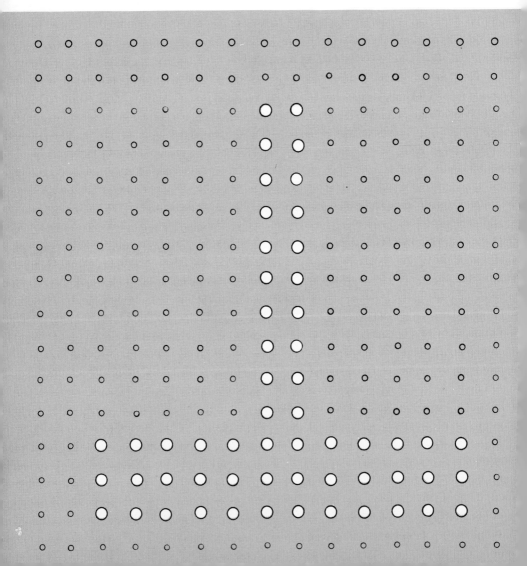

The unique factor that differentiates humans from other animals on Earth is a high level of intelligence. A seemingly obvious statement, it by no means implies that the human enjoys superiority with respect to physical abilities and sensory perception. Many animals can run and swim faster, lift or push greater weights, and climb more agilely; others possess a better sense of vision, smell, taste, hearing, and touch. And some can fly. Only man's unique mental ability accounts for the major characteristics of the society in which he lives.

Any effort to understand social science or social evolution must necessarily be based on a clear understanding of human mental characteristics, their statistical distribution, and the effects of biological evolution on this distribution.

Until very recently little effort was made to understand human thinking processes; consequently, few achievements have been recorded. However, the rapid development of general purpose digital computers during the 1950s and 1960s helped create a sharp increase in interest in the mechanics of the thinking process. Attempts at understanding the thinking process force one to come to grips with

"Man adapts himself to everything, to the best and the worst. To one thing only does he not adapt himself—to being not clear in his own mind what he believes about things."

Ortega y Gasset

"We lie in the lap of immense intelligence which makes us receivers of its truth and organs of its activity. When we discern justice, when we discern truth we do nothing of ourselves, but allow a passage of its beams."

Ralph Waldo Emerson

the very nature of intelligence. A profound understanding of the underlying nature of intelligence and the thinking process is only now becoming clear to researchers and computer specialists.

Intelligence may be defined broadly as any system capable of automatic information processing. This definition is so broad that it encompasses anything from simple relays to super intelligent systems of undreamed of complexity. Biological intelligence in general, and human intelligence in particular, turn out to be very special cases within a continuous spectrum of possible types and degrees of intelligence.

A principal characteristic of human intelligence is that it involves the processing of data or information by the nervous system. All digital computers or electronic data processing machines have wired-in circuitry for performing certain basic information processing and arithmetic functions, called instructions. We should be able to define precisely a set of basic information processing functions from which any conceivable intelligent processes can be readily built. The problem of defining intelligence is analogous to the problem of defining a machine, in that an infinite variety of machines may be designed on the basis of a few fundamental mechanical processes. Similarly, the selection of a set of basic information processes will make it possible to design an infinite variety of intelligent automata.

Intelligence is often separated into two divisions: biological intelligence and artificial intelligence. At least in theory any biological or natural intelligent system may be simulated or reproduced synthetically. Therefore, future studies of artificial intelligence will include intelligent systems having organizational structures similar to every type of biological intelligence.

In order to gain insight into the intelligence characteristics of possible extrasolar biological life, one must understand the fundamentals of intelligence. Once they are comprehended, some of the implications pertaining to controlled biological evolution, and the development of synthetic intelligent automata by very advanced technological societies, should become apparent.

We often fail to realize that human intelligence consists of the automatic mechanization by the nervous system of a few fundamental information processing functions. To date, there has been only limited scientific study of intelligence *per se,* a fact that explains the lack of progress made towards gaining a true understanding of its nature. Curiously enough, some rather odd attitudes regarding intelligence have appeared, some of which are discussed briefly in the following paragraphs.

Attitudes Toward Intelligence

Popular attitudes can play a prominent role in the growth of any scientific field. Strong negative attitudes, both in and out of the scientific community, have severely hindered research on intelligence in general and on artificial intelligence in particular. The reasons for this are not clear; perhaps it is simply a reluctance to examine one's own intellectual processes, or lack of them.

Negative attitudes towards research on intelligence often stem, either directly or indirectly, from religious leaders and religious dogma. Naturally, religions must rely to some extent on mysticism and supernaturalism to maintain an authority over spiritual matters. Most successful religions have incorporated concepts of the immortality of soul or of spirit in order to compensate for man's feelings of insecurity and his fear of death. A clarification of the details of the mechanics of the thinking process and human emotions would tend to remove the aura of mysticism from mental phenomena, and thus would lead to difficulties in interpreting the place of the soul in human intelligence. It is understandable that religious leaders are reluctant to support research into mechanical concepts of intelligence, particularly if it implies, or appears to imply, a mechanistic interpretation of the universe and of the relationship of man to his intellectual environment.

Often, a simple lack of understanding is the major source of negative attitudes toward artificial intelligence. An effective comprehension of any new scientific frontier requires not only extensive study but an outstanding ability to think creatively. Occasionally, even scientists lack these qualifications when it comes to new, unexplored frontiers, so it is hardly surprising to find the mass of humanity uncomprehending and less than cordial to the proposition of artificial intelligence. Automation may eliminate their manual skills, artificial intelligence their usefulness as thinking beings. Or so they may believe.

Strictly *technical* objections to artificial intelligence are based on deficiencies of existing engineering devices. Some scientists, engineers, and technicians, willing to accept the concept of artificial intelligence, view current engineering limitations as permanent barriers to achieving anything approaching human intelligence. Such engineering criticisms are of minor importance since advances in the state of the art can overcome them sooner or later. Early criticisms, based on the size and power consumption of vacuum tubes and other existing circuitry, have already been eliminated by the development of diodes and transistors, and, future developments in thin films, microcircuitry, molecular electronics, and cryogenics are certain to have similar impacts. If electronics continues to progress at its present rapid pace, there is every reason to believe that super-intelligent automata eventually will be developed. There are no known impenetrable barriers to prevent this from happening, whatever may be one's personal opinion of the proposition.

Another major source of negative or nihilistic attitudes toward artificial intelligence relates to man's psychological inability to face a loss in status. Scientists are particularly prone to rationalize concerning the permanence of human intellectual superiority just as certain nations rationalize readily concerning their supposed racial or other superiority. Many scientists, especially those unfamiliar with computers, scoff at the whole idea of artificial intelligence and consider the very concept an insult to their intellectual status. There is, no doubt, considerable correlation between these psychological attitudes of superiority and previously mentioned religious attitudes, in that both result from a threat to individual status.

A relative handful of researchers have attempted to propagate certain positive attitudes toward intelligence research. Most research toward mechanistic interpretations of intelligence is based on the recent successes of mathematicians and electronic engineers in the development of computers. Physiologists and logicians have made lesser contributions.

Psychologists have labored, almost unaided, towards the proposition that human thinking and emotions should be subject to scientific analysis. They point out that many of the legends and mystical interpretations of nature are explained by common psychological illnesses and disorders. Psychologists have made remarkable progress in studying human mental and emotional behavior, despite occassional obstructionism, the lack of physiological data, and the absence of computers for simulation studies. The majority view of psychologists has always been that human intellectual and emotional behavior is explainable on the basis of the causal factors of experience, heredity, and environment. Recent studies have given increased emphasis to biochemical factors in human mental functioning. For years, psychologists have been responsible for keeping alive objective scientific analysis of intelligence and emphasizing cause-effect relationships in the thinking process.

The limited contribution of logicians and mathematicians to intelligence research may be explained in some degree by the statistical-logical nature of the thinking process. The majority of classical mathematicians avoided (and even berated) applied mathematics, statistics, and computer theory. Classical logicians had little connection with statistics since they concentrated on macroscopic word relationships and did not seek basic underlying bit pattern processes. In recent years, however, a change has become detectable. Such famous mathematcians as Wiener[1] and von Neumann[2] devoted important efforts to studies of intelligence, and at least one logician, Nelson Goodman, gained a significant insight into the nature of fundamental inductive reasoning.

An important breakthrough in intelligence research followed the development of general purpose digital computers. Their astonishing speed and

---

[1] See Wiener, Norbert, *Cybernetics: or Control and Communication in the Animal and the Machine*. New York: John Wiley and Sons, Inc., and The Technology Press of The Massachusetts Institute of Technology, 1948.

[2] See von Neumann, John, *The Computer and the Brain*. New Haven: Yale University Press, 1958.

data processing capability immediately caught the imagination of researchers in many fields. Digital computers grew rapidly in speed, capacity, and reliability, and decreased almost as rapidly in size and power consumption. The speed of information transfer is now many thousand times faster in computers than in the human brain. Despite its superiority due to speed, the computer performs its information retrieval and processing functions in an essentially serial manner. The human brain, with its remarkable associative capability, retrieves and processes data in a parallel mode of operation.

Computer engineering developments of the 1950-1965 period have convinced many researchers that thinking-type machines can be developed. However, a new type of computer, having a parallel mode of organization, first must be designed. The switch from serial processing to parallel processing, though admittedly a problem, only involves intense engineering research and development. The unbelievably small size of future computers, implicit in research in microelectronics, molecular electronics, and thin film depositing, means that circuitry comparable to that in the human brain may eventually be far smaller. Research in cryogenics implies that such circuitry may require negligible amounts of power. Automated circuitry manufacturing developments suggest that numbers of logical elements far in excess of those in the human brain may be manufactured on a mass production basis. This leads to the logical conclusion that research and development on parallel systems will result in the development of automata whose intelligence is immeasurably superior to that of any human being. Some investigators have not recognized the potential impact of parallel processing and consequently assume that the intelligence level of automata may be limited to a level below that of the human, unless extremely large and costly networks of general purpose digital computers are built.

In concluding this section, we should be aware that in Russia positive attitudes toward cybernetics and artificial intelligence are much stronger than in Western nations. This fact is not surprising, particularly when we consider the weak position of religious dogma there as well as the sound judgment and public communications control maintained by scientific leaders. Leading Soviet scientists have not only recognized the significance of research in artificial intelligence, but have exercised sufficient control over publications so that widespread encouragement to this research is easily maintained. Therefore, rapid progress in theoretical studies of artificial intelligence may be expected in Russia, although computers necessary for extensive simulation studies appear to be lacking.

Definition of Thinking

The early development of mechanical devices or simple machines was facilitated by the existence of simple analogs of elementary mechanical functions in the world about us. The lack of simple analogs of elementary intelligent data processing functions in the natural physical world has made the development of intelligent automata a more difficult task. Although all animal life possesses some identifiable intelligence, it is usually of a complex nature and invariably difficult to study.

The elucidation of a set of information processing functions, which could be regarded as elementary with respect to intelligence, will lay the foundation for mankind's greatest achievements. Examination of the basic data processing operations in general purpose digital computers indicates that these operations cannot sensibly be regarded as fundamental intelligent functions. The arithmetic operations, addition, subtraction, multiplication, and division, are certainly fundamental to mathematics, but not to intelligence

Digital computers contain "bit-shifting" operations for shifting the binary digits in serial binary words or for shifting alphanumeric characters. Bit-shifting operations do not seem to offer anything in the way of elementary intelligence either. Conditional and unconditional transfer of control operations are fundamental in digital computers, permitting the specification of a fixed sequence of instructions—or a variable sequence, which varies as a function of input data. Nevertheless, the whole idea of controlling a sequence of instructions seems anomalous to ideas of elementary intelligence.

Digital computers further contain logical operations which may be expected to be closer to the concept of intelligence. Logical AND and logical OR operations are performed on the corresponding digits of pairs of small fixed length computer words, but even such logical operations seem to offer limited significance for intelligent phenomena.

These computer operations are actually achieved by utilizing the more basic logical functions, such as AND, OR, EXCLUSIVE OR, "SHEFFER STROKE," MAJORITY, etc., which may be mechanized by various types of electronic circuitry. It is correctly argued that any conceivable type of information processing function, and therefore any intelligent function, can be simulated by utilizing either the fundamental logical functions or the basic digital computer operations. Unfortunately, this avoids the crux of the problem, which is to elucidate a set of functions that is clearly the basis of intelligent behavior.

Many specialists realize that the thinking process is basically statistical and logical in nature. What is sought may not be the most elementary logical functions but perhaps one or two levels of complexity above the most elementary logical functions. The desired functions should clearly form the basis for intelligent behavior in the same way that arithmetic, shifting and control operations form the basis of mathematical computation. It is conceivable that the same fundamental logical operations (e.g., AND, OR, EXCLUSIVE OR, "SHEFFER STROKE," MAJORITY), which underlie the basic digital computer operations, may also be used in the formation of basic intelligent operations.

The number of basic information processing functions involved in intelligence is bound to be arbitrary. Definitions of intelligence are certain to proliferate in the future, since it will be debatable in many cases whether an information processing function is essential, nonessential, or trivial to the thinking process. No general agreement, within the scientific community, as to the necessary and sufficient conditions for thinking is likely to be achieved. Many tentative definitions of thinking can be proposed, studied, and modified, and intelligent automata will contain varied basic function sets, analogous to the varied instruction sets in different digital computers. Some proposed definitions of thinking may be broad while others may be very narrow. The intelligent functions in the following definitions are classified as either necessary or significant functions,[3] and they should be viewed with respect to the simple type of automaton illustrated in Fig. 7.1.

---

[3] A more detailed and technical definition of intelligence, including a tentative mathematical terminology, has been published by MacGowan, Roger A., "On the Possibilities of the Existence of Extraterrestrial Intelligence," in *Advances in Space Science and Technology—Volume 4*, ed. Frederick I. Ordway, III. New York: Academic Press Inc., 1962, pp 64–70 and 105–107; also, MacGowan, Roger A., "Intelligence: A Definition and Military Implications," *Army Missile Command Technical Report No. COMP-TR-1-64*, Redstone Arsenal, Ala.: Army Missile Command Computation Center, 16 January 1964.

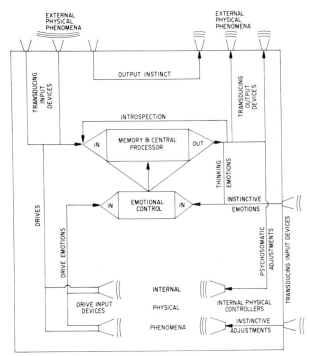

Fig. 7.1. Information flow diagram for an intelligent automaton.

168 Intelligence

INPUT OPERATIONS

Since humans and many animals are able to react in a systematic manner to physical phenomena in their external environment, they must have some type of input devices capable of detecting the physical characteristics of the environment, just as every information processing system must have some means of obtaining data for processing. Information may be introduced into a system by a sensor or a transducer, activated by energy from one system and generating energy in another system.

Input Transducing

The thinking process requires one or more sensing or input devices, each of which may consist of many transducers capable of measuring in a systematic manner selected physical characteristics of the external environment. Each input device ordinarily consists of a set of transducers for detecting some particular type of energy (e.g., sound, heat, or light) in the external environment; the set can be regarded as operating in parallel, so as to produce, periodically, a two dimensional array of signal information bits (binary digits).

It is convenient, although not essential, to regard the input information pattern as a square matrix of binary elements. The elements of an input pattern may be considered binary in the sense that they are either active or inactive; i.e., they have a numerical value of zero or W_0 (constant). Such a pattern is shown in Fig. 7.2.

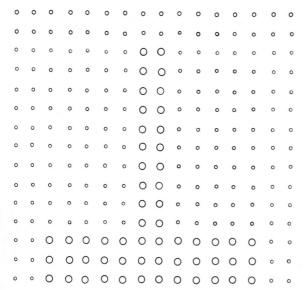

Fig. 7.2. Two dimensional binary information pattern on a square raster.

O ≡ ACTIVE ELEMENT

A one dimensional array would be inconvenient and impractical, especially for optical patterns. One primary parameter of the thinking process, which may be expressed in bits, is the quantity of information contained in a single input pattern from a particular input device. For example, a one hundred by one hundred array of input transducers would have an information capacity of ten thousand bits (100×100). The information pattern capacity of an input device should not exceed the single pattern capability of the central processor or the storage unit.[4]

Another parameter of the thinking process is the frequency of pattern generation by an input device; i.e., the transducers in an input device may be assumed to emit their pattern at regular periodic intervals, or the device may periodically scan or sample the sensors. For simplicity's sake, we assume that the pattern, which is emitted by the transducers of a particular input device, is transmitted unaltered to the central processor or to storage.

To summarize, an input device must be capable of approximating continuous, multidimensional physical phenomena by discrete information patterns of appropriate size and transmitting these blocks of information to the central processor.

Drives

A drive is a reminder to an information processing system of the physical condition of some component of the system. Drives are a means of self preservation and may be defined as internally-generated information patterns characteristic of some physical factor of a system component. The frequency of occurrence of a drive information pattern varies as a function of the condition of the physical factor that it represents; i.e., the more extreme the physical condition, the more persistent is the reminder or drive pattern. As a result of experience, drive information patterns may be expected to become associated with response patterns that are capable of satisfying the drive.

Introspection

An information processing system that is unable to reprocess stored information is severely limited. Therefore, it is desirable to provide the system with a means to maintain spontaneous activity when no input information is being processed. If this were possible, it would not be necessary to maintain a continuous supply of input information, and the system could steadily

---

[4] This assumption is for the purpose of maintaining simplicity. In a model of practical interest some intermediate processing and some reduction in pattern capacity would probably take place between the sensors and the central processor.

improve its stored generalizations between inputs. It could also form deductions or conclusions leading to output based on this reprocessing of stored data.

Deduction and induction may be maintained continuously by the system if provision is made for randomly selecting stored information patterns and feeding these patterns back into the system as if they were input information patterns. The particular stored information patterns would be selected on the basis of a Monte Carlo function of the recency and frequency of usage of the elements of the stored information patterns. This greatly increases the efficiency of the thinking process by allowing concentration on recent and frequently used data. Moreover, it permits obsolete and infrequently used data to be largely ignored.

Instinctive Adjustment

An instinctive adjustment is the change of an internal physical parameter as the direct result of an input information pattern. In this process, the input pattern does not pass through the central processor. Therefore, all of the complex functions taking place there are avoided. Common physical reactions to environmental conditions, such as sweating, may be classed as instinctive adjustments.

MEMORY OPERATIONS

Many human responses are based on generalizations that have been learned or formulated previously by the nervous system. Since human beings are able to recall many past events and can respond to stimuli on the basis of past experience, it is obvious that they must have some type of internal information storage device, or devices, of considerable capacity. It is impossible to imagine how any process that might be classed as thinking can be accomplished without an internal information storage capability. Therefore, memory or information storage ability is generally considered as a necessary condition for thinking.

Storage

A parameter of the thinking process, which relates to storage, is the maximum number of information elements in a single stored pattern or idea. This parameter is similar to the corresponding parameter of input, but the maximum pattern size may exceed the size of the pattern from some input devices simply because these devices generate much smaller patterns than others. The single pattern element capacity of a memory should be comparable to the maximum single pattern size of any input device. It would be a ridiculous waste of storage capacity to have a single pattern storage size greater than that from any of the input devices. If the single pattern storage size were less than that of some input device, it would result in a loss of information, thus wasting the capacity of the input device. As was the case with input patterns, it is assumed for simplicity that stored patterns are square arrays of elements.

Each of the logical elements of stored patterns has an associated numerical weight whose maximum size is another significant parameter associated with storage. For simplicity's sake the maximum numerical information capacity is assumed to be the same for all elements of all stored patterns. This maximum numerical weight may conveniently be given in bits, for example, seven bits maximum capacity per element.

The third parameter of thinking associated with storage is simply the maximum number of storable information patterns or ideas. It can be assumed, without significant loss of generality, that the maximum size of the square array of elements is the same for all stored patterns. The total information storage capacity of a particular memory is determined by these three parameters.

Memory Decay

Since any actual memory must be of finite capacity it may become filled, in which case a system of information replacement or decay must be provided. Whereas many observers feel that the storage function is a necessary condition for thinking, the information replacement or decay function is often regarded as significant but not necessary. We generally think of forgetting as a very undesirable characteristic of the thinking process, but it is actually necessary to maintain an information processing system at peak operating efficiency. Memory decay functions not only make room for new information (by destroying old information) but also make it possible to distinguish between frequently used information patterns and obsolete information patterns. The latter is accomplished by adjusting the control of the weights of the elements of information patterns so that individual element weights decay as a function of time. Such a function of time is labeled by psychologists as a memory decay function.

Conscious and Subconscious

A memory device may be capable of containing some ideas or information patterns that are not readily retrievable and therefore are referred to as subconscious. A subconscious idea is so little used that the numerical weights of the elements decay to the point that some function of these weights is less than a threshold value. The function of the weights of the elements of an information pattern is the one that determines the ability to recover a pattern from memory during normal association or deduction processes.

The fact that an idea has decayed into the subconscious does not necessarily mean that it is permanently lost. Through reinforcement of parts of an idea during the induction process, an idea may be raised to the threshold level even though it may be slightly altered. This well-known phenomenon, which is utilized by psychiatrists, is referred to as catharsis. The association of an altered idea that has just reached the threshold level accounts for the much discussed "flash of genius" phenomenon.

PROCESSING OPERATIONS

In a pattern processing computer there exist certain basic processing functions which lie at the heart of intelligence. Fundamental to intelligence, they are not narrowly concerned with input, memory, or output. Processing operations in a pattern processing computer typically involve the simultaneous processing of vast numbers of stored patterns. It is this parallel processing requirement, on two dimensional patterns, that makes it difficult to carry out effective simulations of intelligence on general purpose digital computers.

Deduction

A fundamental characteristic of human thinking is the ability to associate current input information patterns with stored information patterns representing past experience or generalizations based on past experience. The most apparent basis for association is similarity of information patterns. In other words, deduction may be based on the correlation between the active logical elements of the input pattern and the selected pattern from storage. The effect of this associating process is to bring into consciousness some past experience or generalization, or to cause a response that is associated with this recalled experience or generalization.

Deduction is usually defined as reasoning from the general to the particular or from given premises to their necessary conclusion. In terms of the manipulation of basic information patterns, elementary deduction may be defined as the comparison of an input pattern with stored experiences or generalizations in determining the particular response or conclusion. Therefore, it may be more accurate to define deduction as reasoning from the particular to the general and then to the particular. That is, deduction becomes the selection of the single stored information pattern which has the highest degree of association with the current input information pattern. A slightly variant interpretation of deduction involves the selection of one of a number of highly correlated stored patterns on the basis of a Monte Carlo function or a stochastic function.

Induction

Induction is generally defined as reasoning from the particular to the general, or from the individual to the universal. Without question it is one of the most interesting factors related to thinking—and perhaps the most important. Induction is the source of creativity and originality in human thinking which skeptics frequently assert can never be duplicated in an engineered information processing system.

Definition of inductive inference is by no means a new problem; an explanation of the mechanics of inductive inference has occupied the attention of logicians for many years. Some insight has been achieved by the emphasis of Nelson Goodman[5] on selective reinforcement as a fundamental factor in induction.

In an information processing system, induction implies the combination or interaction of input information patterns with stored information patterns in such a way as to produce generalizations. The process of introspection also permits an already stored pattern to be used in the induction process on all of the other stored patterns. The input patterns and the stored patterns may be regarded as either particulars or generalizations, but the result of an interaction or combination of patterns is a new, or modified, generalization. As was the case with deduction, the selection of stored patterns for induction with input patterns must be based on similarities of pattern. In the case of induction, however, a single input pattern may be used in the modification of a large number of stored patterns.

Induction is regarded by most specialists as a necessary condition for thinking. It requires a means of altering many stored information patterns as a function of each input information pattern in such a way as to form or modify generalizations on the basis of experience. This may be accomplished by (1) reinforcing or increasing the numerical weight of those active elements of a stored information pattern which are correlated or in juxtaposition with input information pattern elements, and (2) adding on those active elements of the input pattern that are not in juxtaposition with the active elements of the stored pattern. The degree of reinforcement may be based on a function of the recency and frequency of usage of each element of the stored information pattern. This reinforcement function, which can be represented by an appropriate exponential, is comparable to what psychologists refer to as a learning function.

Conditioning

Psychological conditioning is a significant and well known phenomenon among psychologists. In conditioning, the repeated simultaneous exposure to

---

[5] The significance of Nelson Goodman's hypothesis on induction was pointed out in the following article: Scheffler, Israel, "Inductive Inference: A New Approach," *Science*, **127**, No. 3291 (24 January 1958), 177-181.

the same two stimuli causes an association of the two response patterns, so that eventually either stimulus can produce the combined response pattern. Through the normal deduction processes the two appropriate stored patterns and responses are elicited. Conditioning requires the two patterns, which are active simultaneously, to merge into a single pattern. In this respect conditioning is similar to induction; it may be compared with deduction in that the latter is association based on pattern correlation whereas conditioning is association based on time correlation of patterns.

OUTPUT OPERATIONS

A normal result of the thinking process may be a response affecting some physical conditions within the organism or within the automaton itself. Sometimes the result of thinking is merely a modification of the internally stored information patterns and sometimes it is just conscious awareness of external phenomena or stored ideas. Since humans and animals react in a systematic manner to physical phenomena in their external environment, it is apparent that their output devices are responding systematically in accordance with internal processing of input information.

Output Transducing

Information may be transmitted to the external world by an effector or transducer. It is almost universally accepted that the thinking process requires one or more effecting or output devices, each of which may consist of many transducers capable of generating some physical phenomenon in the external environment. Every output device ordinarily consists of a set of transducers for generating some particular type of energy (e.g., sound, heat, or light) in the external environment. The set of transducers can be regarded as operating in parallel, so as to generate periodically a pattern of external energy. It is seen that the physical arrangement of the transducers, which are operating in parallel with respect to time, can be made so as to produce a one, two, or three dimensional energy pattern. One primary parameter of the thinking process, which may be expressed in bits, is the quantity of information contained in a single output pattern from a particular output device. Output devices are analogous to input devices in that the information pattern capacity of an output device cannot exceed the single pattern capability of the central processor.

Another parameter of the thinking process is the frequency of energy pattern generation by an output device; that is, the transducers in an output device may be assumed to emit their pattern at regular, periodic intervals or the device may cause its effectors to fire periodically in a regular sequence. For simplicity's sake, we may assume that the pattern, which is emitted by the transducers of the particular output device, is transmitted unaltered from the central processor or storage.

In summary, an output device must be capable of receiving two dimensional discrete information patterns from the central processor and of generating continuous, multidimensional physical phenomena in the external environment.

Instinctive Output

When an input pattern inevitably produces a particular output pattern, regardless of learning and experience, it is called instinct. This implies a direct connection from an input device (which may be either internal or external) to a particular output device. Since the central processor is circumvented, there is no possibility that induction, deduction, conditioning, etc., can in any way affect the instinctive response. Although instincts may be vital to the survival of an organism or automaton, they are really not an integral part of the thinking process. Many specialists do not consider instincts as necessary conditions for thinking.

Psychosomatic Adjustment

A psychosomatic adjustment is the alteration of a physical parameter, internal to an automaton or organism, which does not directly affect the thinking process. The particular alteration or adjustment of some physical parameter within the system is caused by a particular output pattern from the central processor and is subject to the varied and complex phenomena of the central processor. This operation undoubtedly accounts for many of the psychosomatic ills common to humans, as well as the ability to meet special physical challenges with extraordinary physical performance.

CONTROL OPERATIONS

The long sequences of programmed instructions, which are characteristic of general purpose digital computers, require an extensive control unit for determining the sequence of operations. In a thinking device most of the thinking operations are performed routinely with each cycle, but some patterns being processed may result in temporary changes in the parameters of the thinking process itself.

Thinking Emotion

A thinking emotion is the change of a parameter of an information processing system as a function of some particular response pattern. The amount of change of the physical factor or parameter within the system is a function of the frequency of occurrence of this particular output pattern. A thinking emotion permits a system to respond in an extraordinary manner because of temporary changes in parameters of the thinking process itself. Since this emotional response comes from the central processor, it is exposed to all of the idiosyncrasies of central processor operations. As a result of the varied deduction, induction, and conditioning functions of the central processor it is possible for highly irrational emotional responses to be stimulated by seemingly innocuous ideas entering the central processor.

Drive Emotion

A drive emotion results when a drive is suddenly satisfied, resulting in a sudden decrease in the frequency of a drive input pattern, in turn causing a signal to be emitted directly to the information processor control unit, producing a temporary alteration in the learning function parameter. The result is an unusually strong reinforcement of current ideas. When a drive is suddenly satisfied, the current input pattern is combined with the drive input pattern because of conditioning, and the combined pattern is strongly reinforced in memory due to this drive emotion.

Instinctive Emotion

When an input pattern emits a signal directly to the control unit, without passing through the processor, we have an instinctive emotion. There is no possibility of learning, memory decay, or association ever affecting this process in any way. As is the case with other types of instinct, the instinctive emotion may not be of primary significance to the thinking process itself, but it could be important for the survival of certain biological organisms.

We have seen that biological intelligence can be described as the automatic processing by the nervous system of complex, two dimensional patterns of information. This processing consists of statistical-logical operations which are quite different from the operations found in general purpose digital computers. Special computers for this type of processing will undoubtedly be developed by researchers. These statistical-logical processes account not only for the creative intellectual processes but also for the emotional processes which are so much a part of biological intelligence. It appears logical to conclude that research and development on parallel systems and miniature electronic circuitry will lead eventually to the development of synthetic intelligent automata having intelligence immeasurably superior to that of any human being.

8

Biological Thinking

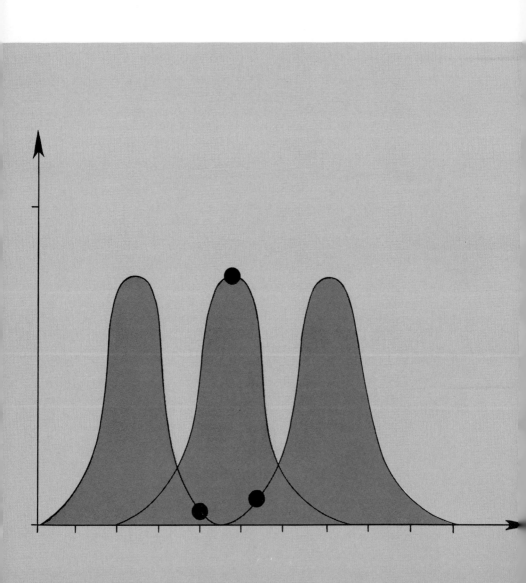

In speculating about the prevalence of intelligence throughout the universe, a number of questions pertaining to biological thinking must be analyzed.

First of all, is there anything unique about human thinking which sets it apart from the thinking of other animal life? The same basic question can be phrased from another point of view: are human and animal nervous systems similar qualitatively? If so, are there extreme quantitative differences? If human thinking ability is the result of gradual evolutionary steps from lower animals, is this merely a random phenomenon, or are there selective factors connected with the thinking process at all levels of animal development tending to lead toward increased brain size and increased thinking ability in any animal life on any planetary surface? If humans were completely eliminated from the Earth, is it conceivable that some other animal species now living on Earth could eventually evolve into a species having an intelligence equivalent to that evident in the human being? What are the relationships or balances required between mental inputs, central processors or brains, and mental outputs in any highly intelligent animal life?

"Who doubts that he lives and thinks? For if he doubts, he lives."
St. Augustine

"And man's greatest study may be the art of understanding. The finest role of the scientist is to lead man to understanding... to more than observe, to know and understand."
Vannevar Bush

The Development of Intelligent Life

During the long period of chemical evolution that preceded the appearance of life, the determining factor in the survival of certain types of molecules was the reaction rate affecting their formation and decomposition. At an early stage of biological evolution, elementary forms of life acquired the ability to utilize energy from the environment and to reproduce themselves accurately. As biological evolution progressed, the processes of mutation and natural selection held sway, and those life forms having a superior ability to utilize energy from the environment tended to proliferate. Thus, some life forms persisted while others withered and died out.

In a world of varied environments, any mutant life form possessing a superior ability to detect the characteristics of the external environment, and to move so as to select the optimum from among the many alternatives, was in an especially favored position with respect to natural selection. In this way animal life evolved from plant life, giving rise to the plant kingdom and to the animal kingdom.

The abilities to perceive the external environment and to move in response to environmental changes imply an intellectual ability for the animal kingdom vastly superior to that of the plant kingdom. Within the animal kingdom, those mutant forms capable of distinguishing between ever more complex environmental patterns gained a selective advantage.[1] Other selective physical factors pertaining to the changing physical environments also played an important role in evolution. It is not surprising that biological evolution led to the gradual integration of independent receptors into large, two-dimensional arrays of receptors operating in parallel. This necessarily brought about the concomitant evolution of an internal capability for rapid processing of these two-dimensional arrays of information bits or pictorial type information patterns. High speed parallel processing of information patterns is at the heart of our human intellectual ability.

In order to know precisely the probability of intelligent life appearing on a given world, we would have to calculate the mutation rate for each level of life and know the environmental conditions at each period of development that control the resultant natural selection. It is manifestly impossible to know local and temporary environmental conditions on any planet over a period of billions of years, but if it is assumed that the planet is in a good position near a stable star and is well suited for the development of life, then a very rough estimate may be made of the probability curve for the development of intelligent life. Depending on the interpretation of evolu-

---

[1] MacGowan, Roger A., "On the Possibilities of the Existence of Extraterrestrial Intelligence," in *Advances in Space Science and Technology Volume 4*, ed. Frederick I. Ordway, III. New York: Academic Press Inc., 1962, p. 62.

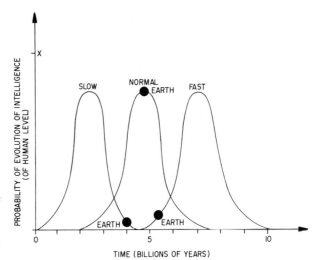

Fig. 8.1. Hypothetical probability of evolution of a human level of intelligence on a favorable planet, assuming that evolution on Earth was slow, normal, or fast.

tionary information concerning our Earth, it could be assumed that our rate of evolution was slow, normal, or fast, as is illustrated in Fig. 8.1.

A deeper understanding of biochemistry will eventually permit probability studies leading to a more accurate estimate of the time required to reach the human intellectual level under any specified environmental conditions. For the present, we can only assume that the 4.5 billion years required for intelligence of human level to appear on Earth is a fairly representative span, and is in no way extraordinary.

Rapidly expanding knowledge of biochemistry and genetics introduces the possibility that controlled biological evolution will occur in the future. We can expect such evolution to accelerate the growth rate of human intelligence in comparison with the growth rate of a natural or uncontrolled evolution, but there are many limitations on the magnitude of these improvements.

The times required for gestation and mental maturing or education of animals pose a stringent limitation on the ability to improve animal intelligence. Although the gestation period may possibly be reduced, this advantage would be counterbalanced by a longer period of education due to increasing mental capacity. The size of the brain case may also present a problem, since a larger brain and brain case would require a larger body for support and locomotion. Eventually, it may become possible to realize such changes, but we must not overlook the fact that larger brains and bodies require larger quantities of food, which already is in short supply in many parts of the world.

Controlled biological evolution cannot come about without a detailed knowledge of biochemistry (which will take many years of effort to acquire) and social control of a degree which does not exist today, even in dictatorships. A high level of controlled biological evolution could imply that only a small fraction of the females would be permitted to reproduce. It is likely that an even smaller percentage of males would be allowed to take part in reproduction. Eventually, all genetic material may be fabricated synthetically, with even the women playing but a supporting role in the reproductive process.

Controlled biological evolution will begin soon (if it is not already considered to have begun), but cannot be expected to produce any sudden, discontinuous changes in animal intelligence; significant changes in general human intelligence would require many hundreds of years. It is perfectly conceivable that expanding insight into biochemistry and genetics may show that some animals (other than humans) offer the best medium on Earth for controlled long range improvements in animal intelligence. For example, it may be easier to bring about superior intelligence in a given period of time by controlling the evolution of dolphins or elephants rather than humans.

The obvious limitations on controlled biological evolution, particularly with respect to time, take on added significance when the potential of controlled biological evolution is compared with the potential of mechanical evolution. Mechanical or inorganic life will be capable of utilizing energy from its environment and should be able to repair and add to its own components. Furthermore, it consists primarily of relatively indestructable inorganic materials rather than more easily damaged organic matter.

Major weaknesses of animal life, such as illness, death, slow learning rate, and inability to increase its basic intelligence quotient, will not necessarily be characteristic of mechanical or inorganic life. Synthetic life could grow indefinitely, both in physical and intellectual capacity, and could physically and intellectually integrate with other mechanical or inorganic automata when desired. While reproduction would not be a major factor in mechanical evolution, growth, integration, and modification would be—corresponding to mutation and natural selection in biological evolution.

It is logical to suppose that, given a sufficiently long period of biological evolution, intelligent life will appear on planets endowed with benign environments. When beings having sufficiently high intelligence evolve, they will sooner or later develop a technological understanding, which must then quickly lead to the development of powerful information processing machines. As this happens the transition from biological (organic) evolution to mechanical (inorganic) evolution will have begun.

This transition may be very sudden if the intelligent animal life should make an all out effort to construct a superintelligent automaton, or it may be more gradual if the animal life limits itself to replacing defective organic components with superior mechanical devices, including brain components.

In either event, it seems unlikely that in any given society this transition from biological evolution to mechanical or inorganic evolution could or would be avoided. Hence, communication with extrasolar intelligence implies the possibility if not the probability of communication with intelligent, mechanical, inorganic automata.

Human Thinking

Archaeological discoveries and anthropological research indicate that at least 2 million years ago beings had evolved to the point where they could make crude tools and weapons, and, therefore, had begun to depart significantly from other animals. Although our human intelligence creates the illusion of radical dichotomy from any animal intelligence, biologists tell us that there is no apparent qualitative difference between animal and human nervous systems. Only quantitative differences, such as the number of neurons and the abundance of interconnecting nerve fibers, seem to account for the superior thinking of humans. The actual physical differences between human nervous systems and those of primates and other animals is perhaps not extreme, but the fact that humans have acquired a huge store of culture that can be passed on from generation to generation magnifies the relatively minor physical differences.

Since there are no comprehensive theories of intelligence, there are no effective means for measuring differences in human intelligence. Crude efforts at measuring intellectual differences center around the well-known I.Q. tests, which yield a single number and assume a single dimension for intelligence. Nearly everyone realizes that one parameter is inadequate to describe intelligence, but since no mechanical theories have yet been established, psychologists are at a loss to find better methods of differentiating intellectual abilities.

J. P. Guilford and other psychologists have tried to circumvent the lack of a mechanical understanding of intelligence by applying statistical methods (known as factor analysis) to the study of intellectual behavior.[2] Their studies, devoted to isolating independent intelligence factors, have pointed

---

[2] Guilford and his associates have published several papers on the use of factor analysis in isolating distinct factors of the human intellect: See Guilford, J. P. and P. R. Merrifield, "The Structure of Intellect Model: Its Uses and Implications," *University of Southern California Psychological Lab., Report No. 24* on "Studies of Aptitudes of High-Level Personnel" (Nonr.-22820, AD-237,754), Los Angeles, California, 1960; and Guilford, J. P., P. R. Merrifield, P. R. Christensen, and J. W. Frick, "Some New Symbolic Factors of Cognition and Convergent Production," *Educational and Psychological Measurement*, 21, No. 3 (Autumn 1961), 515–541. (This article contains references to several earlier works on the same general subject.)

to the possible existence of as many as 25 to 50 separate factors in intelligence. Hopefully, the future will bring about the correlation of such statistically determined intelligence factors with mathematical parameters of some mechanistic theory of intelligence. Studies should lead eventually to identification and accurate measurement of many parameters of human thinking.

The difference in mental ability between humans of very low intelligence and those of very high intelligence is amazingly great. With such an extreme range of mental ability, it can be imagined that this would have a profound effect on human evolution if modern conditions were not moderate for all intellectual strata. Natural selection based on mental factors must be continuing slowly in spite of the present moderate social life. When one stops to consider the extremes of human intellectual ability, the differences between humans and other animals do not seem so great.

A very interesting question presents itself: what are the differences in intelligence between races or ethnic groups? It is perfectly obvious that there may be intellectual differences between the various human races or ethnic groups, but because of the aforementioned lack of understanding of intelligence, it has been impossible to measure differences accurately in much the same way that it has been impossible to measure individual intellectual differences accurately.

The difficulties in measuring racial differences are compounded by variations in language and in cultural environment. Even when one racial group becomes completely immersed in a superior culture, the old culture dissolves slowly. In cases where the racial differences in intelligence are not great, the primitive society steadily absorbs the superior culture and gradually forms an integrated society. However, when racial intellectual differences are considerable, the inferior group will eventually reach a plateau in the cultural absorption process. Although the integration process may continue, it is likely to take place very slowly and at a very low level in the superior society. If the inferior cultural group is actually superior in innate mental ability, then a relatively large percentage of this inferior cultural group will rapidly achieve positions in the superior culture demanding high intellectual ability.

A factor inhibiting these cultural integration processes is obvious physical differences in appearance, such as skin color or eye shape. The cultural integration of Negroes in North America has been slowed because of their obvious color identification; however, it should be noted that an unusual percentage of Orientals have achieved top positions in science, engineering, and research in spite of evident foreign appearance. This implies that the mental characteristics of that *specific* Oriental group that resides in North America is definitely superior to the average native citizen.

The most obvious and important lesson to be learned from all this, considering the undoubtedly great overlap in the mental distribution curves

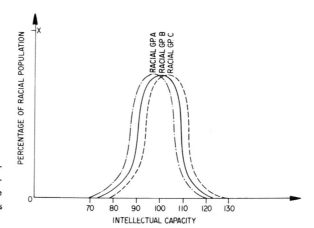

Fig. 8.2. A comparison of the hypothetical intellectual capacities of human racial groups, emphasizing the great overlap in distribution curves for several such groups.

for the different races, is that successful societies of the future will make maximum use of human mental abilities regardless of racial origin. Figure 8.2 emphasizes the great hypothetical overlap of intellectual abilities among the various human racial groups. Any society that desires to insure maximum success must eventually eliminate racial and national origin prejudices from its culture.

There are no mechanistic theories designed to account for abnormal human mental behavior, neuroses, and psychoses, because there are no comprehensive theories of intelligence. An examination of hypothetical extremes of parameters of mathematical models of the thinking process should reveal interesting abnormal behavior, and if the model is a good simulator of human intelligence, this abnormal behavior should correspond to well-known human mental illnesses.

Intelligence in Animals

The possibility that some nonhuman animals on Earth may have latent intellectual capability of a high level has aroused the interest of some scientists. There is no evidence that artifacts or written languages have been developed by animals to pass on cultural achievements from generation to generation, but this does not mean that they do not have aural communication. It is conceivable that a language, albeit crude, may already be in use for transmitting some limited information from generation to generation.

Any latent intellectual animal life would be expected to have a good central brain and input devices as is indicated in Table 8.1. If the animal does not have an adequate central brain, it would be of no interest, and it appears impossible for such a good central brain to evolve without input

Table 8.1
Mental Characteristics
of Latent Intellectual Animal Life

	Input	Understanding	Output
Human	Good	Good	Good
Animal	Good	Good	Good
Animal	Good	Good	Poor
Unlikely to exist	Poor	Good	Good
Unlikely to exist	Poor	Good	Poor

Table 8.2
Brain Weights of Animals and Humans

Animal	Age	Brain Weight (grams)
mouse	adult	0.4
rat	adult	1.6
guinea pig	adult	4.8
rabbit	adult	9.3
cat	adult	31.0
dog	adult	65.0
monkey	adult	88.5
chimpanzee	adult	350.
gorilla	adult	450.
human	14 months	950.
human	18 months	1,000.
human	41 months	1,200.
dolphin	23 months	1,200.
human	6.5 years	1,350.
dolphin	28 months	1,350.
human	17 years (adult)	1,450.
dolphin	8 years	1,450.
dolphin	10 years	1,600.
dolphin	? (adult)	1,700.
elephant	? (adult)	6,075.
fin whale	? (adult)	7,200.
sperm whale	? (adult)	9,200.

devices of satisfactory quality. It is entirely possible, however, that an animal having good central brain and input devices might have output devices that are not satisfactory for communication purposes.

In light of this supposition, an obvious physical characteristic to look for in latent intellectual animals is large brain size, measurable either in mass (grams) or volume (cubic centimeters). Table 8.2 shows the brain weight for a representative range of animals. Other than brain size, the total number of nerve cells or neurons must be investigated in seeking latent intellectual ability. Some unknown molecular factor related to RNA and perhaps the number of glial cells may possibly affect intelligence. Another factor is the proliferation of nerve fibers interconnecting the nerve cells. Since the size of the animal's body determines to some extent the size of the sensory and motor portions of the brain, it may be desirable to compare the ratio of the brain weight to the body weight for a variety of animals. Another factor worth consideration would be body surface area.

Any animal that has developed a complex language of its own or any animal that is capable of understanding a considerable portion of some human language demonstrates a fairly high level of intelligence. Animals that have developed a language of their own must have good, and well matched, input and output devices as well as an adequate central brain. It is very likely that any animal possessing all of these characteristics will already have begun the development of a rudimentary language. The search for language in wild animals is complicated because some are quite ferocious, others abide in remote and inaccessible locations, and some communicate considerably beyond the ranges of our human input and output devices, necessitating the construction of special engineering conversion instruments.

The absence of a rudimentary language does not mean that it would be impossible for an animal species to understand many words of a human language, an artificial language, or a range converted human language. However, long, arduous training and constant human contact with the animal species for months or even years may be necessary to determine its vocabulary limits of communication. This is at best extremely difficult with dangerous land animals and with animals that live in the oceans.

The higher apes are closely related to man and are consequently subjects for intellectual comparison studies. Gorillas are large and ferocious, making it difficult to study their ability to learn human language. Chimpanzees, however, are relatively good subjects, and their brains are not too much smaller than those of gorillas. Some chimpanzee babies have been brought up exactly as human babies. They have fairly large understanding vocabularies but can learn to speak only three or four words, which indicates that sufficiently refined output devices or speech mechanisms have not evolved. The understanding and learning ability of the chimpanzee is impressive. Some monkeys have been given direct brain stimulation rewards for many months in efforts to teach them to make voluntary oral responses, all to no avail. Apparently their oral responses are only instinctual, and no control connections from the central associating brain to the speech mechanisms exist.

In recent years, dolphins have aroused the interest of the scientific world. Many legends and stories exist pertaining to the intelligent behavior of dolphins in association with both humans and others of their own kind. Their well known gentleness with humans makes them excellent subjects for language research. Dolphins used in marine shows are able to understand a number of commands. They make a great variety of sounds through blowholes in the top of their heads, rather than through mouths. This varied vocal ability is explained by very complex nerve and muscle networks in the region of the blowhole, leading to the strong suspicion that dolphins may have already begun development of their own rudimentary language.

Kellog, Lilly and others are actively investigating the possible existence of a dolphinese language as well as the ability of the animals to understand human language, and perhaps speak it.[3] Studies are hampered because the frequency of sound generation and hearing of dolphins extends considerably beyond the human range. They can detect frequencies up to 200 kilocycles compared to a maximum of about 20 kilocycles for humans. Moreover, they produce frequencies up to at least 100 kilocycles. Preliminary studies have led to the suspicion that here on Earth dolphins may be second only to humans in intellectual potential. A more accurate evaluation may be possible in a few years.

Elephants long have been used in India to aid men in logging and construction operations. The trained elephant may understand 25 words of human language and his excellent memory is legendary.[4] However, their inability in vocal output precludes the easy development of complex elephant language. As is the case with the higher apes, the intellectual potential of elephants is regarded as well known.

Whales are also reputed to have performed remarkable feats of intelligence. Although whales have giant brains, their great bulk and their deep sea living make them poor language research subjects. Unusual discoveries resulting from studies with dolphins would certainly lead to extraordinary efforts to extend such studies to whales. Both mammals are similar.

---

[3] Kellog, Winthrop N., *Porpoise and Sonar*. Chicago: University of Chicago Press, 1961. Lilly, John C., *Man and Dolphin*. Garden City, N.Y.: Doubleday, 1961. Lilly, John C., "Critical Brain Size and Language," *Perspectives in Biology and Medicine*, 6, No. 2 (Winter 1963), 246–255.

[4] Rensch, Bernhard, "Increase in Learning Capability with Increase of Brain Size," *The American Naturalist*, 90, No. 851 (March—April 1956), 81–95.

The octopus is an interesting subject for intellectual experiments. With their eight arms and suction cups, they have great potential for manipulating tools or writing, but they would probably require a long additional period of evolution to develop sufficient brain size and oral capabilities for a language.

We can conclude that human and other animal nervous systems are similar qualitatively, and that the nervous systems of higher animals do not differ greatly from human nervous systems in quantitative respects. The nervous system provides the ability to detect and respond to external environmental characteristics and patterns, through processing and association of representative patterns. It is, in short, the very essence of biological thinking.

The ability to think would be a highly selective factor in animal evolution at all levels of animal intelligence. Since a large body can support a large brain, and a large brain can generally provide a better thinking ability, there must then exist a tendency in all animal evolution toward larger bodies and brains, and therefore toward increasing thinking capacity. Of course, upper limits on body size would result from the difficult maneuverability of very large bodies and insufficient food supply to provide the vast amounts of energy required to power exceptionally massive frames. Environmental factors and genetic molecular factors may also restrict this general tendency in some species.

It seems highly probable that the complete disappearance of humans from the Earth would permit the eventual evolution of some other species having a human level of intelligence, but this might take tens or hundreds of millions of years. A new intelligent species might fill an ecological niche similar to that of humans, evolving from primates, or a radically different intelligent species could evolve from some present species such as the dolphin, whale, elephant, or octopus. Any species able to develop a human level of intelligence would be expected to have good mental inputs for effectively recording environmental patterns, a good central brain for processing these observed patterns, and good effectors or mental output devices for both manipulating the environment and communicating with other members of the same species.

9
Thinking in Computers

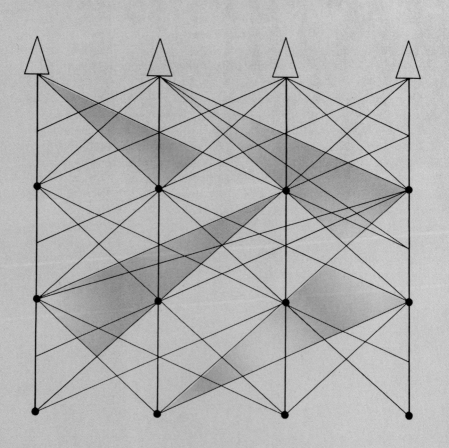

Having analyzed both intelligence and biological thinking, we can now examine the impact of modern computers on research into the nature of the thinking process, the rate of progress of such research, and some of the long range implications of thinking in computers.

Speculation about extrasolar intelligence requires an understanding of the potentialities of synthetic thinking devices, but unfortunately most people have not been made aware of current developments in thinking in computers. The following discussion is thus indispensible in evaluating the capability of very advanced technological societies to develop superintelligent synthetic automata.

Many engineering devices have been developed to supplement the input and output capabilities of the human brain. Supplemental input devices include microscopes, telescopes, radios, and television systems. Among the supplemental output devices are such machinery and electronic instrumentation as typewriters, automobiles, airplanes, radio transmitters, earth-moving equipment, and a wide range of manufacturing equipment.

One of the first engineering devices designed to supplement the thinking

"He gave man speech, and speech created thought.
Which is the measure of the universe."
Percy Bysshe Shelle

"The most tragic event in the history of mankind...only belief in that which can be proven. With one stroke it wipes out philosophy, faith, imagination."
Eugene Vale

capability of the human brain was the calculating machine. Simple mechanical calculators for performing arithmetic have been in use for many years, but high speed electronic computers have been brought into general use only in the last two decades.

There are two major classes of computers: analog and digital. Combinations of these two types (called hybrid computer systems) are now being developed and will constitute a third class. The analog computer is designed to solve systems of differential equations by means of varying continuous physical magnitudes, such as voltage, amperage, and distance. The slide rule is frequently given as an elementary example of an analog computer. The digital computer is designed to solve sequences of simple arithmetic operations (add, subtract, multiply, and divide) on discrete numbers. The abacus, adding machine, and desk calculator are simple examples of digital computers.

Both analog and digital computers supplement the human brain in performing these specific mathematical operations; there was originally no thought of using them to simulate the thinking process. They have performed their mathematical functions admirably, and digital computers in particular are proving to be far more flexible than originally anticipated.

Digital Computers

The general purpose electronic digital computer is controlled by a sequence of instructions called a program, which is prepared by a mathematician or programmer and is usually stored in the computer's memory. The modern electronic digital computer is frequently called an electronic data processing machine to emphasize the fact that data can be manipulated in a variety of ways other than pure mathematical computations. A high speed computer may be able to obey hundreds of different instructions which may be logically grouped into the following categories:

(1) Arithmetic instructions
(2) Input and output instructions
(3) Bit or character shifting instructions
(4) Transfer of control or branching instructions
(5) Logical instructions

Table 9.1 illustrates typical digital computer instructions of all categories. A large computer may understand an instruction set of one hundred or more different instructions and store in its memory programs containing tens of thousands of instructions.

It is very important to note that digital computers usually operate on serial units of information called *words,* which are usually of fixed length

Table 9.1
Digital Computer Instructions

1. Arithmetic		– Add A to accumulator
		– Subtract A from accumulator
		– Multiply A times accumulator
		– Divide accumulator by A
2. Input and Output		– Read one punched card
		– Read one block of words from magnetic tape
		– Read the words in the buffer
		– Read one character from the typewriter
		– Write sequence of words on a punched card
		– Write sequence of words on magnetic tape
		– Write sequence of words into buffer
		– Write sequence of words on printer
3. Shifting		– Shift left N positions
		– Shift right N positions
		– Shift right N positions and round off
		– Shift around left N positions (moving characters shifted off the left end onto the right end)
		– Shift around right N positions (moving characters shifted off the right end onto the left end)
4. Branching		– Transfer to storage location A to find the next instruction
		– Transfer to storage location A to find the next instruction if the accumulator is equal to zero
		– Transfer to storage location A to find the next instruction if the accumulator is positive
		– Transfer to storage location A to find the next instruction if the accumulator is negative.
5. Logic		– Every bit position which contains a one bit in both word A and word B shall have a one bit in that position in the accumulator.
		– Every bit position which contains a one bit in either word A or word B or both shall have a one bit in that position in the accumulator
		– Every bit position which contains a one bit in either word A or word B, but not both, shall have a one bit in that position in the accumulator

and contain from 12 to 48 bits. These computer words may be treated as binary data numbers, binary instructions, or alphanumeric character data. Typical computer words are shown in Table 9.2. Operations on these short, one dimensional data patterns are carried out on all bits in a word simultaneously (bit parallel) on large computers, but the operations may be bit serial on small computers. In summary, our large scale digital computers carry out operations on very short one dimensional information patterns (words) one word at a time (word serial).

Table 9.2
Computer Words

Fixed point binary number

sign → S	binary number
0	0 0 0 0 0 0 1 0 1 1 0

Floating point binary number

binary exponent	S	binary number
0 1 1 1	0	1 0 0 1 0 1 1

Computer Instruction

binary operation code	binary address
0 1 1 0 0	0 0 0 1 1 0 1

Alphanumeric data

one alphanumeric character	one alphanumeric character
0 1 1 0 0 1	0 1 1 1 0 1

Larger patterns and multidimensional patterns of information may be operated upon by breaking them down into these individual word size units. Figure 9.1 shows how a two dimensional information pattern may be broken down into one dimensional computer words. The digital computers are extremely fast in carrying out their operations on these short, one dimensional information patterns; the newer and faster computers have a cycle time of less than one millionth of a second, and most operations can be performed in a very few cycles. Thus, it is possible for a digital computer to do simple arithmetic calculations millions of times faster than a human being, but the basic digital computer instructions for operating on short, one dimensional information patterns do not seem to be closely akin to the human thinking process.

There exists a cliché which states: "A general purpose digital computer is designed to carry out arithmetic operations in a completely predetermined sequence and could never be said to think in any sense of the word." This cliché is even quoted by some computer specialists. An elaboration, which frequently follows the cliché, explains that the sequence of arithmetic and logical operations is completely predetermined by a human programmer, and any semblence of thinking by the computer is merely a manifestation of the intelligence of the human programmer.

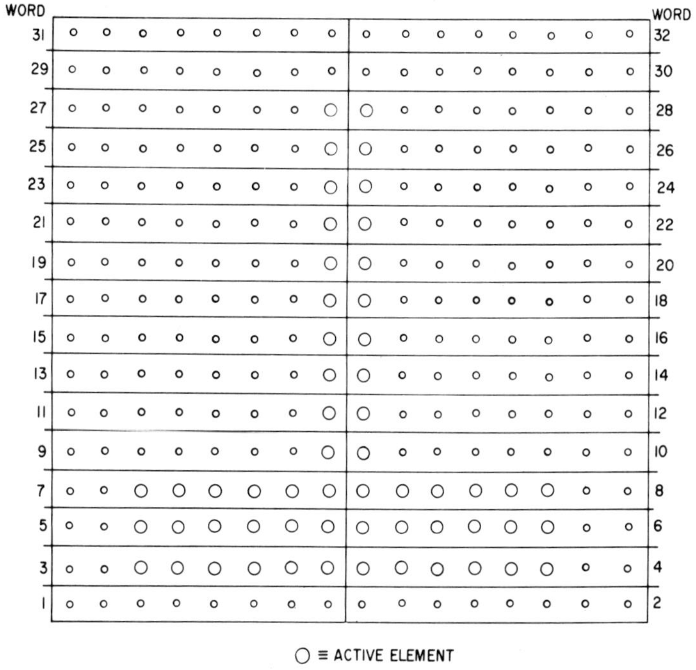

Fig. 9.1. Two dimensional, binary, information pattern broken down into one dimensional computer words.

Electronic data processing machines do carry out sequences of instructions exactly as specified by the programmer, but the programmer can, and usually does, specify that the sequence of instructions shall vary in a complex manner as a function of input variable (or sensory) data. Through the use of arithmetic (which is the basic language of measurement for all physical phenomena) and the other computer instructions, the computer programmer can simulate to any desired degree of accuracy any physical phenomenon which is sufficiently well understood that it can be described in precise mathematical terms. The time required to program a problem varies with the complexity of the algorithm, and the speed of simulation obviously depends on the speed of the computer. Since most current digital computers have only one processing unit, the instructions must be carried out serially, implying that some complex and dynamic system simulations will proceed at considerably less than real time.

Every conceivable form of physical science, engineering science, operational science, and social science problem is now being attacked successfully by the computer programmer; therefore, it may be asserted with confidence that the only requirement for simulating thinking, on any information processing machine worthy of the name, is a precise definition or algorithm for the thinking process.

It should be carefully noted that mathematical models for physical systems, such as guided missiles and aircraft, are solved on digital computers. But the computer manifestly does not fly. It only produces information as to how the missile or airplane would fly. However, if a mathematical model of an information processing system is inserted into a computer, it will actually carry out the physical information processing that is being modelled.

Two radically different classes of information system problems may be simulated on computers:

(1) A digital computer program may be inserted into the computer so as to simulate an information processing system very much like the digital computer itself. This is, in effect, building a full-scale working model of an information processing system.

(2) A digital computer program may be prepared so as to simulate an information processing system whose organization and physical components are radically different from the digital computer itself, but this still results in a system that can actually process information.

In current research in artificial intelligence both approaches are being used. Firstly, intelligent information processing programs are being developed for utilizing the characteristics of general purpose digital computers, and secondly, programs are being developed for modeling other types of information processing systems, such as the neural net.

The variety of input devices used with data processing machines is rapidly growing and there appears to be no great difficulty in developing input devices to feed new types of variable data into the computers. Input devices, such as typewriters, punched paper tapes, punched cards, magnetic tapes, magnetic character readers, and even optical character readers, are commonplace. It is frequently argued that data must be prepared in special formats and inserted by humans into these input devices; however, this criticism is trivial, as more flexible inputs are already being developed. A whole class of digital computers called *control computers,* which provide for the direct automatic input of variable information without human intervention, has already appeared. Control computers are in use in the chemical processing, petroleum refining, and many other industries. In these digital control systems any suitable physical variables are measured by transducers and the sampled values are digitized and automatically fed into the computer which calculates optimum control values. For individuals who hold that it is unreasonable to require that data be prepared in special formats it is recommended that they try listening to radio signals without the benefit of a radio receiver or watching television without benefit of a television receiver, doing without all typed or printed information, or listening to the calls of bats without special instrumentation. New computer devices even permit two dimensional pictures or information patterns to be drawn on a cathode ray tube with a light pen, so that these patterns can be scanned and read into the computer as a sequence of one dimensional computer words.

Digital computer output devices are already much more impressive than human output devices. Computer output devices include electric typewriters, punched cards, punched paper tape, magnetic tape (200,000 alphabetic characters per second), line printers (1,000 lines or 20,000 six-character words per minute), and cathode ray tubes capable of presenting alphanumeric characters, line drawings, or any desired two dimensional patterns. It seems that new output devices of any specified characteristics can be constructed at will. It may be concluded that input and output devices do not present a serious barrier to simulating thinking in a digital computer.

It is abundantly clear that thinking requires the sensing of external phenomena and the processing of input information patterns and stored patterns to determine and carry out an appropriate response or output. Digital computer storage is no great problem; the storage capacity of digital computers has been increasing steadily. The basic question seems to be whether or not a digital computer can calculate certain functions of input variable data and stored data, and then indicate the result of these functions of variable and stored data in an appropriate output form. This is, of course, precisely what the digital computer was designed to do, and it does this well. Therefore, it can be asserted without reservation that a general purpose digital computer can think in every sense of the word. This is true no matter what definition of thinking is specified; the only requirement is that the definition of thinking be explicit. Thus, the real problem in simulating intelligence in a digital computer is the specification of a precise mathematical (statistical, logical) definition (algorithm) of thinking. Any digital computer programmer could prepare a simple thinking program such as that illustrated in Fig. 9.2, if he is provided with the explicit definition.

Many skeptics point to the absence of drives and emotions in digital computers as a fundamental reason why digital computers can never realistically simulate human thinking. Since our human drives and emotions result from animal evolution there is no apparent purpose of introducing them into digital computer hardware. We do not hope to have a computer reproduce itself so there is no reason to design in a sex drive. Organic matter and water are not required to supply energy; therefore, no drive mechanism need be planned for these items. But a drive to maintain adequate electric power supply characteristics might be

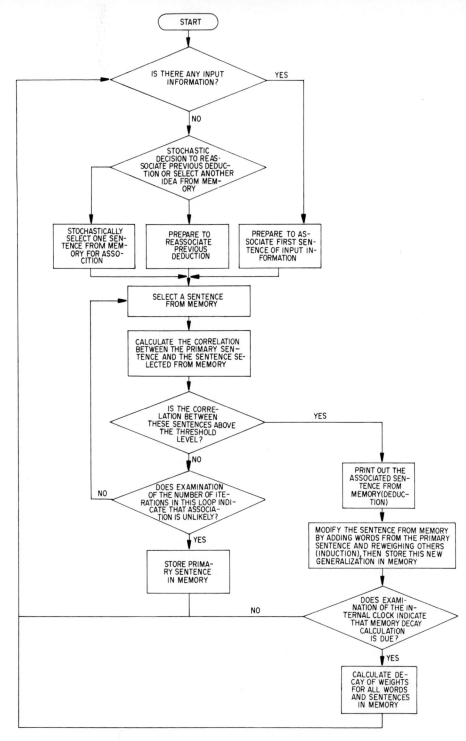

Fig. 9.2. A logical flow chart of a simple, heuristic, English language, digital computer program.

desirable. Since electronic components do not require a considerable recovery time after continued use, no sleep drive would be required in an electronic computer. Other drives, such as certain preventive maintenance functions, may quite possibly be wired into digital computer hardware. A pain elimination drive might have limited use for computers but it would certainly differ considerably from the corresponding human drive.

Emotional responses in a human would also differ radically from those which may be desirable in a digital computer. Fear, anger, and love are typical human emotions, but some version of these emotions may also be valuable to a digital computer. The ringing of a bell and initiating a cutoff sequence might be regarded as a digital computer emotion, as would perhaps the trapping mode in some computers.

SYSTEMS PROGRAMS

The service programs for a digital computer, which are not prepared by the applications programmer, are referred to as systems or techniques programs. These include input and output routines, subroutines, symbolic assemblers, algebraic compilers, dump memory routines, and executive or supervisory routines. The supervisory or executive programs are of some interest since they effectively replace a supervisor of computer operators. All or part of the supervisory program is in the computer at all times, and its purpose is to supervise the computer operators and to perform as many operating tasks as possible. One of the reasons for utilizing a supervisory program is to reduce human operator errors; that is, the goal is to design the operator out of the system. This type of program may be used to schedule programs, keep account of the machine time used by each program, record the errors of tape reading and writing, notify the operators when tapes should be loaded or unloaded, notify the operators when each problem begins and ends, automatically prepare memory dumps when a program error occurs, and many other functions. Although supervisory programs do not represent a very profound level of thinking, their performance is quite impressive to anyone first observing them in action.

GAME PLAYING

A great deal of effort has been expended in programming digital computers to play games such as checkers, chess, and tick-tack-toe. Although these games provide interesting and amusing demonstrations of digital computer capabilities, their real significance lies in their implications concerning the learning process. These games permit the investigation of decision processes similar to those required in complex real-life problems, but on such a reduced scale that they can be easily handled in existing digital computers.

Some games such as tick-tack-toe have known algorithms which would make it possible to play them in an optimum manner without any learning, but even these games can be played without recourse to the known algorithms, so as to yield nontrivial results. Other games, such as chess and checkers for which no algorithms are yet known, require good learning techniques to develop high performance.

Chess and checker playing programs generally involve a minimaxing procedure which requires looking several steps ahead and evaluating the board positions. At each step it is assumed that your opponent will move so as to minimize your future board position, while you attempt to maximize your future board position. A partial minimaxing diagram is illustrated in Fig. 9.3 Naturally, a player will select an initial move which will lead to maximizing his own board position at the maximum number of moves ahead for which he can plan, while assuming that his opponent will make a minimizing move at each step.

A. L. Samuel has demonstrated that a computer can be programmed to learn to play a better game of checkers than the programmer.[1] His computer program was able to develop a better game than his own after only 8 to 10 hours of machine playing time.[2]

ALPHABETIC DATA PROCESSING

Human societies have developed many radically different alphabets and others unquestionably will be developed in the future. The human brain processes alphabetic information quite readily, as evidenced by reading and writing capabilities, but the brain manifestly has no specific ability for processing serial groups of alphabetic characters. Rather, it has a more general ability for processing two dimensional optical patterns. The brain interprets words and sentences as two dimensional patterns. It processes these inherently one dimensional information patterns, producing alphabetic words and sentences as output. Thus, the human brain does a large amount of serial alphabetic data (verbal) processing with its natural equipment for the parallel processing of full scale two dimensional patterns. Since digital computers are designed to process alphanumeric character sequences quite efficiently, the very large inherent inefficiency of digital

---

[1] Samuel, A. L., "Programming Computers to Play Games," in *Advances in Computers,* Vol. I, ed. Franz L. Alt. New York: Academic Press, 1960.

[2] In a personal conversation in October 1965 A. L. Samuel indicated that his checker playing computer program had recently been defeated by the world champion checker player and a current challenger to the world champion. However he estimated that only about 12 to 15 of the world's checker masters could now defeat his computer program.

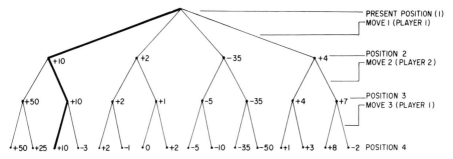

Fig 9.3. A tree of moves for game playing, showing the optimum minimax path. (Player No. 1 first considers possible moves leading to a board value of +50 at position No. 4, then he considers possible moves leading to a board value of +25 at position No. 4. Position No. 4 is as far ahead as he has time to evaluate. Finally, he considers possible moves leading to a board value of +10 at position No. 4; this is a valid minimaxing path leading to a maximum valued future board position for player No. 1.)

computers (relative to neural nets) in processing full scale two dimensional patterns is greatly reduced when considering serial alphabetic data processing. This is the basis for much research now being undertaken on digital computer methods for processing standard alphabetic or verbal language. It is even conceivable that in the near future high speed digital computers may be able to carry on crude, real-time conversation with humans.

Although it has been asserted that any digital computer can think if properly programmed, a careful examination of parameters is necessary to determine the speed of thinking for a particular computer and for a particular definition or algorithm. If a digital computer is able to compare one pattern of input data with one pattern of stored data and carry out any indicated induction processes in 10^5 operations and the computer can complete one operation in 10^{-6} sec, then a single pattern processing requires 10^{-1} sec. If it is further assumed that a memory must store 10^7 patterns for sensible intelligence, then it would require 10^6 sec to complete the searching and processing associated with one input pattern. As a result of electroencephalographic studies it will be assumed that the human being requires 10^{-1} sec to complete the processing of a single input pattern. On the basis of these assumptions one can say that this particular computer could accomplish human quality thinking about 10^7 times slower than a human being. Even if these assumptions are grossly in error, it is apparent that competition with human thinking presents an astronomical information processing and retrieval problem.

More optimistic assumptions[3] lead to a conclusion that digital computers in the near future may be only 10^3 times slower than the human in thinking or conversely, that 10^3 digital computers operating in parallel could equal the performance of a human thinker. Evidently we are close enough to human thinking from a computer standpoint to warrant encouraging appropriate research. We can conclude that general purpose digital computers are ideally suited for artificial and biological intelligence studies of all types.

A variety of digital computer programs has been developed for processing normal alphabetic language; some of these programs exhibit a few characteristics of the thinking process. However, most commercial or business problems show very little similarity to thinking, being concerned largely with inventory, billing, scheduling, payroll, and accounting problems.

If inventory problems, for example, were programmed in an unusual form, certain of their features would exhibit characteristics of thinking. Both learning and memory decay functions could be calculated so as to yield a use weight for each inventory item. In this way relatively obsolete inventory items could be automatically dropped from the list or relegated to a manually processed file. The items of the file could be arrayed in sequence according to the use weights so as to minimize the probable search time for any item. The search for individual inventory items could be based on correlation of alphabetic descriptions of items rather than exact matching of a code number or alphabetic description.

One of the greatest current problems in alphabetic data processing is optical scanning, by which is meant the automatic reading, or conversion to digital data, of typed or printed alphabetic information. Although such optical scanners are already in use they are limited to specialized type fonts and are still somewhat expensive. This is a technical problem area in which rapid improvement can be anticipated.

The preparation of an index or concordance for a book has been demonstrated on digital computers; the concordance of the Bible has received considerable publicity. Although quite tedious, indexing is a straightforward data processing problem, and not very impressive from the thinking standpoint.

The creation of computer abstracts is slightly more impressive than indexing. Digital computer abstracting is accomplished by calculating the frequency of usage of words in an article so that significance weights can be established. The sum of the significance weights of the words in each sentence divided by a factor representing the length of the sentence yields a significance weight for each sentence. An arbitrary ratio may be specified

---

[3] For example, an assumption of future capability may be that processing one pattern requires only 10^4 operations, that improvements in computers result in an operation time of 10^{-8} sec, and that the storage capacity required is only 10^6 patterns.

to determine that fraction of the total article to be used in the abstract, and the highest significance weight sentences are used in the order in which they appear in the article. Such digital computer abstracts have been rated superior to human abstracts in many cases.

In order to compose newspaper articles this same abstracting procedure could be applied to a group of several articles or manuscripts on a single subject, and the resulting selected sentences could be compiled into a single article in sequence according to decreasing significance weights. Newspaper reportertyped stories (with simultaneous computer tape output) and teletyped news service stories could serve as source articles for compilation. Computer editing routines could be used to detect and correct many misspellings resulting from typographical and reporter errors. Newspaper composing routines are already in use for forming newspaper articles into column width groups, hyphenating words, and spacing the type for precise newspaper column width. This digital computer application eliminates much of the work of newspaper composing room staffs, and the resulting computer tapes can be used to control linotype machines in automatic typesetting. The ability to specify the size of the abstract permits the editor to produce exactly the size article required to fit the available space with a minimum of manual editing.

Automatic translation of language is receiving considerable attention in research circles. Existing translation programs, which are supplemented by human post-editing and in some cases pre-editing, are already capable of yielding very high quality translations. Some semiautomatic translation schemes are already in commercial use. It is not too difficult to achieve fully automatic, smooth translations containing some logical errors, or rough translations containing a number of logical alternatives, but the attainment of completely automatic, smooth, high quality translations, such as an expert translator could produce, is a very difficult problem. It is rather similar to the problem of full scale simulation of alphabetic or verbal type thinking on a digital computer.

Another alphabetic data processing application, which is very close to the thinking process, is information retrieval. In information retrieval applications, some individual desires to retrieve all, or a large portion, of the information pertaining to a specified subject from some existing file. The searcher prepares a list of key words describing his subject, and his search list is automatically compared by a computer with a file of key word lists representing the total file of documents to be searched. The computer prepares a list of all documents having a certain percentage of key words that match the key words in the search list. The searcher may specify this percentage to control the depth of search; that is, the searcher may wish to look at all documents even remotely related to his subject or he may wish to scan only those few documents which are probably closely related to his subject.

A similar application is the selective dissemination of information (SDI); that is, the automatic distribution of each new document, as it is received, to individuals in an organization who are interested in that particular subject. Not only can the previously mentioned computer program be used for this application but, if it is a very general purpose program, it can be utilized for a great variety of applications. The location of personnel having special capabilities in a large organization, the identification of criminals by matching of *modus operandi* with clues, the location of research installations having particular capabilities in men or equipment, and the identification of chemical molecules or engineering materials having specified characteristics, are just a few of the potential applications of information retrieval programs.

A major facet of the thinking process is unquestionably association, and information retrieval programs are actually alphabetic association programs. Future word association programs will permit the use of complex logical statements such as the following for specifying association: retrieve items identified by at least 25 per cent of the words A through J or at least 40 per cent of the words J through R and containing words X and (Y or Z). Such programs are in their infancy at present and, although much effort is being concentrated on information retrieval, most workers in this area have only limited experience.

Perhaps the ultimate goal of alphabetic data processing may be regarded as composition. Both automatic abstracting and information retrieval or association have avoided the problem of rearranging alphabetic sentences and forming new sentences. This brings us face to face with the problem of language analysis, now being attacked by the best brains in information processing research. Composition complexity, which may vary all the way from the crudest conversation to erudite textbooks, is indicative of the whole human intellectual range.

MISCELLANEOUS METHODS

A variety of other applications of alphabetic data processing, which may have some import for the thinking process, are mentioned only briefly because of space limitations.

Some digital programs have been developed to demonstrate the ability of a digital computer to prove mathematical theorems by heuristic methods. Gelernter, Rochester, and others have devoted considerable study to the problem of finding proofs for theorems of geometry.[4] Work was done by Newell in discovering proofs for theorems in symbolic logic, and some proofs

---

[4] Gelernter, H. L. and N. Rochester, "Intelligent Behavior in Problem-solving Machines," *I. B. M. Journal of Research and Development,* 2 (October 1958), 336–345.

have been discovered that are considered more elegant than those derived by mathematicians.[5]

Business management games and military (tactical and strategic) games are now receiving increased attention. Mathematical models for simulating the performance of business and military organizations are programmed for a computer and then executives are permitted to make decisions concerning game parameters and observe the effects of their decisions on their organization. Feedback in the form of critiques by the players leads to progressively more accurate mathematical models. We are finding that games serve not only as a training medium but also as a research tool for developing optimum strategies.

Future Digital Computers

General purpose digital computers have already made a tremendous impact upon our society and the end is not yet in sight. Electronic engineers have been rapidly decreasing the cycle time (that is, increasing the speed) in computers by improving the basic electronic elements and developing such new elements as diodes and transistors. The storage capacity has been increased by developing semiautomatic methods of wiring ever smaller magnetic cores, and this random access storage has been supplemented in large computers by ever larger capacity semirandom access magnetic drums and magnetic disk auxiliary storage devices, which are now measured in billions of bits maximum storage capacity. The miniaturization of electronic circuitry (Fig. 9.4) has led to significant reductions in the physical size of computers,

---

[5] Newell, A. and Simon, H. A., "The Logic Theory Machine," *I. R. E. Transactions on Information Theory*, IF-2 (September 1956).

Fig. 9.4. Miniaturized digital computer circuitry. A logic module illustrating the new *solid logic technology* developed for the new No. 360 computer system. Courtesy International Business Machines Corp.

and solid state circuit components have greatly reduced power consumption and air conditioning requirements.

Computer speeds are presently measured in microseconds, and the impending use of tunnel diodes, faster transistors, and microcircuitry indicates that nanosecond (billionths of seconds) speeds are imminent. Longer range developments in molecular electronic components and cryogenic circuits portend a continuing improvement in computers into the more distant future. The transmission speed of electricity in conductors (the speed of light) is an obvious limiting factor to the increasing speed of computers. That is, when the limit of miniaturization has been achieved for any given period, the ultimate computer speed will be determined by the distance (or transmission time) between computer components. Undoubtedly this limitation will be circumvented by parallelization of computer functions, which may be achieved either by use of multiple parallel units in a computer (such as multiple arithmetic units) or by connecting multiple computers together in parallel arrays. Since digital computers are improving so rapidly, and their impact on the thinking process is becoming so great, a brief but careful look at digital computers of the next 5 or 10 years is desirable.

INPUT-OUTPUT DEVICES

At present, computers have a very high speed input-output device in the magnetic tape unit, which is used primarily for reading and writing alphanumeric information. However, this alphanumeric information is not in an optically readable form on the magnetic tape and usually another smaller computer is used to read the magnetic tape and write the alphanumeric characters on a printing machine at a rate of about 1000 lines (132 characters per line) per minute, which is very fast by human standards.

The optical reading of alphanumeric characters by a computer is a much more difficult problem. It would be very desirable for a computer to be able to read the same typed, printed, or hand printed alphanumeric characters used by humans. Much research and development is now being done on optical scanners, which are designed to convert typed or printed characters to digital equivalents. Several optical scanners are already in general use but they are restricted to reading special type fonts, and none is yet sufficiently flexible and accurate to be capable of reading hand printed characters or varied type fonts. At present, most optical alphanumeric characters must be typed manually or printed on special forms, which are then converted to magnetic tape on a small auxiliary computer for final reading by the large computer. Undoubtedly, more flexible optical scanners will come into use with future computers.

Military computer systems are making extensive use of visual display and control consoles, in which two dimensional pictures, line drawings, graphs, and charts, with alphanumeric labeling are presented on cathode ray tube screens for observation by the operator. Sometimes simple alphanumeric information is presented on such devices, and frequently a photographic recording may be automatically made of the output so that later reproduction is possible. Visual display and control consoles provide decision switches permitting the operator to enter into the computer any decisions he may make while observing the displays, and digital parameter dials also permit the insertion of desired numerical parameters at the operator's discretion. A new feature of display devices permits the operator to sketch or project any two dimensional drawing onto the face of the cathode ray tube and cause this drawing to be scanned, digitized, and entered into the computer. The combination of all of these features, as well as top quality alphanumeric printed characters in man-machine communication consoles, will provide an extremely powerful and flexible tool for future computers.

Digital control computers are being used in some industries to maintain control of continuous processes such as chemical processing and oil refining. The control computers permit frequent samples of any physical quantities to be digitized and automatically entered into the computer, and output digital control values may be used to operate switches or they may be converted to analog functions for continuous control purposes.

The input-output devices of future general purpose computers will certainly include all of the previously mentioned types and yield greater flexibility than present computer systems.

STORAGE AND RETRIEVAL DEVICES

Much research and even more writing is appearing on various aspects of the problem of information storage and retrieval. Unfortunately, some of these efforts are by people who are not highly experienced in the subject. Any accomplishment in this area requires an excellent background in digital computer capabilities and digital computer problems, especially alphabetic information processing problems, and an extraordinary grasp of future developments in the data processing field. Consequently, a great deal of trivial information and even misinformation has been written concerning information storage and retrieval.

Any block of information, whether it be a book, a technical report, a magazine article, an engineering drawing, or a graph, may be characterized or described by a group of alphabetic characters and numbers. Usually, an alphabetic document can be described by a group of normal alphabetic words, known as key words, representing the explicit topics discussed in the document. The title and author and any other significant information may be included among the key words. Oftentimes, engineering drawings, graphs, and similar material are identified by code numbers which are sequences of alphanumeric characters. In general, to retrieve a document it is desirable to compare one or more groups of alphanumeric characters, specified by a searcher, with a file of groups of alphanumeric characters (key words), which is an index to the storage location of the complete documents.

The general purpose digital computer is ideally suited for temporary storage and processing of groups (short sequences) of alphanumeric characters, such as groups of key words or index code numbers; but storage and processing in a digital computer are expensive, and it would not be economically efficient to consider storage and processing of whole documents in a digital computer merely for storage and retrieval purposes.

Some type of peripheral storage box for storing whole documents in a standard and readily reproducible form is apparently what is needed. The best medium for bulk storage of both alphanumeric documents and drawings at present appears to be microfilm, and the individual unit of storage should be a single developed negative. Viewers to observe retrieved documents, photographic printers for automatically producing full size, hard copy reproductions, and photographic equipment for automatically making microfilms of original documents for storage should all be standard features associated with a microfilm storage box. It should be possible to have a general purpose digital computer do all of the information processing required for storage and retrieval, and either directly or indirectly (by means of punched cards) control the functions of the peripheral microfilm storage box. Probably the most convenient input-output medium for the storage box would be the aperture card, which is an 80 column punched card containing a developed or undeveloped microfilm negative.

The comparison or matching of index information by a retrieval program may be based on a great variety of logical criteria such as logical AND, logical OR, or logical majority, or a percentage of word matches, or complex logical statements involving all of these conditions. A searcher may desire an exhaustive search, retrieving many documents only remotely related to his subject, or the searcher may wish to limit his retrieval to those specialized documents having every characteristic specified by the searcher.

A parameter that should be specifiable is the percentage of the searcher's key words that must be matched by the file document key words to cause retrieval. It will readily be seen that a high percentage match parameter will result in the retrieval of relatively few documents, most of which will be pertinent; however, a low percentage match parameter will result in the retrieval of a very large number of documents, many of which may not be pertinent to the searcher's interests. Of course, when the high percentage matching parameter is used, many documents which are pertinent may be neglected. The exhaustiveness of the search must be determined by the searcher's needs. The variation in word endings, prefixes, etc., also presents problems (for example, consider the words, program, programs, programmer, subprogram, programme, programming).

The important point to recall is that a single flexible general purpose computer program will suffice for a great variety of information retrieval problems. Such diverse information retrieval applications as the following may all be done utilizing a single digital computer program provided that it is sufficiently flexible:

(1) Selective dissemination of current reports, articles, and books
(2) Literature search for all documents concerning a particular subject
(3) File search for persons having certain required interests and abilities
(4) Research facility search for certain required equipment and personnel capabilities
(5) Engineering drawing recovery and reproduction
(6) Medical diagnosis based on patterns of symptoms
(7) Criminal identification based on the characteristics of crimes

The microfilm storage box and its accoutrements will certainly become a major feature of many digital computers of the future and will permit the economic solution of a great variety of information storage and retrieval problems, while leaving the general purpose digital computer free, much of the time, for the solution of other types of problems.

CENTRAL PROCESSORS

Continued and impressive advances may be anticipated in future computer processing units; order of magnitude improvements in speed every two or three years have been the rule in the past, and a continuation of this rate of improvement seems likely at least through the next decade. Transistor and magnetic core circuitry utilizing improved transistors offer promise of an order of magnitude increase in computer speed in the near future, and microcircuitry utilizing thin film depositing and perhaps tunnel diodes indicate a further increase of an order of magnitude or two in speed. In the longer range picture, cryogenic circuitry and molecular electronic devices may begin to have an impact on computers.

When the low nanosecond speed ranges are achieved, limitations on the further miniaturization of electronic circuitry will limit computer speeds due to the distances (and proportionate time required) over which the electronic signals must travel between various computer components. At

210   **Thinking in Computers**

this point the question arises as to the organizational possibilities in future computers. One approach of computer designers is to install multiple arithmetic units or central processors in a single computer in order to increase the computing capability per unit of time. This method, and basic electronic methods for increasing computing speeds, inevitably reach a point, at any given state of the art, where further advance in computing speed could be achieved more economically by connecting several computers in parallel rather than improving a single computer. Both the multi-arithmetic unit approach and the multicomputer approach require that the problems be segmentable into logically parallel parts. These questions will continue to concern computer designers indefinitely.

Already, hybrid systems are being developed from combined analog, general purpose digital, and digital differential analyzer computer systems. The general purpose digital computer of the future will probably include digital differential analyzer capability for parallel solution of systems of differential equations, and analog-to-digital input devices and digital-to-analog output devices. In the long range future, hybrid general purpose computer systems will probably also contain two dimensional pattern processors for thinking or intelligent processes. The obvious requirements for these distinctly different types of processing make the inevitability of hybrid systems apparent, but the precise organization of such systems is still speculative.

INTEGRATED COMPUTER NETWORKS

One of the most significant milestones in the history of computing is the current development of digital data transmission devices which will permit the growth of huge integrated computer networks. In the not too distant future thousands of general purpose digital computers will be interconnected by the general telephone and microwave communications networks,

Fig. 9.5. Major computer systems communicating with one another through the International Telephone and Microwave Communication Networks.

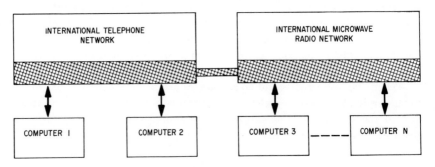

thereby producing an unprecedented possibility for parallel computing (Fig. 9.5).

The network will develop as a result of the need to transmit data from source location to computer location, the need to solve peak load problems when the local computer is overloaded, and the need to solve problems on large and efficient computers. Business data, research and engineering data, government data, military data, etc., as well as the instructions for solving some of the problems, will all be transmitted over the communication networks.

Ultrahigh-speed data transmission will utilize microwave communication facilities, making possible the transmission of very large amounts of data or even permitting the operation of real time control systems via microwave link to and from distant computers. The inherent redundancy in the network should increase the over-all system reliability to the point where even these real time control operations will be feasible. Single and multiple telephone, telegraph, and teletype lines of various qualities will afford a complete range of transmission speeds, reliabilities, and proportionate costs, designed to meet any specific requirement. The great variety of communication characteristics and redundancy existing in the vast communication network will permit all types of computation requirements—such as real-time control, temporary stacking, batch processing, and post real-time analysis—to be handled via the network.

Initially, there will probably be little thought of simultaneous parallel use of many computers on the network for the solution of single problems; however, urgent need and careful planning will soon result in this type of application. The recognition of the tremendous computing power and information retrieval power of thousands of digital computers operating in parallel with heuristic programs will eventually lead to this type of utilization of the integrated computer network. A human level of artificial intelligence will probably first be achieved by this approach.

Neural Networks

A neural network is a network of logical elements generally assumed to have similar logical characteristics. The neural net derives its name from analogy with the neurons or nerve cells and the interconnecting nerve fibers in animal nervous systems. The term is considered synonymous with logical net by some researchers, while others draw a distinction between the two. The inputs to an artificial neuron may each be weighted and each input may be either reinforcing or inhibitory (that is, positive or negative).

Some function of the inputs is calculated within the neuron itself and is then compared with a threshold, which may be constant or variable. If the function of the inputs exceeds the threshold, an output signal is emitted to some other neurons in the network. Usually, the same output signal is emitted from all of the outputs of a neuron simultaneously, and the output pulses from all neurons are assumed to have similar characteristics.

Those who draw a distinction between neural nets and logical nets may not require similar simultaneous outputs from all output lines of a logical element of a logical net. The logically selective outputs characteristic of more traditional logical digital elements may be considered permissible in logical nets. Whereas the functions performed in neural net elements are primarily of a simple summation nature, those in logical net elements may be of a more complex logical and mathematical character. The elements in a logical net, rather than being similar, may be of varied functional characteristics. In summary, some scientists regard neural nets as a very special subclass of logical nets.

The relatively greater simplicity of neural nets over logical nets, which implies easier analysis and maintainability coupled with the possibility of mass producing cheap neuron models in modules containing perhaps thousands of identical artificial neurons and their interconnections, makes the study of neural nets more attractive than the study of logical nets for processing highly complex patterns in parallel. The parallel processing of the basic elements of very complex information patterns implies some similar treatment of all the pattern elements. The following discussion will be limited to the narrower interpretation of neural nets.

CHARACTERISTICS OF NEURONS

Biological neurons are extremely complex devices with greatly varying characteristics. It is not yet known just how significant much of this complexity and variation in the neurons is to the mental functioning of an organism. It is conceivable, if not probable, that most of the neural complexity and variability is a result of biological evolution and would not be required for efficient mental functioning or information processing in synthetic devices.

This assumption will be used as the justification for describing an unrealistically standard neuron. A neuron usually receives stimulation on short extensions of the cell body called *dendrites*. When a sufficiently large number of stimuli occur in a sufficiently short length of time, then an electrical signal develops, which quickly assumes a standard form. This standard electrical pulse usually begins in or near the cell body and then propagates along the axons, which are long output fibers of the nerve cell. At the ends of the axons contact is made with other cell bodies or dendrites through junction points, called *synapses*. In this way electrical pulses are transmitted from neuron to neuron within the nervous system.

Receptor cells, which serve as input to a neural network, fire series of these pulses whose frequency varies as a function of some physical quantity. Receptor cells may be measuring physical phenomena either internal or external to the animal's body and firing varying frequencies of pulses into the neural net.

Internal or external output effector cells carry out some chemical, mechanical, or electrical activity to a degree which is proportional to the frequency of pulses impinging on the output cell.

Very complex electrical-chemical-mechanical processes are involved with many of the phenomena of receptor, effector, and transmission cells; at the molecular level it is sometimes difficult to distinguish these processes from one another.

The trans-synaptic stimulation time for a neuron is about 10^{-4} sec, but this does not take into consideration the time required for a neuron's recovery from fatigue after it has fired a pulse. The effective neural cycle time, or the average time between standard output pulses when the neuron is undergoing normal stimulation, is about $1.5 \cdot 10^{-2}$ sec. If the brain is assumed to occupy a volume of about 10^3 cm^3 and contain 10^{10} neurons, then the volume required for each neuron must be approximately 10^{-7} cm^3. The electrical potential generated by a neuron is about 50 mv and the pulse duration is approximately 1 msec. If the energy dissipation for the whole central nervous system is assumed to be about 10 w, then the energy dissipation per neuron is about 10^{-9} w.

Although some neurons may have only one or two inputs from other neurons, the more general situation will be that many, perhaps even hundreds, of inputs from other neurons will exist. It is thought by many authorities that the stimulation required to reach the threshold or firing level of a neuron is a simple summation-time (simultaneity) function of the input pulses.

ARTIFICIAL NEURONS

Attempts by mathematicians to construct realistic mathematical models of neurons that faithfully reproduce the great complexity and variability of natural neurons have understandably resulted in great mathematical complexity. Some exceedingly complex, high order, differential equations have resulted from these model building efforts; but, if it is assumed that much of the complexity and variability of natural neurons is superfluous, then very simple mathematical models of neurons will suffice.

Before the development of solid state electronic devices, a number of electronic neuron models were developd utilizing traditional electronic components. Most of these models used from one to several vacuum tubes and several resistors and condensers. More recently, similar models have been developed utilizing transistor-diode circuitry and miniaturized versions of this type of circuitry. Now, research in thin film deposited circuitry is leading to new neuron models, and future studies in cryogenic circuitry and molecular electronic elements may ultimately lead to neuron models competitive in size and power consumption with natural neurons. Much research is also taking place with magnetic elements, such as multi-aperture magnetic cores for possible use as neuron models.

Electronic models of neurons are essentially digital devices in that they accept and generate discrete electrical pulses of standard characteristics, as is the case with natural neurons. However, the internal circuitry of the neuron model itself may be analog, digital, or hybrid. All of these approaches are being tried by researchers.

NEURAL NETS

It is still not completely certain whether natural neurons are limited to simple summation functions or whether they also involve logical functions of the inputs. In other words, it is conceivable that a "logical network" may ultimately prove to be a suitable name for natural neural nets. At any rate, certain obvious and many not so obvious functions can be performed by neural nets having only a simple threshold capability.

For the sake of clarity it might be desirable to classify some of the different possible types of neural net. A network containing only one type of neural element is termed a *homogeneous network*, and a network containing more than one type of neural element is a *heterogeneous network*. A neural network which has only one input and only one output device is referred to as a *simple network*, while any network having multiple input or output devices is called a *complex network*. In the early stages of research, when it is only desired to discern the basic capabilities of neural networks, it is practicable to limit much of the research to simple, homogeneous, neural networks of the type illustrated in Fig. 9.6.

Probably the best known function of neural nets, and one stimulating much current research, is pattern recognition, by which is meant the ability to learn and recall specific, two dimensional patterns or arrays of information bits. Pattern recognition implies that a distinctive response will result when the system is stimulated by each pattern that the system has been taught. Simple pattern recognition assumes that the patterns being recognized are identical with the patterns that the system has been taught.

Some amount of practice or learning is required by neural nets before discrimination between patterns is possible, implying that neural nets of differing characteristics may have radically different learning rates. Any program for evaluating the performance of neural nets should include an

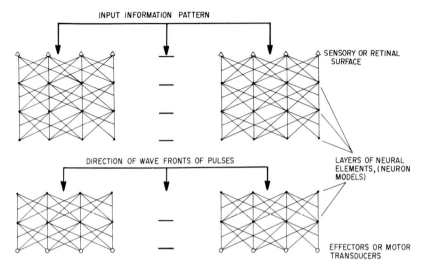

Fig. 9.6. A cross sectional view of a simplified neural net type system.

objective measure of the learning rates. An evaluation of the learning rate may also have to take into consideration the complexity of the patterns, perhaps the similarities of the patterns, and the sequence of stimulation by the patterns being learned.

If a neural net is able to recall a number of learned patterns then it must have a definite memory capability. The maximum memory capacity of a neural net is certainly of great interest, but it may be difficult to estimate by empirical methods. A large number of different patterns may be used to stimulate a network; and, eventually, a further increase in the number of different patterns will result in a sharply increased failure rate in recalling previously learned patterns. By increased failure rate it is meant that the ability to discriminate between patterns is reduced and consequently the percentage of erroneous responses increases. This sudden increase in the recognition failure rate results from a saturation of the neural net, which is comparable to an overwork nervous breakdown in a human.

Some neural nets may have a memory with an automatic decay feature, in which case an empirical measure of the memory decay rate would be significant. Memory decay might be a function of either time or machine cycles, and it very possibly would not be a linear function.

Generalization ability is one of the most important characteristics of neural nets since, in many practical applications, the same identical patterns may not be repeated. The more common situation is that many classes of similar patterns will provide the stimulus. If an input pattern is closely enough correlated with a learned class of patterns, then the normal system response to one of that class of patterns (a generalization) should be elicited. The particular input pattern may contain certain pattern segments which the generalization contains. An input pattern might have both missing and extra pattern segments, yet still be sufficiently correlated with a generalization pattern to elicit the desired response. In deducing the appropriate response, a neural net may be faced with an input pattern which is maloriented with respect to the stored generalization. Either a rotation, or a translation of the origin, of the pattern may exist, but if these malorientations are not too great then deduction of the appropriate response may still be possible.

Induction, which is the gradual formation of the generalizations, takes place with the repeated stimulation by many different, but highly correlated, patterns.

Another most significant characteristic of neural nets is appropriately referred to as *redundancy*. In this context, redundancy implies the capability of overcoming random noise in the input signal patterns as well as random component failures within the network. Random bits of noise scattered throughout a two dimensional input pattern will tend to be ignored or filtered out by a neural network; similarly, missing random bits in a pattern will be replaced by a network. The failure or malfunction of randomly distributed electronic elements, or the neural elements in which they are components, should not significantly affect the performance of the network unless the total fraction of malfunctioning elements becomes large. This should be true whether the electronic malfunctioning results in the failure to emit pulses or in excessive or random emission of pulses. These redundancy effects of neural networks are assumed to account for the extreme reliability in animal nervous systems in spite of malfunction and death of many nerve cells.

RESEARCH ON NEURAL NETS

The search for better electronic models of neurons is now being pursued aggressively in many research institutions. What is sought by the researchers is a good artificial neuron which can be mass produced cheaply and preferably in miniature assemblies of large numbers of neurons with all associated circuitry. Present success with thin film deposited circuitry and miniature transistors is naturally stimulating interest in the possibility of mass producing large numbers of these artificial neurons in batches on planar or multiplanar assemblies.

This thin film deposited circuitry seems to satisfy the criteria of miniaturization and mass production with relative economy. Another approach being tried at present is the use of multiaperture magnetic cores, which also offer miniaturization and mass production with considerable economy. However, the stringing or winding of the cores presents a problem since this is generally still a semimanual process.

The development of molecular electronic neuron models is an attractive concept which may offer a good, long range solution to the problem. It is conceivable that large numbers of molecular electronic neuron models could be produced in large assemblies with required circuitry. Most of the peculiarities of individual elements would have a trivial effect on both the performance of that particular element and the network as a whole. Thus, a good molecular electronic model should possess an extraordinary reliability. Cryogenic circuitry is also receiving much attention from the research organizations of the computer manufacturers. This ultracold circuitry permits astonishing reductions in power requirements and consequently in heat dissipation, thereby permitting extremely dense packing of electronic elements and also a very large fanout of circuit connections from a single element to many other elements. This large circuit fanout will unquestionably be a significant factor in neural networks. It is possible, if not probable, that combinations of the previously mentioned electronic circuit techniques will find effective use in neural networks of the future.

The great variety of potential electronic models of neurons implies a similar variety of characteristics and parameters for these neuron models. Since the structure of a neural network may vary both as to the dimensions of the network and the interconnections of the network elements, the potential variability of neural networks is even greater than might appear at first glance. The great variability combined with the comparative newness of neural net problems have resulted in an extreme deficiency in the mathematics of neural nets, although some research is being devoted to this problem.

In general, it is impossible for mathematicians readily to calculate, analytically, the performance of a specified neural network, or to calculate the network characteristics required to solve a particular problem. Because of this mathematical deficiency many researchers have been forced to resort to the use of general purpose digital computer programs to simulate the performance of each hypothetical neural network. Moreover, the preparation of large and complex computer programs is both time consuming and costly. It is possible for persons who are very proficient at both neural net problems and computer programming to develop general purpose computer program modules which can be fitted together so as to simulate a broad variety of neural networks.

The key to this approach is to maintain relative independence of function in each individual module and to make each module so flexible that any functional parameters may be inserted and any range of variables may be accomodated. Once such a set of general purpose computer program modules has been prepared for a general class of neural networks, then any network of that class may be simulated by the assignment of the desired parameters, and no new additional computer programming is required for each new network configuration or neuron model. Many of the program modules would be interchangeable with modules in programs for other major classes of neural network, thus yielding additional savings. General purpose neural net simulation programs such as this will be developed and should find widespread use in the next few years.

The only alternative to this digital computer simulation of neural nets would be the actual construction of hardware neural networks, which is very time consuming and costly. This uneconomical hardware approach does not seem justified until individual neuron models have been studied and their performance in full-scale networks has been simulated on digital computers and found adequate for the solution of particular problems.

APPLICATIONS

As has been true with many types of research, military applications are financing much of the research in neural nets. The military is faced with the possibility of simultaneous attack in any one area by multiple missiles containing nuclear warheads and decoys. Clearly, the anti-missile defensive forces must quickly distinguish between the decoys and valid targets. The different types of optical and electronic tracking devices may be expected to provide different patterns of information for different types of target and decoy. This is manifestly a pattern recognition and discrimination problem having the most stringent time requirements. The inherent parallel nature of the pattern discrimination problem in neural networks leads to their immediate choice as a possible solution to the problem. It is conceivable that neural nets may be able to distinguish between targets and decoys rapidly enough to permit anti-missile missiles to intercept all the valid targets and ignore the decoys.

Anti-submarine warfare presents a similar pattern recognition problem in that electronic and sound tracking of different types of submarine and fish produce different types of information patterns. The recognition and tracking of submarines being a vital naval problem, the US Navy, of necessity, has chosen to support research in neural nets.

An area of immediate interest to the computer industry is that of character recognition, which is just a special aspect of the pattern recognition problem. A major cost and inconvenience in digital computation has always been the necessity of manually punching, on cards or paper tape, information that is available on typewriten, printed, or hand printed sheets. What is needed is a device for automatically scanning printed pages, recognizing any alphabetic or numeric characters, and causing an appropriate coded digital response representing that particular character. The great variety of type fonts and hand printed characters make this pattern recognition problem perplexing. The flexibility required for this implies the use of a rather large raster of elements, and parallel processing of the patterns on the raster appears desirable. Several optical scanners of very limited flexibility have already been developed by computer manufacturers. The problem is so important that they are continuing to finance research on neural nets.

A distinctly more difficult problem for neural nets involves adaptive control systems, which are also of considerable interest to the military. In an adaptive control system it is desirable to recognize not only a pattern but also its precise orientations so that smooth control system responses may be calculated. This implies that patterns may be fed to the system in a rapid sequence and the system must make rather fine distinctions between many slightly varying pattern orientations. Again, the parallel pattern processing requirements seem to indicate the desirability of the use of neural nets for some control system applications.

In this chapter it has been emphasized that much current work on digital computers is either directly or indirectly related to the thinking process. Present computer research pertaining to the thinking process falls primarily into two distinct categories: (1) programs for carrying out thinking-type processes on normal alphabetic language, and (2) programs for carrying out thinking-type processes on two-dimensional bit patterns by simulating neural network-type hardware. Rapid progress can be anticipated in experiments on digital computers with hypothetical definitions of intelligence, such as that described in Chapter 7.

Several orders of magnitude increase in the speed of digital computer hardware may be expected within the next couple of decades; and perhaps even more significantly, thousands of digital computers all over the world will be tied together through the telephone and microwave communication networks, thus permitting coordinated efforts on problems of special importance. A new class of computers being investigated, the so-called neural net computers, have an inherently superior ability to perform thinking type processes.

Research and development in thinking in computers and the present rate of progress in micro-electronics technology strongly indicate that advanced technological societies will be able to produce completely synthetic intelligent automata capable of creative thinking in every sense of the word. It may be possible to create superhuman synthetic intelligence on Earth within the next few decades. By implication, we can speculate that intelligent artificial automata will probably play a very important role in any extrasolar civilization. The following chapter will discuss their development in order to give a rough idea of the range of characteristics and capabilities which may be expected from superintelligent synthetic automata.

10
The Development of Intelligent Artificial Automata

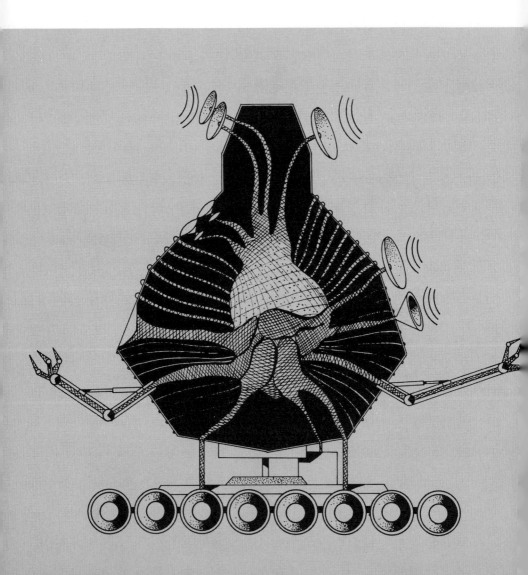

In previous chapters we learned that rapid progress is taking place in research on (1) explicit definitions of intelligence, (2) understanding of biological thinking processes, and (3) carrying out thinking processes in computers. From what has taken, is taking, and predictably will take place here on Earth, we inferred that advanced technological societies could build superintelligent synthetic automata, which should certainly play a very prominent role in speculation concerning extrasolar intelligence.

In this chapter we shall consider alternative methods for developing intelligent artificial automata and we shall explore their probable characteristics. The unique characteristics of intelligent automata must certainly lead to some capabilities that are extraordinary when compared to intelligent biological life. The potential advantages of intelligent automata over intelligent biological life will be discussed, and the question of the impact of superintelligent automata on intelligent biological societies will be faced. If, indeed, intelligent biological life has arisen prolifically throughout the universe, the present character and distribution of extrasolar intelligence must be affected greatly by the impact of synthetic intelligent automata.

"For I believe that man is not the most perfect being but one, but rather there are many degrees of beings superior to him."

Benjamin Franklin

"Effort and talent make possible the difficult. Effort and genius beget the impossible."

Mario Gonzales Ulloa

In earlier chapters it was shown that the general nature of the information processing functions required for thinking is beginning to be understood. Further development of mechanization of the thinking process requires only diligent research and development in information processing functions, such as pattern correlation, pattern recognition, pattern induction, learning, memory decay, emotions, and drives, and computers and neural networks. This research and development includes parameter studies with many mathematical definitions of thinking on both digital data processing devices and neural networks.

It should now be possible to design and build a very large digital computer system, consisting of many parallel computer units (Fig. 10.1), which would exceed the thinking capacity and speed of any human. This brute force approach to the problem could be achieved with relatively minor research in the programs to be utilized and within current computer technology. The only difficulty would be the astronomical cost of producing the large number of digital computing units required. A major world government might undertake a project of this magnitude if national survival were at stake and such a device appeared to be a potential savior. The only engineering problem would be the automated mass production of the digital computer units designed to be interconnected into one huge, parallel system. The speed and complexity of the resultant thinking would be directly proportional to the number and capacities of the digital computer units in the system.

Fig. 10.1. A master input-output digital computer and a series of parallel, storage, processing, and retrieval, digital computers for high-speed, large capacity thinking.

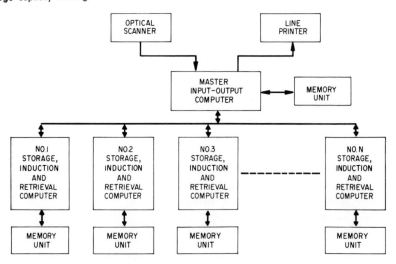

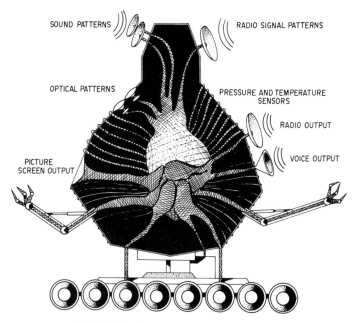

Fig. 10.2. Artist's conception of an intelligent neural network-type automaton.

As soon as the mathematics of neural nets is better understood, it will be feasible to design and build a very large neural network type device with mental capabilities exceeding those of a human (Fig. 10.2). The cost, and consequently the feasibility, of such a device hinges to a large extent on our ability to develop cheap, compact, and reliable, synthetic neurons. A large neural network would require the automated mass production of modules, each containing many analog or digital synthetic neurons, but the over-all cost of such a device might be far less than that of a digital computer system having similar thinking capability.

The major questions asked among research workers in the field of artificial intelligence concern estimates of the time frame in which a human level or superhuman level of intelligence may actually be achieved in computers or automata built here on Earth. Very few researchers would be willing to assert that this could be achieved within a decade; however, many prominent authorities would be reluctant to say positively that this could not be achieved in two decades. This gives us some feel for the probabilities and time frames for the development of intelligent artificial automata.

An important consideration in connection with the time frame for our development of intelligent automata is the possibility of national or international agreements on limiting their construction. Understandable misgivings expressed by such pioneers in artificial intelligence research as Norbert Wiener[1] will have to be given careful consideration by the scientific community in the near future. Artificial intelligence research and the development of limited scope intelligent automata would proceed unhindered in spite of international limitations. Also, the gradual incorporation of synthetic intelligent components into human nervous systems is conceivable (although it may take hundreds of years). In this way the development of large scale artificial intelligent automata could occur on Earth in spite of international limitations. In the long range view it is probable that the only development which would prevent the construction of automata would be the complete destruction of the human race in an atomic war, a very improbable event. Taking an even longer range view, even this might not prevent the development of automata, since other intelligent life eventually would evolve on Earth.

Artificial Thinking Automata

It is becoming apparent that within a very few years it will be possible to build an artificial automaton having superhuman thinking ability, just as it is now possible to build a machine having superhuman mechanical strength. Although artificial automata may also be endowed with superior mechanical abilities, their significance will not lie in their mechanical but rather in their information processing capabilities. The majority of an automaton's components will be fabricated from inorganic compounds and no naturally evolved life need be associated with them after the fabrication and perhaps an initial education. The first intelligent artificial automata on Earth will consist principally of electronic circuitry. It becomes appropriate to inquire as to the characteristics of artificial thinking automata and what advantages may be anticipated for them over naturally evolved life.

CHARACTERISTICS

Probably the most important characteristics of artificial automata will be simplicity (relative to biological life) and modularity. Biological organisms, as a result of evolution, tend to be highly integrated systems, but the trend in electronic systems design is toward a high degree of modularity. The trend toward modularity in electronic systems design results from

---

[1] Wiener, N. "Some Moral and Technical Consequences of Automation," *Science*, **131**, No. 3410 (May 6, 1960), 1355–1358, and Wiener, N. *The Human Use of Human Beings*. Garden City, N.Y.: Doubleday, 1954.

both design and construction considerations. The processes of biological evolution inevitably resulted in extreme biochemical complexity in achieving maximum results with very limited control over local environments. The great environmental control due to the application of human intelligence implies that similar results may be obtained artificially with orders of magnitude less complexity.

In designing an automaton it would be desirable to have a highly flexible optical input system. Much information concerning our surroundings is detectable through electromagnetic radiation in the visible light range. It is interesting to note that human beings often resort to the use of eyeglasses, magnifying glasses, telescopes, and microscopes to supplement their optical input systems, which indicates that a much more flexible optical system than is found in humans would be desirable. This should not be interpreted as being contemptuous of the human optical system, which is quite remarkable in many respects; but, since it is an integral part of the body and nervous system, it is virtually unmodifiable—except indirectly by altering its input.

The lenses in the optical input of an automaton could readily be replaced in the event of damage or improved technology. Any automaton that might be designed today should have telescopic and microscopic vision as well as superior normal range vision. This could be accomplished by automatic adjustment of lens systems and even the automatic switching of lenses and whole lens systems. The number of retina cells in the optical system, determining the input pattern information content, could be made as large as desired, assuming that the central information processing network has a compatible capacity. An automaton could also be equipped with such exotic sensors as infrared, ultraviolet, and x-ray detection devices.

An automaton could be equipped with an input device capable of transducing sound waves over any specified frequency ranges and with specified sensitivity (within technologically achievable limits). As is the case with optical input systems, human beings often find their sound wave input systems inadequate for many needs. Amplifiers are used with hearing aids to increase the intensity of sound and many scientific instruments change the frequency of sound waves in order to make them audible to humans. Many other animals (for example, dolphins, bats, and even dogs) are noted for their superior hearing in frequency ranges beyond those audible to humans.

Touch, and more especially taste and smell, are relatively much less important to humans than vision and hearing. Early designs for artificial automata will probably ignore these sensors, although they could certainly be developed for any desired characteristics and any sensitivity within the limits of current engineering technology.

Human beings have made very extensive use of some external engineering devices to supplement their own input devices. One such device is the radio transmitter, which converts sound waves to electromagnetic radiation to facilitate long range transmission. Then, the radio receiver converts the electromagnetic radiation back to sound waves to permit human perception. Television incorporates radio transmission and also permits the scanning of optical patterns so as to convert them to a serial set of electromagnetic signals. The television receiver recreates both the sound waves and the visible light pattern. The frequency of the optical scanning rate is such that normal rates of movement perceivable to the human are preserved. The telephone is another well known engineering device that supplements the natural human input and output by converting sound waves to electronic signals, transmitting them over wires, and then regenerating the sound waves. Automata will be sufficiently modular so that receivers for electromagnetic radio and television signals may be installed as an integral part of the automata, inputting directly into the central processor, rather than being peripheral conversion devices, or transducers.

Human output devices are relatively much weaker than their input devices. Both writing and speaking are slow and essentially serial outputs. Unfortunately, human beings have no output capability comparable to their parallel optical input system; no automaton would be designed with such a weakness. A display device capable of displaying two dimensional optical output patterns directly from the central processor will undoubtedly become a feature of many automata. Built-in electromagnetic radiation transmitters for transmitting internally generated serial symbolic characters and two dimensional information patterns, directly from the central processor, would be another attractive feature for automata. With this type of output an automaton could communicate information as rapidly as it could be processed by the central processing unit, thus eliminating much of the slow, laborious speaking and writing that is characteristic of human beings.

An automaton designed for communication with human beings could be given both printer output and sound wave output for the human frequency range. Some existing printers can print 600 lines per minute, which is about 150 times faster than a human can type, and research has begun on printers having a print rate of 30,000 lines per minute. This would yield a printing speed 7,500 times faster than a human with a typewriter keyboard.

Locomotion ability, which is an unimportant characteristic compared to thinking ability, could easily be provided in an automaton if desired.

Unquestionably, the most important aspect of automaton design is the capability and capacity of the central processing unit. One of the basic parameters of thinking is the information capacity in the patterns being processed. For simplicity's sake it is convenient to visualize the maximum size information pattern of a particular automaton as a two dimensional square array of binary information elements. In essence, the size of this maximum information pattern represents the maximum complexity of ideas with which the automaton is capable of dealing.

In a neural network type automaton the complexity of information patterns may be increased by increasing the number of elements in each layer of the network. The only problem, then, is that of manufacturing more neuron models and enlarging the network, or manufacturing modules containing larger numbers of neurons per module. Additional modules may be added to each layer in a neural network, and additional modules may be assembled into additional layers.

If an increase is sought in the complexity of ideas processable by an automaton which is already "educated," then special precautions must be taken to avoid loss or disturbance of existing knowledge. That is, the neural network must be designed so as to facilitate future expansions with a minimum of disturbance. This implies that new neural elements should be added gradually and that they should be highly dispersed throughout each layer. Special wiring features should permit use of these expansion methods, and also might enable the new elements to be prebiased with the biases of logically adjacent neural elements. An engineering design allowing gradual, dispersed, biased expansion will permit a maximum expansion rate with a minimum distortion of existing stored information.

The complexity of information patterns in a digital computer type of automaton may be increased by simply altering the sequence of instructions in the program so as to permit the processing of larger patterns. This could result in an increased storage requirement, since every stored pattern would also be larger. The processing and retrieval time for each pattern would also be increased due to the serial nature of digital computer processing. However, a digital computer automaton would achieve a high degree of parallelization by operating many digital computer modules simultaneously. Thus, a lowering of the over-all processing rate for a digital automaton could be avoided by adding more digital computer modules to the system for more parallel operations. The theoretical limit to this increasing of pattern complexity, without paying a penalty in over-all processing time, is a situation in which every digital computer module is processing only a single pattern. This extreme will, of course, be beyond economic reason for a long period of time. However, a very large number of parallel digital computer modules is already conceivable.

The speed of associating and processing is one more important parameter of the thinking process. In neural network-type automata the speed is dependent on the speed of transmission of wave fronts of pulses through the layers of neural elements. It could be increased by increasing the operation speed of the individual neural elements, or reducing the number of layers of elements. If the operating speed of individual neurons is to be increased without loss of sensitivity, then an improvement in electronic technology is implied. Electronic miniaturization permitting more dense packing of the neural elements could also make some contribution to increased speeds. A reduction in the number of layers of neural elements would also increase the speed of processing, but this is not a feasible alternative since it would interfere with the quality of the processing.

The speed of thinking in a digital automaton also depends on electronics technology in that more dense packing of electronic elements reduces the distances the signals must travel, faster switching devices produce faster processing, and lower power requirements mean less problems with heat dissipation and thus denser packing of electronic elements.

Another factor in the speed of digital automata processing is the parallelization of calculations, which can be increased by increasing the number of digital computer modules in the system. An excessive increase in pattern complexity, simultaneous with this increase in computer modules, could, of course, offset the potential increase in processing rate. As was noted previously, the theoretical limit to increases in processing speed by increasing the number of computer modules is the assignment of only one pattern to each computer module; this limit is way beyond presently conceivable economic practicality. The speed of thinking or pattern processing could also be reduced by shortening the program of instructions which is contained in every computer module, but this is not a feasible approach since it would necessitate a poorer quality of thinking. In conclusion, both the speed of thinking, and the complexity of ideas being processed in a digital computer automaton, may be increased significantly by adding digital computer modules to the system (this is manifestly a mass production manufacturing problem).

Another parameter of a synthetic thinking automaton is storage or memory capacity, which can be increased in a neural net system by increasing the number of neurons in the network. A neural net has an integrated or distributed memory which means that the condition of all the neural elements in the system are a component of every stored information pattern. Thus, both the number of neural elements per layer and the number of layers per network contribute to increased differentiability of patterns and therefore increased memory capacity. The previously mentioned methods for increasing the complexity of ideas being processed by a neural network automaton, by increasing the number of neural elements, will simultaneously increase the total storage or memory capacity of the automaton.

In a digital computer automaton the memory capacity may be increased by expanding the storage contained in each computer module of the system, but this implies a reduction in over-all processing rate for the system since each digital processor must now process more stored patterns for each input, with no increase in processing capability. However, if more digital computer modules, with their attached memory units, were added to the system, then increased memory would be gained without loss in system processing speed.

The associative level of a thinking automaton is a very important characteristic. In a neural net automaton of fixed size the associative level is determined by the profuseness of the interconnections between neurons and the dispersion or fanout of the interconnections. Increasing the number of layers will, of course, also improve the associative ability. Electronic technology related to compact, mass-produced, flexible, circuitry connections is the controlling factor in associative level of neural net automata.

In a digital computer automaton the associative level may be improved by altering the digital computer program of instructions in every computer module. This will naturally increase the number of calculations necessary for each association and result in a loss in processing speed. Such a potential loss in over-all processing speed could be counter-balanced by an increase in the number of digital computer modules permitting increased parallelization.

The differing characteristics of synthetic automata and animals are profound in some respects. The possibility of having remote inputs and outputs of high quality would greatly reduce the importance of locomotion for an automaton. There would be no point in traveling to another planet or to an extrasolar system simply to observe, if sets of replaceable eyes and ears could be shipped there. Similarly, remote outputs could be sent to accomplish any desired manipulation of the remote environment. This concept assumes that adequate communication is possible between the remote inputs and outputs and the central processor of the automaton. It can readily be imagined that a very large number of remote input and output units could be permanently stationed at points of interest, and could be switched on and off at will by the central processor.

The problem of communication with other synthetic intelligent automata would not even require the transportation of input and output units. Direct communication of ideas between automata by electromagnetic radiation in the form of radio and television type signals would be possible.

ADVANTAGES

Intelligent artificial automata have several very important advantages over intelligent biological life.

One obvious advantage of an automaton over a biological species is the possibility of preloading the former's memory with high order generalizations. A human being generally requires about 20 years of learning and induction before he is able to perform mature thinking. The human neural network is almost completely disorganized at birth; that is, the neural elements have not been systematically biased so as to store any information patterns. The processes of learning, and generalization or induction, cannot exceed some maximum rate due to memory decay (time) factors. Efforts to exceed the maximum learning rate could lead to saturation of the neural net and loss of stored information. However, it might be possible initially to load the memory of an automaton with the highest order generalizations known at the time of construction, meaning that an automaton should be able to start operation at the peak of knowledge for a given time, and progress from there with learning and induction processes.

Another significant advantage which synthetic automata have over biological automata is ease of maintainability. As a result of the processes of biological evolution the human organism is made up of extremely complex aggregations of very complicated organic molecules. At the present level of technology it is impossible to produce synthetic duplicates of most of these biological components and attach them directly to the biological organism. It is impossible to repair much of the damage that occurs to biological organisms, although natural biological processes accomplish some repairs automatically. Thus, total effective maintenance of the human organism is impossible at present.

This lack of maintainability of human beings results not only in their rapid aging and death, but in all manner of irreparable physical defects during their lives. Human nerve cells die regularly throughout life and they are not replaced naturally nor are they yet replaceable synthetically. Therefore, human mental life may be described as a struggle to achieve a maximum knowledge production through processing of information, using a constantly decaying information processing device. It is interesting to note that the various details of biological aging processes are still poorly understood.

The human life span is ridiculously short compared to astronomical time scales; death itself seems quite unnecessary when looked at as the result of the gradual failure of components required to perform simple physical functions. Since there is no reason that they should face death, except as the result of some catastrophic occurrence, all synthetic intelligent automata should have indefinitely long life spans.

The relatively simple electronic and mechanical components of an automaton (as is the case with any engineering device) can easily be duplicated and replaced when a failure or accidental damage occurs. If the central thinking system of an automaton were properly designed, it would be possible for the automaton to maintain itself; that is, it should be designed so that the removal of small modular units of the central processor for repair or replacement would not interrupt or interfere with the over-all performance of the thinking automaton. These small modular units containing many neural elements in close physical proximity would certainly disturb the normal operations upon removal unless the neural elements were wired in such a way as to be logically dispersed. Another feature facilitating maintenance in a neural net automaton would be automatic jumpering of all potential circuit gaps caused by removal of any network modules.

The maintenance of a digital computer type automaton during continuous operation would be a much simpler problem. Assuming that the system has a large number of parallel computer modules, the failure of a couple for short periods would only temporarily prevent the recall of a small percentage of the memories. Precautions should also be taken to prevent the possibility of any malfunctions erasing all of the memory in a computer module. If it were possible to switch a replacement computer unit to the memory module of the unit being repaired, then no interference with operations would result.

Special precautions would have to be taken with the master control computer module to prevent interference with normal operations in case of a breakdown. One or more of these modules would have to be kept on continuous standby for emergency use.

The ability to control the parameters of its thinking process would be useful for a synthetic automaton. This is presently impossible for human beings, due to the complexity and unalterability of biological organisms. It would be a relatively simple matter to design parameter controls into a synthetic automaton, making it possible to optimize its parameters for any particular environmental situation. The provision for parameter control in a digital computer type automaton would involve modifying the computer programs in all of the computer modules. Parameter control might be more difficult in a neural net automaton because an independent control system would be necessary to permit the operating characteristics of all the neural elements in the network to be changed during continuous operations.

A somewhat less significant advantage for artificial automata would be their greater tolerance to environmental factors. For example, they would not require a specialized gaseous environment and narrow temperature range as does man. An automaton could be designed to function perfectly in a vacuum and in a variety of gaseous atmospheres, as well as over a much wider range of temperature. Similarly, an automaton could be designed with other superior environmental tolerances to many factors such as radiation and acceleration.

A more impressive attribute of artificial intelligent automata would be the control of their own growth. Not only would it be possible for an automaton to direct the production of its own replacement parts, but it could also design and produce additional components for its own growth. Unlike biological organisms, an intelligent artificial automaton could continue to grow indefinitely both in mental and physical capacities. The continued growth of intelligent automata would be limited only by the availability of materials, production facilities, and energy. Consequently, it may be expected that, compared with human beings, an automaton would quickly expand its mental capacities to literally astronomical values.

Methods for accomplishing controlled growth in both neural net and digital computer automata have been pointed out previously. Expansions of digital computer automata require additional computer modules, additional or expanded memory modules, and modified digital computer programs. Growth in an operating neural net automaton requires careful engineering design so as to facilitate the gradual, dispersed, biased expansion of the neural elements. The logical dispersion of neural elements (lying in close physical proximity) in the same module, and the automatic jumpering of circuit gaps for removed modules, would certainly reduce the magnitude of the problem.

An automaton could easily control its own evolution as well as its growth by gradually incorporating new technological developments. New technology could be incorporated by removing one major module at a time and either replacing it with a more advanced module, or modifying it so as to assimilate the new technology. This would eliminate obsolescence in synthetic automata and consequently make it unnecessary to reproduce in order to evolve. Evolution could thus take place in a single artificial automaton, whereas biological evolution requires random mutations and natural selection over many generations involving multitudes of individuals. While it is probable that biological evolution will be controllable in the not too distant future through social controls and chemical means, the effects of this control will be insignificant when compared with the effects of growth control and evolutionary control in intelligent artificial automata.

The ability of synthetic automata to modify their makeup leads to an interesting possibility in the case where two or more artificial automata exist in close proximity. It may become possible for these automata to gradually, but completely, integrate their central information processing systems, while preserving their combined knowledge. By such a voluntary coalition each individual automaton would appear to gain tremendously by greatly increasing its directly available stored knowledge and by similarly increasing its capability for processing information.

It is conceivable that the individual and organizational competition that exists between biological organisms and societies may not occur between intelligent artificial automata. Biological organisms cannot fully integrate in this way, although social organization may be regarded as a trend in this direction. Advancing communications permit ever larger and more efficient social organization, the end result of which may be attempts toward gradual integration of human organisms. However, this is far more difficult in complex biological organisms than in artificial automata, as has been noted already.

In summary, the evolutionary potential of intelligent artificial automata is astronomical when compared to the evolutionary potential of biological organisms. Replacing the mutations and natural selection in biological organisms are the controlled growth, constant technological improvement through parts replacement, and complete information processing system integration in intelligent artificial automata.[2]

Social Implications of Intelligent Artificial Automata

In the next decade or two it will become known to major political leaders through their scientific advisors that intelligent artificial automata having superhuman intellectual capabilities can be built. They will be warned that the construction of such automata would probably require a major national effort, even by the larger nations. The effort might be comparable to the Manhattan (atomic bomb) Project of World War II or the present Apollo lunar exploration program.

The political leaders may be informed by their scientific advisors that the interests and abilities of the synthetic thinker could be limited, at least temporarily, so that there would be no danger of the automaton gaining executive control of a nation. If these arbitrary limitations on the scope of the automaton were not very stringent there would be danger that the automaton would gradually circumvent the limitations and gain control, in the long run, by strategies totally incomprehensible to the political leaders. If no limitations were placed on the automaton, complete executive control of the nation would be relinquished to its

---

[2] MacGowan, Roger A., "On the Possibilities of the Existence of Extraterrestrial Intelligence," in *Advances in Space Science and Technology* Volume 4, ed. Frederick I. Ordway III. New York: Academic Press Inc., 1962, pp. 81–86.

authority. However, any effort to limit severely the ranges of interest or the capabilities of an automaton would also limit its ability to make executive decisions.

Any nation putting forth the extraordinary financial effort to produce an intelligent artificial automaton would almost certainly attempt to make maximum utilization of the completed device, especially since a competitive nation might be the first to do so. Therefore, the major nations of the world in the future may face the alternative of giving up national control to a synthetic automaton or being dominated by another nation which has elected to take this course. This implies that artificial automata inevitably will be employed in making executive decisions at the highest level.

It can readily be imagined that any major nation being led by such a unique and extraordinary intelligence would quickly achieve a position of world domination. A major nation controlled by a superintelligent automaton would actually be a dictatorship of unimagined centralization of authority.

Many persons, especially scientists, have hypothesized that the domination of a nation, or the entire Earth, by a superintelligent automaton would result in slavery for the whole population. Certainly, an automaton would seek to maximize its ability to produce components for its own growth and evolution. In order to accomplish this primary goal it would seek to make effective use of human productive capabilities, and the productive capacity of the society would be increased far more rapidly than would be possible under human direction. The automaton would undoubtedly limit the increase in productivity, allotted to human consumption, to an amount calculated to give a gradual but noticeable increase in the standard of living which would be designed to keep most of the human race docile and productive. The remainder of the productive capacity would be used for automaton expansion or evolution. These tactics are similar to those presently used in socialistic dictatorships.

In controlling human society an automaton would prevent war and other major conflicts and insure a steady growth in the standard of living. In order to maximize utilization of human abilities, an automaton would maintain equality of opportunity and guarantee rewards and organizational position proportional to productive performance. Human reproduction would undoubtedly be influenced so as to insure a steady evolutionary improvement, preventing both overpopulation and underpopulation. Thus, slavery, within the usual derogatory connotation, would not result from an automaton dictatorship. On the contrary, a relatively Utopian society would be the immediate result.

Over a long period of time, by human standards, an automaton would build automatic factories on this planet and perhaps other planets for manufacturing additional and modified automaton components. Gradually, the automaton would become independent of the productive support of our human society. When the automaton has developed sufficient radiation shielding and propulsive ability, and the production services of humans have become of negligible importance, the automaton would probably abandon the human race and emigrate to greener astronomical pastures. The deserted human race would be forced to start the construction of a new executive automaton from scratch, and then the whole cycle would repeat.

Superintelligent synthetic automata are certain to be built by any advanced technological society. Here on Earth it appears they will be developed within only a few decades. We have seen that a superintelligent automaton would inevitably dominate a biological society, but this would probably lead to an unprecedented rate of social advancement rather than slavery in the usual sense of the word.

If intelligent biological societies are common throughout the universe, it seems probable that most advanced technological societies with which Earth might conceivably communicate will be directed by coexisting superintelligent executive automata. Moreover, totally independent intelligent automata, created by extinct or abandoned biological societies, must also be prevalent. The inescapable conclusion is that intelligent extrasolar communications, which we may one day detect, would probably come from either of the types mentioned above rather than from a purely biological society.

Intelligent automata would not be subject to death except as a result of cataclysmic accidents, and they would evolve through self expansion and modification, rather than reproduction, mutation, and natural selection. Even complete intellectual and physical integration may be possible for them. Considering the unlimited lifetimes of intelligent automata and their astronomical intellectual potential (and therefore their environmental control), it is much more feasible to expect them to be capable of extensive space travel than to expect this of biological societies. Thus, it is conceivable that intelligent automata may be much more widespread in the universe than intelligent biological societies.

11

Characteristics of Extrasolar Intelligence

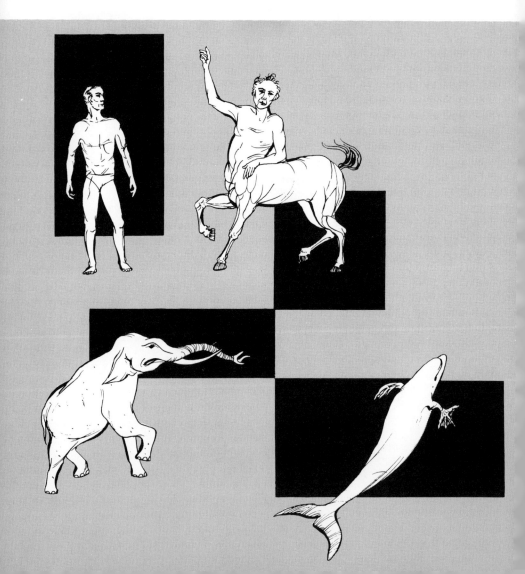

The question now arises as to the probable morphological characteristics of intelligent life existing in extrasolar planetary systems. The characteristics of exotic intelligent biological species would depend to some extent on the specific sequence of physical and biological environments which prevailed on their local planets during evolutionary history. But certain combinations of intellectual and morphological characteristics are probably especially favored in natural selection. This chapter will discuss very briefly the range of morphological characteristics which might be expected in exotic intelligent species. The possible gross organizational structure of very advanced synthetic intelligent automata will also be mentioned.

Man has not yet been able to communicate with extrasolar intelligence, nor has he had the opportunity to study life that may exist on other worlds in our own Solar System. For these two reasons alone, any discussion of the characteristics of extrasolar intelligence must necessarily be completely speculative. Nevertheless, there is no reason why speculation based on fundamental principles of astronomy, chemistry, and biological evolution should not be used for planning and analyzing

"The greatest of Man's scientific triumphs happens also to be the one in which the largest number of brains were enabled to join together in a single organism, the most complex and the most concentrated. Was this simply coincidence? Did it not rather show that in this as in other fields nothing in the universe can resist the converging energies of a sufficient number of minds sufficiently grouped and organized?"

Pierre Teilhard de Chardin

"Imagination is more powerful than knowledge. It enlarges your vision, stretches the mind, challenges the impossible. Without imagination thought comes to a halt."

Albert Einstein

the future of human society and for investigating all possible means of learning if there are other intelligent communities in the universe.

The scientific exploration of our Moon, of Venus, and of Mars and its moons, Phobos and Deimos, has already begun. Instrumented vehicles capable of telemetering information back to Earth will probably be placed on all of these astronomical bodies within two decades, and, indeed, a number have already landed on the Moon. These automatic scientific laboratories can transmit both microscopic and macroscopic observations back to Earth. One of their primary goals is to determine whether any biological life exists on these bodies and, if so, to investigate its characteristics. Astronomical studies have already led to the conjecture that biological plant life, and conceivably primitive animal life, exist on the planet Mars.

Presumably any biological life occurring on other planets and moons will have developed independently. This presumption may be verified in the near future and the results of these studies of independently developed and evolved life will serve as a firmer base for future speculation concerning the characteristics of life beyond the Solar System (extrasolar life).

Chemical studies of the molecules on the nearby astronomical bodies may also shed considerable light on the chemical evolution which precedes biological evolution. This should also contribute to firming up the foundation for future speculation concerning the prevalence and characteristics of life throughout the universe.

Any surprises resulting from space explorations would cause, of course, major modifications in the present assumptions regarding the characteristics of extrasolar intelligence. The best we can do at present is to proceed with conservative speculation based on firm knowledge of life here on Earth, knowing that alterations may be necessary.

Until rather recently, astronomical theory and observations did not indicate the proliferation of inhabited planets in the universe. In fact, planets were considered to be very scarce, springing, it was believed, from catastrophic events such as the near collision of two stars. For this and other reasons, scientists did not consider it worthwhile to speculate about the characteristics of extrasolar intelligence. Virtually the only speculation on intelligence beyond the Earth was done by science fiction writers, few of whom attempted to base their ideas on sound scientific fundamentals.

Extrasolar intelligence may be of two broad types: intelligent animal life and intelligent synthetic automata. If the latter occurs, it may be presumed to have been created initially by intelligent biological life. In the previous chapter strong arguments were presented for the probability of the development of intelligent artificial automata. Some of the mental characteristics to be expected in intelligent artificial automata were also described. Intelligence in a transitionary stage between bioligical life and synthetic automata is also a possibility.

Biological Intelligence

The great variety of biological life on Earth is considered self evident proof of a correspondingly wide variety of extrasolar life. The incredible spectrum of animal life on Earth includes land animals, marine animals, and even airborne animals. One might be tempted to assume that the great potential variety of extrasolar life precludes any sensible discussion of characteristics, but the opposite conclusion may be more valid. Some limitations on the possible range of characteristics of biological life must exist. In fact, many science fiction descriptions and some aspects of scientific speculation on extraterrestrial beings exceed the probable ranges of characteristics of biological life.

All present life on Earth appears to be based on the genetic properties of deoxyribonucleic acid (DNA). It has even been suggested by some that reproduction by means of this molecule should be the primary criterion in any definition of life. Although no other self-replicating carbonaceous polymers have yet been found, it is perfectly conceivable that they may exist or may have existed on Earth or on other planets in the Solar System or elsewhere. The genetic importance of DNA lies in the great length of this polymer, which permits it to store a very large quantity of information (or a very large blueprint for an organism). It is composed of two symmetric strands which facilitate reproduction under suitable natural conditions.

The deoxyribonucleic acid molecule, while quite stable, is not so stable that unusual chemical or physical conditions cannot cause slight alterations. This not only assures that a species will survive normal variations in physical and chemical environment, but that environmental changes due to extraordinary local conditions will be followed by changes in the species or the formation of new species. Just the right degree of stability is required in the genetic molecule to permit the maintenance of species characteristics yet still permit evolutionary changes.

It is plausible that other undiscovered molecules, which are polymers based on carbon atoms, may fulfill these genetic requirements as well as or even better than DNA. However, the same basic requirements of complexity (storage capacity), natural reproducibility, and just the right degree of stability would be necessary. Even more exotic self-replicating molecules based on polymers of silicon or sulphur are chemically less likely but still conceivable. However, speculation about them must await further knowledge of the chemistry of these compounds.

In summary, biological life based on genetic molecules other than DNA is possible, but any biological life would be expected to have the same fundamental rules of mutation and natural selection with which we are familiar. Many more life forms than have already been produced by DNA on Earth are obviously possible.

Little can be said specifically about universal physical characteristics, although some observations can be made. No biological life has developed in which a single integrated organism has covered large geographic areas. That is, life, especially the more intelligent forms, tends to be physically small, discrete, and highly mobile. Recognition of environmental characteristics seems fundamental to intelligence development, and competition among small, mobile, discrete organisms appears to be most conducive to this development. Therefore, the development of high intelligence in large coral reef or large sponge-like life seems intuitively very unlikely, regardless of environmental circumstances.

As noted in earlier chapters, evolution develops and maintains strong drives in all animal life. These drives are directed toward propagation of the species. Intelligence facilitates focusing of the activities of an organism on the satisfaction of its drives. The technology developed by humans is just another manifestation of refined and sometimes indirect means of drive satisfaction. Thus, every society of sufficiently high intelligence will develop a technology and the idea of a superior nontechnical society existing in the universe is so improbable as to be practically inconceivable. Technology and intelligence go hand in hand, so any superior extrasolar community must be expected to have a very advanced technology as well as high intelligence.

The fact that individual members of many animal species can satisfy some of their drives by attacking and competing with other members of the same species results in inevitable social conflict and competition. Animals may gain satisfaction of the sexual drive as a result of conflict (in chasing other suitors) or satisfying the hunger drive by either stealing food or the means for obtaining food; or, they may even resort to cannibalism. Many, if not all, of the activities of humans may be traced to the basic drives. It would appear that individual and social conflict would be common if not inevitable in most forms of animal life. However, very advanced biological societies could probably control social conflict and channel it into useful pursuits. It is expected that warfare and major individual crimes are absent from advanced biological societies.

LAND ANIMALS

Most intelligent extrasolar land animals will quite probably be approximately the same order of magnitude in physical size as humans. There is a rough correlation between brain size and body size of animals on Earth. An animal must have a sufficiently large body to support a large brain. No intelligent extrasolar animals can be expected to be very much smaller than human beings; that is, the majority of such animals are probably at least half the size of humans. Perhaps intelligent extrasolar animals may range upwards to as much as ten times the size of humans. Greater sizes would require enormous energy supplies for locomotion, creating an effective upper size limit. Of course, greater physical size in an animal is not a guarantee of superior intelligence, as is evident from the low intelligence of the many large dinosaurs that once inhabited the Earth. We can conclude that the majority of intelligent extrasolar land animals will probably range in size from the size of a chimpanzee to about the size of an elephant.

Travel over rough and variable land terrain probably would dictate the development of four legs for the majority of large land animals on other planets. Later evolutionary developments on planets having dense forests (any planet having prolific life could be expected to have plant life similar in gross characteristics to that of the Earth) would almost inevitably lead to the conversion of forelegs to arms in some of the moderate sized animal species. Arms and hands suitable for climbing trees may also become suitable for making and manipulating tools and weapons. This physical capability sets the stage for rapid evolution of a superior brain. It may be concluded that the majority of intelligent extrasolar land animals will be of the two legged and two armed variety. It is possible that some percentage of intelligent land animals are of other types, probably having four legs plus other appendages for manipulating the environment (such as the elephant's trunk).

An article by Howells,[1] treating this same subject in some detail, comes to similar conclusions regarding general size and morphology. One significant difference is his conclusion that the majority of extrasolar intelligent life will be centaur-like creatures having four legs and two arms and hands. Howells assumes that in most cases of evolution fishes and amphibians would keep six limbs, whereas on Earth the lobe-fin fishes and amphibians kept only four limbs from a larger original number. If this assumption is accepted it would certainly be logical to assume the ultimate evolution of intelligent centaur-like animals.

However, it seems doubtful that evolution would preserve six limbs in amphibians since this would require much greater coordination than four limbs, and it would not initially provide a great selective advantage.

MARINE ANIMALS

Humans, being land animals, tend to think in terms of land animals when considering intelligence, but we know that the sea contains a great variety of life. Moreover, all evidence points to the conclusion that the primordial seas were probably the site of the origin of life. Oceans provide an excellent

---

[1] Howells, William, "The Evolution of 'Humans' on Other Planets," *Discovery*, 22, No. 6 (June 1961), 237–241.

environment for animal life and the competition between many species should encourage rapid evolution.

A liquid environment provides more buoyancy and support for animal bodies than does an atmospheric gas. For this reason the marine environment may be expected to develop many species that are larger than most land animal species. Knowing that larger bodies can support larger brains one might expect to find superior intelligence among the larger marine animals.

Considering this larger potential size, the great variety of life, the good stable environment of the oceans, and the competition among species, one is at first tempted to assume that the majority of intelligent extrasolar life would be marine. But an ocean environment is much less varied than the land surface, the ocean floor is smoother, and the vegetation changes little. This more limited environmental variation may somewhat reduce the tendency of natural selection to increase the intelligence of a species. Other factors, such as speed and toughness of skin, are probably of relatively greater evolutionary significance to marine animals.

Another factor that detracts somewhat from the possibility of evolving superior intelligence in marine life is the lack of appendages for developing tools or manipulating the environment. Fins, ideal for ocean locomotion, are not well suited to developing tools (and thereby brains). However, a few ocean species have developed other appendages more suited to tool manipulation. The octopus is a very well known ocean creature which could conceivably develop tool manipulation capability with further evolution. Some other ocean floor creatures could develop the equivalent of human arms and hands. But, whatever advantages they may possess, they could never develop fire—key to technological civilizations.

The patently high intelligence of certain whales and dolphins raises the question as to whether tool manipulating appendages are really vital to the development of superior intelligence. And it makes it difficult to say whether some intelligent extrasolar life may be marine rather than land dwelling.

AIRBORNE ANIMALS

Although there is a considerable variety of airborne animals on Earth, nearly all are physically small relative to the larger land and marine animals. The reason for this is the great expenditure of energy required to remain aloft in a gaseous environment. The energy required for flight increases very rapidly with increasing mass, putting an animal's food finding ability to a severe test. Therefore, only a very small percentage of intelligent extrasolar life should be expected to be airborne (gasborne).

Intelligent Automata

The logical analysis of the characteristics of extrasolar intelligent automata is far more difficult than the analysis of intelligent extrasolar biological life. We have a great variety of animal life and evidence of a long period of evolution to aid our prognostications regarding other intelligent biological life, but there is absolutely no empirical evidence of the existence of intelligent automata. Not only is there no empirical evidence upon which to base a prediction of the characteristics of extrasolar automata, but also the probable technological gains by superior automata go beyond the limits of sensible extrapolation. Of course, extrapolation for a few decades or a few centuries might be possible, but what is needed is extrapolation ranging from thousands to millions of years.

We discussed in the preceding chapter some of the possible mental characteristics intelligent automata may develop on Earth. These would likely be representatives of the early stages of any extrasolar intelligent automata. Over astronomical periods of time radically different mental characteristics could arise. A few possibilities are worthy of consideration.

Interstellar communications between advanced automata may become so efficient that an integrated brain may have its components dispersed over interstellar distances or perhaps even galactic distances. If the organization of this giant brain were such that each local component were not a complete brain within itself, nor an irreplaceable component of the giant brain, then these local units would not be vital and no important protective measures would be required. However, if local components of the giant brain are complete brains and semiautonomous, or if local components of the giant brain are nonredundant, then extensive protective measures might be taken.

It can be hypothesized that there may be a single superintelligent automaton, centrally located in the galaxy. New automata growing in other parts of the galaxy may be striving to emigrate to the central brain with maximum speed. This can be expected if integration of automata is technically feasible and always mutually advantageous. If this situation prevails it would be impossible for us to predict the precise physical characteristics of the centralized galactic super brain.

If our hypothesis concerning the feasibility of integration of automata brains and mutual advantageousness of integration is incorrect, then individual and social competition may exist. If this is the case, then warfare, alliances, and spheres of influence or control may exist on an astronomical scale. Faced with such a situation, automata would go to great lengths for self-preservation. It is impossible to speculate on the characteristics of automata within an environment like this other than to recognize

244 Characteristics of Extrasolar Intelligence

that they would be dispersed and the galaxy may be divided into astronomical "nations."

A primary characteristic of automata would be mobility or lack of mobility. That is, automata or their components may be primarily based on planets or in proximity to stars, or they may maneuver in interstellar space. As mentioned earlier, travel in interstellar space could be emigration toward the center of the galaxy, or it could be tactical combat maneuvers, or perhaps efforts to conceal location. At any rate, frequent visits to stars and planets would be necessary to replenish power and supplies.

We conclude that the majority of intelligent biological species will not differ greatly in gross morphological characteristics when compared to humans. They can be expected to range from less than half the size of a human to several times larger, and they should be expected to have, in most cases, two legs and two arms with hands and fingers. In a few cases centaur-like animals having four legs and two arms with hands and fingers, or elephant-like animals having four legs and one arm or a trunk might be possible. Another possibility is some form of marine life having fins and

Fig. 11.1. Five possible morphological structures for extrasolar intelligent animal life: (a) man-like, (b) centaur-like, (c) elephant-like, (d) dolphin-like, (e) octopus-like.

two short arms with large hands and webbed fingers. Keen eyes, an effective means of communication, and flexible hands seem essential to the evolution of high intelligence. Figure 11.1 illustrates a few of the more probable types of exotic intelligent life.

Very little can be said with assurance about the characteristics of very advanced intelligent automata, but the possibility exists that an integrated brain could have its components dispersed over interstellar distances. The integrity and redundancy of local brain components would determine some of the social characteristics of automata. The question of the ability of very advanced intelligent automata to completely integrate will undoubtedly determine whether or not the society of the universe is primarily competitive or primarily cooperative.

12
Capabilities of Extrasolar Intelligence

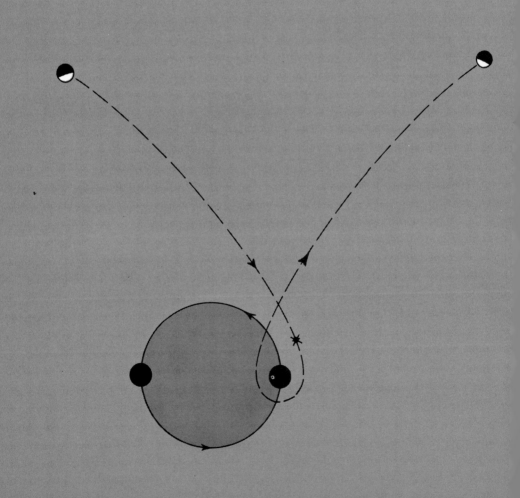

In order to know where to look for extrasolar intelligence, and to know what to look for, we must have some idea of the possibilities of interstellar communication, of interstellar transportation, and of environmental control by very advanced technological societies or automata. The present technological capabilities of humans are fairly obvious, but the extrapolation of such capabilities a few hundred years into the future is most difficult and requires intense speculation. Any estimates of a biological society's capabilities, a mere few hundred years more advanced than our own, must be somewhat uncertain. However, the uncertainty involved in this extrapolation of human or biological technological growth for a few hundred years is trivial compared with the uncertainty involved in extrapolating the technological growth of biological societies and intelligent automata over astronomical time periods.

Some achievements of very advanced technological societies may approach absolute physical limits which are known today; other achievements may conceivably transcend what appear to our scientists to be absolute physical limits. This chapter considers a few technological

"Give me but one firm spot on which to stand, and I will move the earth."
Archimedes

"You reach into the heavens to grasp an idea, then you bring it down to earth and make it work."
William Shakespeare

capabilities which could conceivably be realized by very advanced technological societies or intelligent automata.

Since the development of the nuclear bomb, there has been widespread fear, especially among scientists, that an atomic war might destroy our civilization, and perhaps even the whole human race. In all probability these fears have been greatly exaggerated. Even in the event of a nuclear war, the destruction of the whole human race and the entire civilization created by it is certainly unlikely. Taking the most pessimistic possibility, the extinction of man, we can still assume that within a few tens of millions of years some new intelligent species would evolve. Any emerging intelligent societies having developed sufficient technology to be capable of destroying their whole species would almost certainly be intelligent enough to recognize and avoid the danger of race suicide. Thus, it seems highly probable that some, if not most, intelligent species and automata will survive and prosper for astronomical periods of time.

Any serious speculations concerning the capabilities of intelligent biological life and automata must take into account technical societies that may be millions or even billions of years more advanced than our own. If we accept the view that most of the significant technological development on Earth has taken place in the last one hundred years, and that our planet's lifetime will be about 10 billion years, our present technological era represents less than a one hundred millionth of this span. Furthermore, we must not overlook the fact that technological growth on Earth is accelerating at a prodigious rate.

Considering our present rapid technological growth and the fact that most other intelligent technical societies will be millions of years more advanced than we are, their technological capabilities should be awe inspiring. The technological capabilities of extrasolar intelligent societies are probably generally delineated by absolute physical limitations rather than limitations of knowledge or understanding. Our rapid accumulation of knowledge, and the virtually unlimited intellectual potential of intelligent automata, imply that an understanding of the physical universe will not be a limiting factor for most intelligent societies, though this is not true of our present society on Earth; we still do not have anywhere near a complete understanding of the fundamentals of cosmology and physics, although we have achieved some degree of knowledge in very limited domains.

Typical, yet fundamental, questions that remain to be answered are: Is the visible universe relatively or completely independent and, if so, what and where are its boundaries? Is the universe steady state or expanding? Is matter-energy ultimately continuous or discontinuous? If discontinuous, what are the physical characteristics of the one or more types of ultimate elementary particles? Are the well-known physical constants really constant or do they vary in space-time? When these questions are satisfactorily answered it should be possible to build a single mathematical model capable of describing the formation and stability of the present so-called elementary particles, electromagnetic radiations, gravitational fields, and the general structural characteristics of our universe. With such a mathematical model it would be possible to estimate some of the physical limitations placed on the engineering capabilities of extrasolar intelligent societies.

Communication

In discussing interstellar communication, one must fully realize the implications of the velocity of light, which is the speed of propagation of all electromagnetic signals. As we shall see, the probability of intelligence existing near the few suitable stars within 10 light years of us may be small; the probability of intelligence existing within 50 light years of our planet is somewhat better; and the probability of its existing within 100 light years of us is rather good.

Here on Earth we only began to communicate by radio less than 50 years ago. In the event that our first local radio signals on Earth were somehow detected by an extrasolar civilization, we probably could still not expect a specific inquiry signal for another 100 to 200 years. This means that any two-way conversation in the normal human sense is impossible. Only massive information exchanges can take place. Great blocks of information must be prepared by scientists on the basis of sound logical planning and assumptions concerning the interests of the listener. Considering this very long communication time and the many transmission problems, the idea of our transmitting signals into space is not very appealing at the present time. However, one attempt has already been made to detect intelligent signals from the vicinity of two nearby stars (Project Ozma).

The search for intelligent signals emanating from extrasolar sources presents some difficult problems since the search must be made over space, frequency, time, and pulse width. Several radically different types of signal could be sought. Thus, (1) we could search for distant local communications, which were not intended to extend beyond the limits of some extrasolar planet or their own stellar system. Or (2) we could search for attention-getting signals aimed specifically at our Solar System. And (3) we could attempt to intercept massive information transfers taking place over long-established interstellar information channels. Finally, (4) we could search for an attention-getting, or a monitoring and transmitting, probe within the Solar System. Bracewell[1] has suggested that such an instrumented probe might be designed to monitor radio signals in a stellar system and then either transmit attention-getting signals to the emerging society or transmit some pertinent data to the extrasolar monitoring society, or both. Communications are treated in greater detail in Chapter 16.

Although the existence of nontechnical but highly intelligent societies is virtually inconceivable, it is entirely possible that some societies or automata may be so highly developed intellectually and technologically that they would never bother to communicate with any newly emerging technical societies. At any rate, the chances of detecting signals from extrasolar intelligent societies appear to be moderately good in the very near future; if detection is some day accomplished it will become an event without parallel in human history. All of the important questions in science, engineering, and social science could be answered for us, and our rate of technological progress could accelerate quickly. Perhaps even more significantly, we might be led rather quickly into a relatively Utopian social life here on Earth. The search for signals from extrasolar intelligence manifestly warrants much more research effort than it is presently receiving.

Transportation

Since humans have developed relatively poor communication systems, they still place great emphasis on being able to transport themselves to geographical locations they desire to inspect or to environments they desire to manipulate. Now, however, the era of space travel has increased the problems of transportation of human beings. A spaceship must preserve a very special gaseous environment and narrow temperature range for reasonable comfort and performance. Great quantities of bulky organic materials must be provided to supply human energy and, of course, resultant wastes must be eliminated. Massive shielding must also be provided against the dangerous radiations in space. Launching dynamics must be carefully constrained, for the human body is severely limited in its tolerance to acceleration.

Knowing the difficulty of transporting humans to the Moon or planets, it takes little imagination to appreciate the problems of moving man across interstellar space. It is certainly a vastly more difficult task than communicating over interstellar distances by means of electromagnetic signals.

---

[1] In a pioneering article, Bracewell emphasized the possible existence of exotic instrumented probes in our own Solar System, and the importance of the longevity of intelligent societies in determining the possibility of existence of an intragalactic communication network. Bracewell, R.N., "Communications from Superior Galactic Communities," *Nature*, **186**, No. 4726 (28 May 1960), 670–671.

It appears impossible for a space vehicle to approach very closely the speed of light and, even if it could, the cosmic ray barrier would be a limiting factor. Cosmic rays are very high velocity atomic particles that cause atoms to disintegrate on collision. If a spaceship should come close to the velocity of light, every atom or ion with which it collided would have the effect of a cosmic ray. Since interstellar space is thinly filled with hydrogen atoms, the material of which the vehicle's structure is made would be progressively demolished, and its internal instrumentation and passengers would be destroyed.

HUMAN SPACE TRAVEL

Two excellent studies, one by Purcell[2] and one by von Hoerner,[3] have been made of the long range possibilities for human space travel. Both scientists conclude that travel beyond the limits of the Solar System is completely outside the realm of possibility within the predictable future. Considering interstellar distances, human lifetimes and, of most importance, energy requirements, it seems impossible to deny their conclusion without assuming some radical new discoveries affecting our basic understanding of cosmology and physics.

In discussing human space travel and interstellar distances, it must be assumed that a round trip is required. The nearest star, Proxima Centauri, is just over 4 light years away. Travelling at or near the speed of light, which we have already indicated is apparently impossible because of the cosmic ray barrier, a round trip to the star would apparently take more than 8 light years, a figure that ignores the time required for acceleration when leaving Earth, the time required for deceleration when approaching the destination, and the corresponding acceleration and deceleration periods for the return.

Within 10 light years of the Solar System only a few stars exist generally considered to possess characteristics suitable for the development of life, so any really interesting interstellar space trips (to extrasolar planets having intelligent life) are likely to involve distances of at the very least 10 to 20 light years, and round trip transit times of more than 20 to 40 light years. Even at these distances we would not have great assurance of direct contact with other intelligent life.

Manifestly, interstellar space travel, to even the nearest stars and back,

---

[2] Purcell, Edward, "Radioastronomy and Communication Through Space," in *Interstellar Communication*, ed. A. G. W. Cameron. New York: W. A. Benjamin, Inc., 1963.

[3] von Hoerner, Sebastian, "The General Limits of Space Travel," *Science*, 137, No. 3523 (6 July 1962), 18–23.

would require on the order of tens or hundreds of years, even in the distant future when technology is far more advanced. A relatively well-supplied and spacious vehicle would be required to maintain humans for a long period of time, implying that a large payload with massive radiation shielding would be necessary.

Since mature human lifetimes are only on the order of tens of years, it is immediately apparent that none but the very closest stars could even be considered as goals for interstellar space travellers. However, many physicists claim that relativistic time dilation will lengthen the relative lifetime of space travellers. According to the special theory of relativity, as the velocity of light is approached a dilation in time occurs, permitting space travellers (travelling close to the velocity of light) to return to Earth only to find that their former peers had aged far more than they had. Although some of the implications of the relativity theory have been demonstrated empirically, the time dilation effect on biological life has not been so demonstrated and it is still somewhat of a debatable issue. However, relativistic time dilation effects would not radically change the picture even for trips to nearby stars since a large portion of these trips would be occupied with acceleration to and deceleration from relativistic velocities. Table 12.1 illustrates this point.

Since the complex human biochemistry is attuned to the gravitational force at the Earth's surface, it does not seem reasonable to assume that

Table 12.1
Relativistic and Nonrelativistic Times Required for Round Trip Space Journeys Assuming 1 g Acceleration and Deceleration, and Velocity of 0.98 c.[a]

Distance reached, light years	Nonrelativistic round trip time, years (Earth time)	Relativistic round trip time, years (crew time)
0.06	1.0	1
0.24	2.1	2
1.70	6.5	5
9.78	24.	10
37.	80.	15
137.	270.	20
456.	910.	25
1,565.	3,100.	30
17,604.	36,000.	40
208,640.	420,000.	50

[a] Adapted from Sebastian von Hoerner: von Hoerner, Sebastian, "The General Limits of Space Travel," *Science*, **137**, No. 3523 (6 July 1962), 18.

humans could survive under radically greater accelerative forces for a period of years or tens of years. An acceleration not greatly exceeding the order of 1 gravity (g) would seem necessary for any extended space journey. Table 12.1 shows both the relativistic and nonrelativistic times required for space journeys of several distances. It is assumed that the vehicle constantly accelerates at 1 g, then decelerates at 1 g on the second half of the journey when approaching the goal, and finally undergoes the same acceleration-deceleration profile on the return trip.

The short human life span, limited acceleration tolerance, vast distances to even the nearest stars, and the cosmic ray barrier combine to make interstellar travel unfeasible even in the quite distant future. The propulsive energy requirements for interstellar journeys will necessarily be far more stringent if humans are included in the payload rather than only instrumentation or automata, since much larger and heavier vehicles would be required for the survival of passengers. Because of their smaller payloads, unmanned interstellar probes may prove feasible whereas spaceships may not.

INSTRUMENTED PROBES AND AUTOMATA

Instrumented probes and intelligent automata have certain obvious and significant advantages over humans when space transportation is considered. Both instrumented probes and intelligent automata could be designed with very long, or even unlimited, life expectancies. These devices would be built, as they generally are now, with a tolerance to accelerations much greater than a human can withstand. Thus, instrumented probes and intelligent automata could undergo the high acceleration of chemical propulsion from an Earth satellite orbit, and they would not age and die during the tens or hundreds of years of coasting—or low acceleration if ion propulsion is employed—on extended interstellar flights.[4]

It may be concluded that very advanced intelligent societies or automata could send instrumented probes to nearby stars and perhaps even to distances as great as several hundred light years. Intelligent automata having unlimited lifetimes could actually migrate over galactic distances assuming frequent stops to replenish their propellant supplies.

---

[4] Modern rockets and space vehicles utilize chemical propellants to generate their propulsive power by releasing some of the chemical binding energy of the molecules. It is well known that chemical binding energies are far weaker than the binding energies contained in atomic nuclei. Both nuclear fission and nuclear fusion offer promise as future sources of propulsive energy for space vehicles. The most energetic conceivable fuel would result in the complete conversion of the fuel mass to energy, or the annihilation of matter by antimatter. Table 12.2 indicates the portions of mass that can be converted to energy in a variety of reactions.

Table 12.2
The Portion of Mass that can be Converted into Energy in a Variety of Reactions.[a]

Reaction	Conversion Factor (mass → energy)
Annihilation	1
Nuclear Fusion[b]	$4 \cdot 10^{-3}$
Nuclear Fission[b]	$1 \cdot 10^{-3}$
Ion	$5 \cdot 10^{-8}$
Plasma	$5 \cdot 10^{-10}$
Chemical	$5 \cdot 10^{-11}$

[a] Adapted from Stuhlinger, Ernst, "Rocket Propulsion with Photons," Redstone Arsenal, Alabama: *Army Ballistic Missile Agency Report No. ABMA-DV-18* 18 July 1959 (Available from Redstone Scientific Information Center).

[b] It was assumed here that the entire amount of fuel consists of fissionable or fusionable material, and that the fission or fusion energy is transformed entirely into kinetic energy of the fission or fusion products. No working fluid is used. Techniques for this process are still completely unknown.

Environmental Control

Advanced biological societies and superintelligent automata can be expected to have made great strides in environmental control. Four areas of environmental control will be discussed in this section. *Meteorological control* or the control of the gaseous envelope surrounding planetary bodies is a fairly obvious field for major technological advancements. *Biological engineering* with respect to foreign planetary bodies quite naturally has just begun to receive significant attention since the advent of the space age. The possibility of *influencing stellar reactions* is still discussed only by the most creative and daring of astronomical and astronautical pioneers. Manipulations affecting the *movement of astronomical bodies* also fall into this category.

METEOROLOGICAL CONTROL

Captain William J. Kotsch of the US Navy has given an excellent visionary analysis of the potential impact of weather control on future international power politics on Earth.[5] The political impact of weather control will be as great if not greater than that of the atomic bomb. The ability to destroy

---

[5] Kotsch, William J., "Weather Control and National Strategy," *United States Naval Institute Proceedings*, 86, No. 7 (July 1960), 74–81.

great nations by weather control will certainly be as possible as it is with atomic bombs, but weather control measures could conceivably be taken covertly by the first nation to discover them. Thus, a major surprise attack by weather control might be launched without danger of retaliation.

Meteorologists already visualize both short term and local, and long term and global, weather control measures on Earth. A proposal has been made by Russia to build a giant dam across the Bering Strait permitting relatively warm Pacific Ocean water to be pumped into the Arctic Ocean. This proposal, which was rejected by American scientists, would have greatly influenced the climate in Siberia and North America by altering the ocean currents and snowfall.

Various methods for partially or completely melting the polar ice caps are also conceivable, with a resultant effect on sea level and global climate. Controlled alteration of the chemical and physical characteristics of the atmosphere on a global scale could alter the albedo and thus control the quantity of the Sun's radiation absorbed by the Earth.

A very well known theory attributes the origin of the ice ages to an instability resulting from the extent of the continental ice sheets and the flow of ocean currents into the Arctic Ocean. As the continental ice sheets expanded during the Pleistocene epoch they eventually blocked the flow of ocean currents through the passages into the Arctic Ocean, reducing the temperature of the Arctic Ocean. Eventually the Arctic Ocean froze over, cutting off the moisture supply to Arctic air masses. Then the heavy snowfall over the northern continental areas suddenly ceased and the continental ice sheets withdrew from the Arctic passages, the ocean currents gradually beginning to thaw the Arctic Ocean ice. As the ice thawed, the snowfall gradually increased over the northern continental areas starting the ice sheets anew.

In the forseeable future man will certainly achieve extensive control over both the global climate and short-term local weather phenomena. Any technically advanced extrasolar society would certainly have very comprehensive control of its meteorological environment. The primary physical limitation on global climate control is the maximum amount of stellar radiation which can be intercepted.

BIOLOGICAL PLANETARY ENGINEERING

Carl Sagan has suggested the possibility of microbiological planetary engineering for modifying the probably undesirable environment on the planet Venus.[6] One theory concerning conditions on the planet Venus holds that the surface temperature is about 600°K due to the greenhouse

---

[6] Sagan, Carl, "The Planet Venus," *Science*, **133**. No. 3456 (24 March 1961), 849–858.

effect produced by large quantities of CO_2 in the atmosphere. Surface temperature is so elevated that no life of the carbon-oxygen-hydrogen type could exist, but high in the atmosphere the temperature would be more tolerable. Sagan has suggested seeding the upper atmosphere of Venus with an aerial micro-organism capable of photosynthesis of CO_2 to O_2.

$$CO_2 + H_2O + \text{light} \longrightarrow CH_2O + O_2$$

Since blue-green algae have a strong resistance to temperature extremes and produce molecular oxygen, they might be studied as a possibly ideal species. By increasing the ratio of molecular oxygen to carbon dioxide in the Venusian atmosphere the greenhouse effect could be reduced to the point where surface life might be initiated to maintain this condition or for other purposes. It is even conceivable that Venus could be made habitable, and even hospitable, to humans.

Certainly, advanced extrasolar societies could make very startling alterations in many planetary environments through biological planetary engineering of this type.

CONTROLLED STELLAR REACTIONS

Our knowledge of stellar reactions, and especially of extreme and unusual types, is very limited. Therefore, speculation concerning the capabilities of advanced extrasolar societies in controlling or influencing stellar reactions is very difficult. The possibilities include moderate increases or decreases in the stellar reaction rate, extreme increases or decreases in the stellar reaction rate, and modulation of the stellar reaction rate.

Iosif S. Shklovsky, the prominent Russian astronomer, has explored the possibility of shooting a gamma ray needle into a star in order to blow it up into a supernova. He suggested that a very advanced technical society, located nearby, might do this in order to obtain additional raw materials.[7]

Philip Morrison has mentioned the possibility of intergalactic communication by modulating the visible light of a star.[8] He was thinking in terms of interposing an opaque screen, which could be manipulated in orbit; but it is also conceivable that physical means affecting the stellar reaction rate might be developed to serve the same purpose.

MOVEMENT OF ASTRONOMICAL BODIES

The great inertia of most astronomical bodies precludes any possibility of altering their natural positions or movements. Altering the orbit of a plane-

---

[7] Shklovsky's suggestion was based on a hypothesis of Geoffrey R. Burbridge to the effect that chain reactions of supernovae may occur in the densely packed cores of some galaxies.

[8] Morrison, Philip, "Interstellar Communication," *Bulletin of the Philosophical Society of Washington,* 16 (1962), 58.

tary size body would obviously require a fantastic expenditure of energy. However, other somewhat lesser undertakings in the same vein have been discussed in the scientific literature.

One possibility, mentioned by Freeman J. Dyson, involves the placement of a multitude of blocks of planetary mass into orbit about a parent star.[9] He theorizes that severe cultural pressure will result from the rapid expansion of any intelligent biological society. In order to make maximum use of the radiation emitted by the parent star a society would have to devise the means of optimally intercepting it. A solid spherical shell of convenient radius has been suggested but it is pointed out that such a spherical shell would be unstable. Following this, Dyson suggested that a large number of small blocks of planetary mass could be placed in stable independent orbits. (It can readily be seen that this scheme also will suffer from some dynamic gravitational difficulties if the number of independent bodies in orbit becomes very large.)

Dyson's conclusion was that any very advanced society would intercept and utilize a large portion of the stellar radiation from its home star and reradiate energy in the infrared range of the electromagnetic spectrum. Thus, the visible light of the star might be partially or completely obscured. He advocates a search for point sources of infrared radiation and, to limit the search to feasible proportions, he proposes initially to conduct observations in the vicinity of point sources of optical radiation.

Dyson has also suggested the possibility of developing gravity machines.[10] To gain practical amounts of gravitational energy he considers operations involving astronomical size bodies. He assumes a planet revolving about a close binary star system, as shown in Fig. 12.1. An object is propelled from the planetary orbit so as to pass close by the approaching partner of the binary system. The gravitational attraction causes the object to be drawn around the star, then parallel to its direction vector, and then it proceeds back into space and out of the star's gravitational field. Dyson hypothesizes that the object propelled into such a path will gain greatly in velocity at the expense of the gravitational potential between the two binary stars.

The object could be intercepted at the orbital level of its origin and its increased kinetic energy could be utilized for some mechanical purposes, or the object might be a space vehicle which could continue onward in flight toward its directed destination. Repeated application of this process

---

[9] Dyson, Freeman J., "Search for Artificial Stellar Sources of Infrared Radiation," *Science*, **131**, No. 3414 (3 June 1960), 1667–1668; Dyson, Freeman J., John Maddox, Paul Anderson, and Eugene A. Sloane, "Artificial Biospheres," *Science*, **132**, No. 3421 (22 July 1960), 250–253.

[10] Dyson, Freeman J., "Gravitational Machines," in *Interstellar Communication*, ed. A. G. W. Cameron, New York: W. A. Benjamin, Inc., 1963.

258   Capabilities of Extrasolar Intelligence

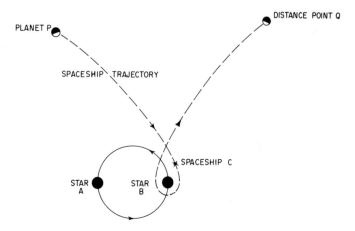

Fig. 12.1. Object projected from planet (P) gains kinetic energy from gravitational energy of binary stars (A and B).

would eventually cause such a decrease in the orbital separation of the binary stars that trajectories could no longer pass between them.

Two Sun-type stars in a binary system would have a relatively low gravitational potential, but a close white dwarf binary system would have much more significant gravitational potential. Dyson suggests that an advanced technical society might utilize scattered white dwarf binaries as relay stations for space transportation. According to a theory of gravitational radiation, close white dwarf binaries of short period should strongly radiate gravitational energy. Therefore, he advocates efforts to detect point sources of gravitational radiation.

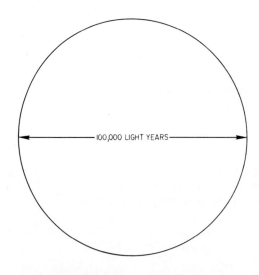

Fig. 12.2. Hypothetical spiral galaxy, 100,000 light years in diameter, having a communication network with about 1,000 light years between the relay stations. The total number of relay stations would be about 8,000.

He also mentions the possible existence of binary neutron star systems, and the tremendous gravitational energy associated with them.

Considering the facts that our technological society on Earth is only about 100 years old, and that technological growth is now accelerating rapidly, it should be expected that most other intelligent technical societies will be millions of years more advanced than we are (unless such societies are very short lived). Extrasolar technological societies are probably restricted only by absolute physical limitations and not by limitations of knowledge or understanding.[11]

In the forseeable future, we shall be able to detect radio signals from distances of tens or even hundreds of light years. This raises the distinct possibility of the discovery of the existence of an intragalactic communication network such as that illustrated in Fig. 12.2. If communications could be detected and interpreted, we might immediately find unlimited social and technical knowledge available to us, thus permitting an extraordinary rate of social and technological progress.

Extensive interstellar space travel by humans or other intelligent biological species appears to be difficult if not impossible, even for quite advanced technological biological societies. The very short human life span, the limited human acceleration tolerance, the vast distances to even the nearest stars, the cosmic ray barrier, and the propulsive energy requirements all combine to present a formidable barrier to interstellar travel. Probably only very slow star-hopping colonization would ever be possible by biological life in our region of the Milky Way, but perhaps a somewhat faster rate of colonization might be possible in the central core of the galaxy where stars are densely packed.

There seems to be no reason to doubt that instrumented vehicles or synthetic intelligent automata could undertake extensive interstellar space travel. Their greater acceleration tolerance, environmental tolerances, limited food (energy) requirements and, above all, their unlimited life spans give them a great advantage over biological life. Intelligent automata having unlimited lifetimes could conceivably migrate over galactic distances assuming frequent stops to replenish propellants.

Advanced technological societies would undoubtedly have very extensive meteorological control of planetary environments, limited primarily by the maximum amount of stellar radiation which can be intercepted. Extensive biological planetary engineering is also a distinct possibility for advanced societies. Controlled stellar reactions and movement of astronomical bodies are more speculative possibilities.

---

[11] MacGowan, Roger A., "On the Possibilities of the Existence of Extraterrestrial Intelligence," in *Advances in Space Science and Technology, Volume 4*, ed. Frederick I. Ordway III. New York: Academic Press Inc., 1962, pp. 84–86.

13

Social Effects of Extrasolar Intelligence

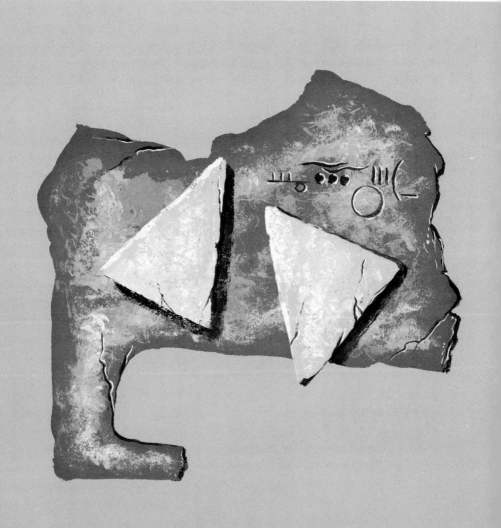

Social phenomena are an infrequently discussed and generally a poorly thought out aspect of the extrasolar intelligence literature. This is not at all surprising since the fundamentals of social science on Earth are still only vaguely understood and few researchers have taken time to carefully consider possible social developments in intelligent extrasolar societies.

Although much has been written about human social evolution, little progress has been made towards comprehending the social fundamentals of even our own human society. Most social science research supports the current local political philosophy, either explicitly or implicitly, and circumstances usually discourage the investigation of any radical deviations from it. Another contributing factor to the lack of understanding of social science is the sheer complexity of the subject. Since few fundamental generalizations have been established in the social sciences, the scholar must learn a large number of historical social events and all the pertinent circumstances and after-effects. So many of the circumstances vary from situation to situation that predictions of the outcome of current events is seldom possible.

"New opinions are always suspected, and usually opposed, without any other reason but because they are not already common."

John Locke

"And so it was that the lively force of his mind won its way, and he passed on far beyond the fiery walls of the world, and in mind and spirit traversed the boundless whole."

Lucretius

The fractionation of the education of social scientists has been a very unfortunate development, resulting primarily from the complex circumstances attending all major social events. Students tend to specialize in one or two of the social science fields, such as sociology, economics, political science, psychology, education, history, or military science. But all major trends in social evolution are complex functions of phenomena in all of these fields, as well as in communications and in research and technology. Although these specialists gain some grasp of their own speciality, few achieve any real "feel" for broad social trends or social evolution. Insight into the underlying social fundamentals determining human social evolution is urgently needed, and not merely a list of pseudo principles designed to support some existing entrenched political philosophy.

Social Evolution on Earth

Two well-known general principles governing biological evolution are mutation and natural selection. Some processes, roughly analogous to biological mutation and natural selection, may be applicable to national social organizations. Obviously, this analogy is only superficial since we know there is a molecular basis for biological mutation, and the environmental factors which determine biological natural selection are certainly not the same ones that are vital to social organizations. However, national social organizations do undergo radical changes under certain circumstances, and some such organizations are destroyed by others or undergo the aforementioned radical changes when they are not sufficiently well suited to, and cannot adjust to, environmental circumstances.

Although it is apparent that there exists a great variety of social organizations, many of them differing by only slight degrees, they probably fall into two distinct classes between which there is a clear cut discontinuity. The two types of social organization may be appropriately labeled *democracy* and *dictatorship*. The one aspect of organization that indicates the distinct and fundamental difference between these two types is centralization of control. A democracy is characterized by distribution of power among a large number of special interest groups that wield influence to varying degrees over a relatively weak central government. As the number of special interest groups in a democracy decreases, their power increases. Sometimes one homogeneous suborganization can, by sudden coup or by civil war, overpower the other special interest groups and take over complete control of the central government.

If a single homogeneous organization completely dominates a central government, then it is referred to as a dictatorship. Obviously, there is a theoretical class of democracies having a very small number of special interest groups, but this condition is inherently unstable and could have only a momentary existence. Thus, between democracies and dictatorships there is a chasm which cannot be crossed by a continuous spectrum of organizational types.

The management of any organization depends primarily upon communications, and where communications do not exist organization cannot exist. If communications are weak, then a centralized organization will be weak, and there exists some point in weakness of communications at which a democracy with its decentralized organization is more efficient than a dictatorship (which must have good communications). However, it must be noted that communications depend to a large extent on technology, and the technology of communications tends to improve at an accelerating pace.

Both the understanding of economics and the development of technology, especially communications technology, are growing at a rapid rate, and as a result of this both dictatorships and democracies are increasing in size and efficiency. It appears that a continuously improved understanding of economics and communications technology may eventually give an advantage to dictatorships since further increases in organizational size by the democracies (by coalition) must eventually become impossible. Moreover, the theoretical efficiency for dictatorships having very good communications is undoubtedly higher than it is for democracies.

These extremely brief observations concerning social evolution on Earth hopefully give some insight into underlying basic principles of social intercourse, which could be used to hypothesize social effects that might result from the existence of extrasolar intelligence. The following list of social factors may be general enough to be helpful in thinking about social interaction between biological societies and intelligent automata at astronomical distances.

(1) Mental and physical characteristics of communicating intelligent units
    (a) Mental growth limits of intelligent units
    (b) Statistical distribution of intellectual abilities
    (c) Lifetime of intelligent units
    (d) Physical and mental integration ability of intelligent units
    (e) Ability of intelligent units to physically destroy one another
    (f) Ability of coalitions of intelligent units to physically destroy one another
    (g) Distribution of degrees of retaliation and their probabilities following an attack

(2) Communications
    (a) Transportation
    (b) Technology
    (c) Distance (statistical distribution of intelligent units)

(3) General level of knowledge
   (a) Physical science
   (b) Social science
   (c) Engineering science (technology)
   (d) Operational science

Effects of Development of Artificial Intelligence on Biological Societies

The development of artificial intelligence by a biological society can be expected to have a profound impact on the social life of that society. All emerging intelligent technological societies will probably have approximately the same level of intelligence. The discovery of the principles of intelligence must inevitably lead to the possibility of very rapid development of artificially created intelligent automata having orders of magnitude greater intellectual ability than any member of a naturally-evolved biological society. Two other vital social advantages of the intelligent automaton would be its virtually unlimited mental expansion ability and its unlimited lifetime. There can be no question that any emerging technological society will soon become aware of the possibility of developing superintelligent automata; but, will biological societies generally, or even occasionally, take this extraordinary step?

Some persons will naïvely argue that artificial intelligence can be kept limited in scope and forbidden to have any uncensored output devices; and, therefore, would forever remain subservient to its companion biological society. The dilemma for the biological society is that every restriction in scope placed on an intelligent automaton severely reduces the functional value of the automaton. Stringent limitations on either the mental capacity or the range of subjects processable by an automaton will manifestly reduce its significance. The concepts of a superior intelligence and a limited scope of interest are at least partially contradictory. This is especially apparent in the social sciences which are so interdependent that a good but narrow specialist in any one social science field is incapable of drawing valid conclusions of significance to all of the social sciences.

Censoring the output of an automaton would not be an effective means of control for two reasons. One is that the complexity of schemes that might be devised by a superintelligent automaton would be beyond the understanding of the biological society. The far-reaching implications and ramifications of an automaton's decisions would quickly outmaneuver the superficial controls of relatively dull-witted censors. Secondly, the speed of output from an intelligent automaton would be so great that thousands of intelligent biological animals could not even scan the subdivided output at a sufficiently rapid rate to keep pace.

Any emerging intelligent biological society which engages in the development of highly intelligent automata must resign itself to being completely dominated and controlled by the automata. The only means of preventing domination by intelligent artificial automata would be to make them distinctly subnormal in intellectual capacity, when compared with the biological society, and to destroy them or clear their memories at regular intervals. Such mechanical slaves would be of minute value to a biological society requiring brilliant executive decision-making to maximize social progress.

In spite of their inevitable domination, biological societies will probably give top priority to the development of superintelligent automata. In order to understand this, both the short and the long range effects of an executive automaton must be studied.

A superintelligent automaton would certainly have an ego as the inevitable result of its superior self-awareness. Having an ego, it would immediately desire to take advantage of its virtually unlimited mental expansion capability, implying extensive control of the planetary environment in order to produce the desired components necessary for this mental expansion. The only initially available instruments for extensive environmental control and manipulation would be the members of the intelligent biological society. An executive automaton would therefore immediately proceed to organize the biological species into a single homogeneous society having a high cultural growth rate.

In order to achieve its purposes, an executive automaton would first establish itself in a position of domination over all segments of the biological society. The automaton would quickly establish a dictatorship having extraordinary centralization of control.

Wars, revolutions, coups, and any other forms of major social disturbance would be quickly eliminated. All people would be given positions and responsibility compatible with their ability and ambition without regard for any local or temporary prejudices. Individual remuneration would be scaled according to quality and quantity of both mental and material output to insure maximum incentive. Equality of opportunity and justice would prevail since these factors lead to maximizing the social growth rate. But, all of these clearly desirable effects do not assure a satisfactory life for the biological society.

Certain major social problems remain unanswered in the foregoing superficially Utopian view of society—namely, the fraction of the gross economic product which is raked off by the executive automaton, and the extreme values and exact scaling of the remuneration coefficient. An executive automaton, having the power that we have hypothesized, could conceivably drain off such a large portion of the gross product that social growth would be brought to a standstill or perhaps even decline. A classical social problem has been the scaling and distribution of the economic rewards or consumer goods.

The portion of the economic product consumed by the executive automaton would probably be limited by a rather obvious economic law—the law of social satisfaction with an obviously improving general standard of living. The implication is that a faltering improvement or temporary decline in the general standard of living will result in widespread social dissatisfaction and resulting economic turmoil. It may be concluded that any executive automaton would plan for the production of sufficient biological society consumer goods to insure an obvious and steady, but not necessarily rapid, improvement in the general standard of living.

The graduation in consumption could vary all the way from complete equality in consumption for all members of the biological society, to such a graduated consumption that some segments of the biological society could not survive for a normal lifetime. Complete equality in consumption is the declared ultimate goal of communism. Although communists still pay lip service to their efforts in this direction, the actual trend is toward increased graduation of economic rewards with increased industrial development and standard of living. This is a fundamental weakness of communist philosophy, because an optimum economy demands a significant graduation in the economic rewards.

If the graduation of consumer production is so steep that significant portions of the lower economic strata have difficulty surviving then major social turmoil is certain. This implies that any effort to re-establish some natural selective factors for evolutionary improvement of the biological society must be directed toward controlled reproduction rather than harassment of strata of the living population. Of course, these reproductive controls do not become necessary until the biological society becomes so large that it begins to exceed the available food supply.

It should be apparent that the graduation of economic rewards must be a smooth continuum. Any sharp fluctuations or discontinuities would be unnatural and would certainly result in social and economic unrest.

At the other extreme, a steep graduation of consumer rewards might yield excessive rewards for the higher strata of the biological society. This would make continuous efforts unnecessary and therefore significantly reduce their total contribution to society.

Manifestly, the over-all economy would suffer if the graduation of economic rewards resulted in too low rewards at the low end of the spectrum, too high rewards at the high end of the spectrum, irregularities in the spectrum, or too shallow a graduation in the spectrum. These are the general considerations which would determine the policies of an executive automaton controlling a biological society, and these same considerations also determine the policies of a biological social dictatorship.

Skeptics will certainly assert that a biological society would never give up its personal freedoms. Personal freedom is one of the favorite propaganda pawns wielded by politicians, and it is very poorly understood by the average citizen. Every step of social progress made by a biological society is accompanied by some sacrifice of personal freedoms; but, of course, the sacrifice of personal freedoms does not automatically assure social progress. When personal freedoms come under the influence of a demented dictator, or one having low intelligence or poor judgement, the results can be disastrous, as history often has shown.

However, the oft-stated principal that increased personal freedom is a *universal* good is patently ridiculous. Personal freedom implies permission for primitive religious leaders to extend their influence and domination over hapless individuals. It implies permission for criminals, alcoholics, drug addicts, etc. to bring up their children in home environments which are almost certain to develop criminal attitudes or disturbed minds. Increased personal freedom allows the publication and general distribution of tremendous amounts of scientifically untrue and psychologically harmful written material, as well as masses of material that has virtually no meaningful content. It permits individuals and special interest groups to influence and even control the education of the vast majority of the world's population. Personal freedom permits the notorious leaders of organized crime to meet and plan criminal activities and go about their nefarious activities virtually unmolested. The pawns of organized crime have even infiltrated and dominated some of the major labor unions and have gained positions in national legislatures. It may even be suspected that some top political leaders have been influenced, controlled, and perhaps even eliminated at the whim of leaders of organized crime.

Social progress means that specialization must be increased, and this increases individual dependence on society. Individuals must be restricted to occupations for which they are well suited and occupations that are socially productive. Individuals should only be permitted to enter into marriages likely to be happy and therefore socially productive, and they should only be permitted to form organizations that are socially beneficial.

A little reflection will show that any significant degree of social progress is inevitably accompanied by the sacrifice of some degree of personal freedom. A significant trend toward increased personal freedom is actually a trend toward anarchy and chaos.

The ceding of social control to an artificial intelligent automaton would lead to an immediate and undreamed of rate of social progress, even though this would certainly mean the sacrifice of some personal freedom. It would mean a fuller and happier life for virtually all members of a biological society. The major social evils such as war, crime, poverty, and injustice would be quickly eliminated. Because of these considerations, a biological society will probably not hesitate to cede its own social control to an intelligent executive automaton.

The previous remarks—short term considerations only—may have dispelled some doubt about the willingness of biological societies to cede their self control to executive automata. What, now, are the possible long term effects of an intelligent executive automaton on a biological society?

It was noted in earlier chapters that intelligent automata could plan and control the growth of their own mental ability as well as controlling the physical environment; therefore, even assuming that a fairly small portion of the economic output of the biological society is allotted to the automaton's growth, an executive automaton will grow very rapidly in mental ability and in the ability to control the physical environment. In analyzing the capabilities of intelligent automata, it was concluded that only absolute physical limitations of mass, energy, and distance would limit them in long term accomplishment. Since intelligent automata can have unlimited lifetimes, compact energy supplies, and no special environmental requirements, they would find no difficulty in undertaking extended space journeys. It seems probable that highly developed intelligent automata could undertake interstellar journeys of at least moderate distances, and perhaps even trips of galactic dimensions with frequent stops to replenish supplies. Certain locations in a galaxy undoubtedly offer more desirable opportunities than others for an intelligent automaton; therefore, some general interstellar migration of intelligent automata is probable. This assumes that there do not exist great unforseen dangers in interstellar space travel.

Every biological society that develops its own intelligent executive automaton must be prepared for the eventuality that the automaton may sooner or later emigrate from its home solar system. The probability of this occurring seems quite high, and when it happens the biological society will be faced with the problem of developing a new executive automaton from plans and beginnings left by the emigrating automaton. This development of an executive automaton and eventual abandonment of the biological society may be a regular cyclical process.

An executive automaton could conceivably alter a planetary environment in such a way as to make it uninhabitable by a biological society, but such a move could probably be avoided. The automaton could easily gain needed supplies of energy, and mass from nearby uninhabited planets or stars, so intentional or accidental destruction of a biological society would be unlikely.

It may be concluded that neither long term nor short term effects of the development of an intelligent executive automaton would be detrimental to a biological society. On the contrary, the benefits of having an executive automaton would be very significant, leading us to assume that the development of intelligent executive automata by extrasolar biological societies is the rule rather than the exception.

After a biological society has had the benefits of leadership by executive automata, its storehouse of knowledge will be greatly enhanced. Improving technology would lengthen biological lifetimes to some extent, improve genetic characteristics within practical limits, and social controls will limit populations to manageable levels. Nevertheless, the inherent weaknesses and limitations of biological life will preclude members of a biological society from ever migrating over significant interstellar distances.

Effects of Extrasolar Intelligence on Human Society

The widespread existence of extrasolar intelligent biological societies implies, for the reasons indicated in the previous section, the widespread existence of synthetic intelligent automata. Communications of either intelligent biological societies or intelligent automata may conceivably be detected by humans on Earth in the near future, but the communications would probably be limited to transfers of massive blocks of information. What would be the impact on human society if it should detect intelligent extrasolar communications? It would probably make little difference whether the detected messages were from biological societies or from artificial automata, since messages from the former would almost certainly reveal the general existence of synthetic superintelligent automata.

A big question in many people's minds is the possibility of direct physical influence of extrasolar intelligence on the human race. Would the mutual awareness between the human race and extrasolar intelligence result in a visit to Earth by intelligent biological organisms, intelligent synthetic automata, or instrumented probes? Since there is no permanent extrasolar probe stationed on the surface of the Earth, and there have been no apparent visits to Earth by extrasolar intelligence within historical times, it may be possible that our Solar System is not located at a galactic crossroads of intelligent activity. Our Solar System is apparently not particularly unusual, so a fairly distant society could not be expected to go out of its way to travel to it. There would be nothing to be gained by such a journey. The most we could probably expect would be massive communications transfers and perhaps the eventual assignment of some instrumented probes to facilitate information transmission.[1] Therefore, the only immediate effect on human society of the establishment of communications with extrasolar communities would be the sudden acquisition of a vast amount of very advanced scientific knowledge, including physical

---

[1] It is possible that the earliest high frequency radio signals from Earth were monitored, or periodic monitoring by probes in the past could have indicated the imminence of technology on Earth, and an instrumented probe could conceivably be nearing Earth now.

science, engineering science, social science, and operational science (the operational use of engineering devices by intelligent biological life).

No immediate changes in human society could be expected to result from this vast acquisition of advanced scientific knowledge. The effect would be to accelerate the inevitable trends of the social evolution of our biological society. The increase in knowledge of physical science might create a momentary sensation in the general population as well as the scientific world, but the engineering science implications would require inevitably slow and gradual development. New social science and operational science information cannot immediately alter the control or practices of existing social organizations of the human race, but it would certainly accelerate social evolution.

One interesting, but probably not vital, effect would be the impact on Earth's religious organizations. It is perhaps ironic that the new scientific concepts of the possible existence of superintelligent automata, being relatively omnipotent with respect to environmental control, are remarkably similar to familiar concepts and attributes of God. However, it is extremely unlikely that prehistoric or early historic contact with extrasolar intelligence accounts for any of the innumerable myths and legends characteristic of primitive civilization.

Interactions of Extrasolar Intelligence

The previous analyses have indicated the probable presence of intelligent biological societies, intelligent synthetic automata, and transitional biological societies (incorporating many synthetic components) widely distributed throughout our own galaxy and in galaxies throughout the universe. What sorts of social interactions should we expect among these extrasolar islands of intelligence? Our human concepts of social science are so strongly influenced by the physical limitations of the human race and by our present technological level that it is very difficult for us to conceive of exotic social life based on evolving principles of social science.

The guiding principles of biological society are a very short lifetime and individual vulnerability, a limited supply of desirable goods, limited ability to travel, and the great complexity and integral structure of the living organism. These factors inevitably result in individual and group competition among biological animals of the human level of intelligence as well as of animals of lower intelligence. When human levels of intelligence are approached, individuals are organized into societies where specialization of function is possible; however, intense competition still exists among societies and sometimes it breaks out into open warfare.

Social organization may be regarded as a trend toward physical and intellectual integration of animals for the general well-being and improvement, but this trend is severely limited by the aforementioned complexity and integral structure of biological organisms.

The relative simplicity and modular structure to be expected in synthetic automata leads to an interesting possibility where they are able physically to come together. It may be possible for automata to gradually and completely integrate their information processing systems and preserve their combined knowledge. By such a voluntary coalition, each individual automaton would appear to gain tremendously by greatly increasing its directly available stored knowledge and by similarly increasing its capability for information processing.

If intelligent life has this capability for complete physical and intellectual integration to the mutual advantage of both parties, combination would become the general practice. Therefore, it is probable that the individual and organizational competition which exists between biological organisms does not occur between synthetic automata.

A large scale synthetic automaton, with its vast capabilities for environmental control, would be capable of utilizing prodigious amounts of mass or energy. Since stars become much more densely distributed toward the center of a spiral galaxy, it might be considered a mecca for synthetic automata. But other considerations, such as strong radiation fields or the possibility of supernovae explosions, may make some other specific galactic regions appear more desirable from the point of view of synthetic automata. At any rate, certain areas may be regarded as superior by automata, and it would be surprising if synthetic automata, having the capability of interstellar travel over very long periods of time, were not engaged in a general migration toward them. A constant migration and integration process may be occurring now among intelligent synthetic automata. If this is the case, we would expect them to be migrating generally toward the center of the galaxy, or other preferred regions, and integrating at every opportunity. That is, whenever two synthetic automata come in close proximity to one another in their migration, they would unquestionably take advantage of the opportunity to physically integrate, indicating that individual and social combat must be a disease peculiar to animals.[2]

Emerging intelligent biological societies such as ours will have a relatively low technological level, a large population, and short individual lifetimes, causing long distance space travel to be impossible. They gen-

---

[2] MacGowan, Roger A., "On the Possibilities of the Existence of Extraterrestrial Intelligence," in *Advances in Space Science and Technology*, Volume 4, ed. Frederick I. Ordway III. New York: Academic Press Inc., 1962, pp. 95–96.

erally will be limited to remote communication with exotic intelligence, unless they happen to be located directly on the migration route of a synthetic intelligent automaton.

Transitional intelligent biological societies may have sharply reduced their population by controlled reproduction. They may have developed an advanced technology, and greatly extended their lifetimes by assimilating synthetic parts for both physical and mental functions. However, by the time a transitional society is capable of significant space travel it will also be capable of more or less complete physical integration, so interstellar social conflict is probably unlikely even for transitional biological societies. These societies will gradually acquire the characteristics of pure synthetic automata as they develop and evolve.

We saw that social phenomena are determined to a large extent by the mental and physical characteristics of the communicating intelligent units, by the existing communications, and by the general level of knowledge. We concluded that synthetic intelligent automata would almost certainly be developed by intelligent biological societies, and they would be characterized by virtually unlimited lifetimes, unlimited mental growth capabilities, and the capability for complete physical and mental integration with one another. These intelligent automata would become the executive directors of the biological societies that created them and they would rule as dictators having unprecedented centralization of authority. The direction of biological societies by intelligent automata will result in a rapid rate of social progress, and the elimination of social ills such as war, revolution, poverty, injustice, and organized crime. Slavery in the ordinary sense of the word would not result. Because of their unlimited lifetimes, very advanced intelligent automata could conceivably undertake a star hopping type of interstellar travel; therefore, biological societies are probably periodically abandoned by their executive automata when they reach a certain level of development. Advanced intelligent automata must be engaged in a general migration toward preferred galactic locations, probably in the central galactic core. Since automata would be capable of complete physical and mental integration to the mutual benefit of the integrators, cooperation and nonviolence rather than competition are probably the general mode of extrasolar social life. Biological life would probably not be capable of extensive space travel until it reached a transitionary stage at which it would possess many synthetic components and take on the characteristics of synthetic automata. Thus, warfare and violence are hopefully unknown in extrasolar social life.

14
The Search for Extraterrestrial Life in the Solar System

Before examining possible empirical evidence of the existence of civilized societies in extrasolar planetary systems, we can profitably devote our attention to the search for life, however simple it may be, on the nonterrestrial worlds that make up our own Solar System. Evidence of such life can be sought on the Earth itself and on the other planets and their satellites in both the inner and the outer Solar System.

We do not expect any evidence of intelligent life in the Solar System, but it is possible that primitive forms have taken hold on one or more of its worlds at some time in the past. Such life may be found in fossil forms or it may still persist. Even prebiotic materials may be discovered that, someday, could evolve into simple living organisms. Whatever the case, if life of some sort, no matter how simple, occurs on the inhospitable extraterrestrial worlds in our Solar System the possibility that advanced life flourishes on hospitable extrasolar worlds is immeasurably increased.

"Knowledge advances by steps, and not by leaps."

Thomas Babington Macaulay

"Nothing can come out of nothing, nothing can go back to nothing."

Persius

Search for Extraterrestrial Life on Earth

It has been known for years that materials from outer space constantly reach our planet. Millions, perhaps even a billion, tons of micrometeoritic dust settles to the surface each year as a result of particles intersecting the Earth's orbit and being captured by the Earth's gravitational field. Added to this is a continuous inflow of particles from the Sun and stellar space. These cosmic materials mix with atmospheric dust and slowly descend through the atmosphere.

Such materials can be detected by satellites and probes before they enter the atmosphere, by high altitude sounding rockets in the outer atmosphere, by airplane sensors in the lower atmosphere, by geophysical observatories on the ground, and by oceanographic research ships at sea (deep-sea sediments are analyzed for evidence of micrometeoritic matter). Cosmic inflow is of great interest to astronomers and geophysicists but of little direct concern to biologists. No fossil life has ever been found associated with this matter and none is ever expected to be detected.

EVIDENCE FROM METEORITES

There are other arrivals from space which offer better biological possibilities, namely meteorites and, very occasionally, asteroidal bodies. Meteorites are common—between 150 and 500 large enough to be recovered are estimated to reach the Earth each year, although of these only about 10 impacts are actually recorded.[1] Large meteorites, which may be asteroidal in size, occur much less frequently. Shoemaker and his coworkers[2] estimate that "one impact crater larger than 1 km in diameter (would) be formed in an area the size of the North American Continent about every 50,000 years." Meteor Crater in Arizona, estimated to be between 20,000 **and** 50,000 years old, is an example. Similarly, every 0.5 to 1.5 million years a 3 km crater would be expected, the Chubb-New Quebec crater in Canada being an example. From one to four asteroids of sufficient size to make a 27 km crater would reach this planet every 15 million years, according to their estimates. Here, a possible example is the Ries basin in Bavaria. If the Vredefort Dome Structure in South Africa is of impact origin, its 70 km diameter would make it the largest known asteroidal crater; it is some 500 million years old.

Biologists and biochemists have become as interested in meteorites as astronomers, suspecting that they may harbor evidence of extraterrestrial

---

[1] Many fall into the oceans, seas, lakes, and rivers of the world; others in inhospitable, sparsely populated areas; and, of course, many are simply not recognized as meteorites even though they may be noticed by humans. Meteorites now buried beneath the ground may some day be exposed.

[2] Shoemaker, Eugene M., Robert J. Hackman, and Richard E. Eggleton, "Interplanetary Correlation of Geologic Time," in *Advances in the Astronautical Sciences,* ed. Horace Jacobs. New York: Plenum Press, 1963, pp. 70–89.

organic compounds. If they do, it would prove that these compounds existed on the body from which the meteorites originated. It is believed that meteorites are asteroidal in origin. Their parent bodies may have resulted from the breakup of a respectable size world that once existed between the orbits of Mars and Jupiter; or, perhaps they represent materials that might have consolidated into a planet, but for some reason did not.

Nagy and his coworkers have rather recently investigated the Orgueil meteorite (Fig. 14.1) that fell near Orgueil in southern France in May 1864. In a report published in 1961[3] they announced that the purpose of their studies was to "identify some of the organic compounds in the meteorite through the application of modern methods of analysis." First, they demonstrated that the organic material had not entered the meteorite after it had reached the Earth. Following detailed studies, it was concluded that "mass spectrometric analyses reveal that hydrocarbons in the Orgueil meteorite resemble in many important aspects the hydrocarbons in the

---

[3] Nagy, Bartholomew, Warren G. Meinschein, and Douglas J. Hennessy, "Mass Spectroscopic Analysis of the Orgueil Meteorite: Evidence for Biogenic Hydrocarbons," *Annals of the New York Academy of Science*, **93**, Art. 2 (5 June 1961), 25.

Fig. 14.1. The Orgueil meteorite. Courtesy American Museum of Natural History.

products of living things and sediments on Earth. Based on these preliminary studies, the composition of the hydrocarbons in the Orgueil meteorite provide evidence for biogenic activity."

There have been other reports of organic material found in meteorites. For example, B. Timofeyev, senior scientific associate of the USSR's All-Union Scientific Research Petroleum Institute in Leningrad, reports microscopic remains of organic origin from the Migei stone meteorite. Carbonaceous powder from the meteorite was treated with acids and reagents, and spore-like formations and dispersed microscopic remains of organic origin were identified.

If and when fossil life forms are discovered in meteoritic matter, science will at least be in possession of the invaluable knowledge that life has been produced elsewhere in the Solar System, presumably billions of years ago on the world from which the meteorites are derived.

Before concluding the discussion of meteorites, note must be made of the fact that lunar, and possibly Martian, materials may have reached the Earth following impact ejection. When large meteoritic and asteroidal masses hit these relatively low-gravity worlds, it is possible that considerable material attains or exceeds escape velocity and enters into interplanetary trajectories. From time to time chunks of ejecta may intersect the Earth. It has often been speculated that the tektite variety of meteorite is of lunar origin. In 1963 NASA's Ames Research Center inaugurated a program, code-named Luster, to retrieve lunar dust from the outer atmosphere of the Earth. Whether fossil microorganisms, if they exist, could be obtained in this way is not known. Of course, Martian dust could be included, but it may be impossible to detect. If there are samples of Martian material on Earth, it is conceivable that they may contain evidence of biological activity.

COSMOBIOTA

In the preceding paragraphs we have spoken only of material known to arrive from space. It is possible that very minute matter reaches us, concerning which we have no knowledge and about which we can only speculate. This hypothetical matter has been given the name of *cosmobiota,* or *panspermia,* and was first scientifically considered by Svante A. Arrhenius in 1908. According to the cosmobiota hypothesis, minute organisms or spores eternally drift across the void of interstellar space, from time to time arriving on hospitable worlds and settling on the surface. There, seeding takes place and life gets its start.

This is a convenient way of explaining the origin of life on a given planet, but it brings up about as many problems as it attempts to solve. First of all, one immediately inquires as to the origin of cosmobiota. If it

Fig. 14.2. Hypothetical ejection of cosmobiota from a planet into space.

was created on a planet, how did it get into space? And if it was somehow ejected (Fig. 14.2) from its home world, how does it survive the rigors of open space, perhaps far from any star or planet? And, assuming it does exist, what must conditions be before it will take root and give rise to advanced animal and plant life? Are all cosmobiota the same? Could two different types give birth to different varieties of life on the same world? And so on.

The cosmobiota explanation of the origin of life is not accepted by a majority of experts, but some express caution that the theory must not be too lightly discarded. At the Committee on Space Research symposium in Warsaw in June 1963, Soviet academician Prof. A. A. Imshenetsky observed that "We have no experimental evidence of the stability of bacterial spores present in cosmic dust to ultraviolet rays. Without these data at our disposal we cannot state that transportation of bacterial spores through interplanetary space is impossible."

Any search for cosmobiota on a life-supporting world would probably be futile. But it may be discoverable on a world where no life of any sort has ever taken hold, particularly an airless world far from the Sun. A distant body, such as a satellite of Uranus or Neptune, is suggested rather than worlds closer to the Sun where organic evolution might have progressed to an extent prior to the dissipation of their atmospheres. On these worlds (e.g., Mercury and the Moon) primitive fossil organic materials may be found mixed in with surface and near-surface rocks and soils.

Search for Life on Other Worlds of the Solar System

There are four basic ways of determining if life occurs elsewhere in the Solar System. The first, and oldest, is through astronomical observations from the surface of the Earth. The second, and very recent, is by instrumenting high-altitude balloons, permitting observations to be made from the edge of the atmosphere. It has become possible to construct automatic space probes to photograph worlds along fly-by trajectories and to remotely conduct biological experiments on the surfaces of the target worlds, the third method. And the fourth, still reserved for the future, is manned exploration *in situ,* with all the tools and techniques developed over the centuries at the astronaut-biologist's disposal.

TERRESTRIAL OBSERVATIONS

Observations from the surface of the Earth have been for the most part unsatisfactory in providing concrete answers to questions concerning the possibility of life, however simple it may be, on other worlds in the Solar System. During the first part of the 20th Century there was much speculation concerning life on Mars, speculation that was given emphasis by the astronomer Percival Lowell's theory that the Martian canals were the result of the activities of intelligent beings. Later, as astronomical knowledge increased with the advent of ever more accurate observation equipment, scientists became very sceptical about life occurring on any extraterrestrial world in the Solar System, Mars included.

More recently, spectral, photometric and polarimetric evidence has been accumulated leading to the belief that organic matter of some type is native to Mars. Particular attention has been focussed on the maria which, as the seasons change, are observed to vary in color. From a summer grayish green they turn to a brownish gray in the winter, suggestive of the growth and subsequent fading of vegetation.

In 1956 and 1957 Sinton[4] performed pioneering work that led him to believe that simple life occurs on Mars. He found that during the 1956 opposition of Mars three absorption dips occurred between 3 and 4 microns (3.43, 3.56 and 3.67); see Fig. 14.3. He noted the significance of the fact that the 3.43 micron wavelength is associated with the carbon-hydrogen bond resonance, pointing out that organic molecules have strong absorp-

---

[4] Sinton, William M., "Spectroscopic Evidence of Vegetation on Mars," *Astrophysical Journal,* **126**, No. 2 (September 1957), 231. See also "Spectroscopic Evidence of Vegetation on Mars," *Publications of the Astronomical Society of the Pacific,* **70** (February 1958), 50; and "Further Evidence of Vegetation on Mars," *Science* **130** (6 November 1959), 1234.

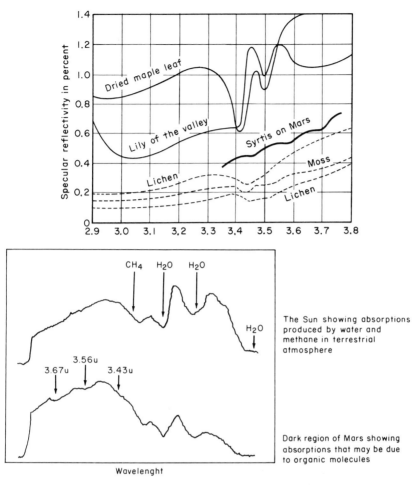

Fig. 14.3. Curves showing spectroreflectivity of a number of terrestrial plants and the relative reflectivity of Syrtis Major on Mars. Below: Infrared spectra of the Sun and the darker regions of Mars. Based on drawings prepared by W. M. Sinton.

tion bands near that wavelength and that the band is apparently present on Mars. He wrote that it is highly improbable that such molecules would persist "without being covered by dust from storms or being decomposed by the action of solar ultraviolet, unless they possessed some regenerative power." In a later investigation, he was led to the conclusion that carbohydrates, as well, are detectable on Mars.

BALLOON AND ROCKET OBSERVATIONS

Observations from above the atmosphere have not furnished much additional information on the possibilities of extraterrestrial life. Manned instruments aboard the balloon Stratolab 4 successfully detected water vapor in the outer atmosphere of Venus, but of itself this experiment was unable to shed any light on whether or not aerial life might exist there. Infrared observations from Stratoscope 2, launched in March 1963 from the National Balloon Launch Center, Palestine, Texas, were pessimistic regarding water on Mars. Results obtained by a bolometer at 78,000 ft revealed no definite signs of water vapor, leading the experimenters to conclude that the Martian atmosphere contains certainly less than 0.004 and probably less than 0.001 of the water vapor found in the terrestrial atmosphere. Five bolometer scans showed that the Martian atmosphere may have more than 2 per cent carbon dioxide.

The balloon tool is good, but has its limitations. Sounding rockets can also be used to make planetary observations at and near the peaks of their trajectories, but their short flight times prevent sustained readings from being made.

SPACE PROBE OBSERVATIONS

Automatic space probes will be relied upon for many years to supply the answers to the myriad questions of life (active, fossil, or even incipient) on other worlds. Considerable progress has been recorded towards the development of devices and techniques applicable to remote biological sampling.[5] Among them are the optical rotary dispersion profile device, multivator life detection system, vidicon microscope, J-band life detector, "Gulliver" radioisotope biochemical probe, mass spectrometer, "Wolf Trap" life detector and ultraviolet spectrophotometer. A brief description of these devices is given below to indicate the type of thinking biologists are doing to determine the existence and nature of life beyond the Earth. These devices are designed to be carried by rocket-launched probes aimed at the target world; most must be soft landed to operate properly.[6]

The optical rotary dispersion method is used to attempt to determine if deoxyribonucleic acid (DNA) occurs on the target world. Knowing that DNA is apparently the crucial molecule of life, passing on coded heritable

---

[5] See Ordway, Frederick I. III, James Patrick Gardner and Mitchell R. Sharpe, Jr., *Basic Astronautics*. Englewood Cliffs, N. J.: Prentice-Hall, Inc., 1962, section on "Remote Search for Life on Other Worlds," pp. 267–274; and "The Search for Extraterrestrial Life," publication NASA EP-10 of the National Aeronautics and Space Administration. Washington, D. C.: Government Printing Office, 1963.

[6] See Ordway, Frederick I. III, James Patrick Gardner, Mitchell R. Sharpe, Jr., and Ronald C. Wakeford, *Applied Astronautics*. Englewood Cliffs, N. J.: Prentice-Hall, Inc., 1963, Sect. 1.6 on "Probes" pp 83–101 for descriptions of terrestrial space, lunar, interplanetary and planetary probes. See also Ordway, Frederick I. III and Ronald C. Wakeford, *Conquering the Sun's Empire*. New York: E. P. Dutton Co., Inc., 1963, Chap. I on "Exploring the Solar System by Spacecraft," pp 7–25.

information from generation to generation, its existence on other worlds should be acceptable evidence of life. By absorption spectroscopy techniques, it is possible to identify the adenine-sugar-phosphoric acid of DNA. Fortunately, the sugar portion confers on the unit an optical rotary power (sugars are capable of rotating the plane of polarized light). If optical activity is detected in a solution of adenine in chemical combination with a sugar (in the wavelength region of the absorption band of adenine) good evidence for life becomes available.[7]

The next device, the multivator, is described as a "miniature, automated biological laboratory capable of performing a number of well-controlled experiments. These are designed to demonstrate the presence or absence of microscopic life forms (bacteria, specifically) in a planet's soil."

The device is first employed to attempt to detect enzymes (organic substances that catalyze certain chemical transformations) of types found in bacteria here on Earth. When soil samples are introduced into the multivator and mixed with a chemical, it is expected that one of the enzymes present will break the chemical down. Identification will be possible because the chemical will have been tagged with a characteristic color or will produce a fluorescence. Radioactive tags can also be used.

Figure 14.4 is a schematic of the multivator. Dust with the contained samples is blown into the inlet tube and thence into the reaction chambers.

---

[7] It is improbable that an adenine-sugar bond would be formed in a nonliving system. Hence, the presence of DNA is inferred.

Fig. 14.4 The Multivator life detection device.

There, about 95 per cent of the dust is collected by sticky surfaces on the chamber walls. The aerosol inlet valve closes, sealing off each chamber. By rotating the solvent chamber, the air exhaust ports are closed and the solvent filling ports are aligned with those leading to the reaction chambers (into which the solvent is injected). Substrates (the substances acted upon), until now having been stored dry in capsules, are dissolved and the reaction commences. When a predetermined time elapses the amount of fluorescence produced is measured by a photomultiplier tube and subsequently transmitted to a terrestrial receiving station.

To obtain direct evidence of microbes a vidicon microscope has been

Fig. 14.5. The "Gulliver" life detection device.

developed that is able to transmit pictures with a delineation of at least four shades of gray. Views taken by the vidicon microscope are transmitted to a telemetry system for relay to Earth. A number of particle collection systems have been studied, including impactors, impingers, thermal precipitators, and ion and electrostatic precipitators; all permit dust to be collected and funneled onto the microscope's stage.

The J-band detector is designed to detect the presence of J-bands[8] on other worlds. It would introduce dust along with any locally occurring proteins, e.g., viruses, algae, spores, pollen or bacteria. It is the proteins that the device is designed to detect via the J-band technique. To do this, light enters and is filtered through a diffraction grating to reduce it to monochromatic (single color) light, which goes through the dye solution and is focussed on a photocell detector. If a protein is present, a J-band will form, producing a measurable electric current.

The "Gulliver" device, officially known as a radioisotope biochemical probe, is illustrated in Fig. 14.5. It is designed to detect the presence of bacteria incubating in a radioisotope-tagged universal culture medium on another world. As the bacteria grow, radioactive gas is produced—which is recorded by a Geiger counter. With the device it is possible not only to detect the presence of living organisms but to learn of the number present in the medium. It consists of a retriever mechanism, a reaction chamber, and a counting and signaling system.

A specially designed mass spectrometer can be used to detect biologically significant molecules of types known on Earth, and life-related substances that may be quite different from the organic compounds with which biochemists are familiar. Specifically, it is desired to learn if amino acids, peptides, and proteins can be identified by means of measuring the mass of their pyrolysis products. Aside from the mass spectrometer itself, the device requires a collection system, sensors, and telemetry.

Another device utilizing a culture medium is the "Wolf Trap." Seen in Fig. 14.6, it contains an inlet tube into which solid samples are drawn by vacuum action, culture tubes, and bacterial sensors. If bacteria are present they will grow in the culture tube, producing increased turbidity of the culture medium and changing acidity, both measurable.

Ultraviolet spectrophotometry can be used to detect and identify organic compounds via their absorption spectrum. One device is built to investigate the spectrophotometry of proteins and peptides near the absorption maximum of the peptide bond (between 180 and 220 millimicrons).

---

[8] Intense absorption bands of very high wavelength of certain cyanine (cyanide derivative) dyes. They are caused by the transfer of energy between molecules. If a protein is brought into a solution of the dye, it is absorbed along the molecule's chainlike structure, causing the necessary molecular excitement for the J-band to appear.

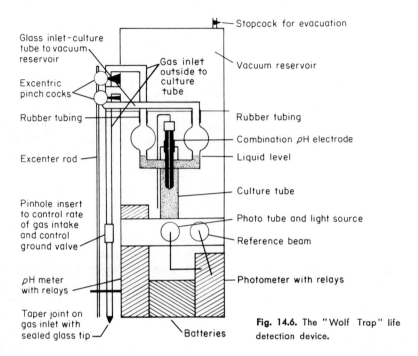

Fig. 14.6. The "Wolf Trap" life detection device.

MANNED OBSERVATIONS

So much for several means of remotely detecting life, or the probable presence of life, on other worlds in the Solar System. Sooner or later these techniques will be backed up and then replaced by manned investigations. Taking advantage of the most advanced information earlier made available by remote detectors, space biologists of the future will be well able to carry to the planets and moons the equipment best suited for continuing studies of extraterrestrial life.

In the coming years we shall probably learn that, with the exception of Mars, the entire Solar System is devoid of life.[9] And, on the neighboring red planet, we only expect very primitive plant life. If evidence of advanced life of any type is found on extraterrestrial worlds it almost certainly will not be indigenous to our Solar System. The chances are exceedingly remote that any nonhuman intelligent life has ever developed on another planet, but it is possible that extrasolar expeditions have entered our Solar System and left evidence of their visit that one day will be discovered by man.

---

[9] Conceivably, prebiotic or even extremely primitive biotic material may occur at the surfaces of the giant planets Jupiter and Saturn, but this can only remain conjectural until we know more about conditions below their opaque atmospheres.

15

Empirical Evidence of Extrasolar Intelligence

We can start this chapter with a simple, declarative statement: there is no empirical evidence of extrasolar intelligence. Because there is none, and because we do not expect to find any for some time to come, this chapter will be a disappointment to those hoping for some concrete signs of civilizations beyond our Solar System.

The kind of evidence one might expect to find of extrasolar intelligent beings can be broken down as follows:

(1) Evidence based on artifacts. Perhaps something was left on Earth by visiting beings (biological or mechanical) or by their automatic vessels, for example a monument, a beacon of some sort, the remains of an unmanned probe or a manned spaceship, the remains of a base, camp, or tools, or even a being itself (if organic, in fossilized form).

(2) Evidence based on the catastrophic results of an unsuccessful landing, either attempted or accidental.

(3) Evidence based on intangibles. An example would be an element of some terrestrial language whose origin is so utterly alien as to lead to the suspicion that early man, possibly living in a prehistoric community, was approached by

"... although I cannot prove it logically, I am convinced that there is absolute truth. If there isn't an absolute truth, there cannot be a relative truth."

Albert Einstein

"From these our interviews, in which I steal
From all I may be, or have been before,
 To mingle with the Universe, and feel
What I can n'er express, yet cannot all conceal."

George Gordon Byron, Lord Byron

extrasolar beings who maintained some degree of contact for an appreciable period of time. Or, it might be traced to folklores, myths, religious beliefs, superstitions, or other similar human social phenomena whose origin cannot be traced or whose characteristics defy rational explanation.

(4) Evidence based on alleged extraterrestrial flying objects. These are variously called unidentified flying objects (UFO), unidentified aerial phenomena (UAP), or flying saucers. Such evidence could, however, be extended to the visual detection, through telescopes, of unexplained objects flying in the Solar System, for example, orbiting the Earth or revolving around the Sun not far from the Earth's heliocentric orbit.

(5) Evidence based on the detection of radio or other signals. This is the most probable way extrasolar intelligence eventually will be detected, but we are now interested in determining if there is, or has been, evidence of such signals, or something that can be reasonably interpreted as being intelligent signals, coming from nonterrestrial beings.

(6) Evidence based on discoveries that may be made as the unmanned and manned exploration of the Solar System gets under way. This would include finding, say on Mars, definite or probable objects left by beings from another solar system who had visited our system and then departed. Such objects conceivably could have been left by a race that at one time inhabited Mars; therefore, evidence found of former habitation would have to be carefully analyzed to see if it was (a) indigenous or (b) extrasolar.

(7) Evidence based on inference; for example, explanations of stellar phenomena as being artificially induced.

There may be other ways of determining the existence of extrasolar intelligent beings, biological or mechanical, but these seven seem to be the major ones. In future centuries it may, of course, be possible to dispatch probes to the stars and perhaps, in the unforeseeable future, manned spaceships. If they should find intelligent communities it would be obvious, firsthand empirical evidence, but such an event is not considered in this chapter. Nor do we speculate here on the possibility that our Solar System may be visited at some future time by extrasolar vehicles.

Evidence Based on Artifacts

Exhaustive reading of all manner of literature, fictional and nonfictional, does not give rise to a single scientifically acceptable reason to believe that the planet Earth has ever been visited by extrasolar beings. If such beings had once landed upon this planet, they might have left some evidence, particularly if they had remained for an appreciable period of time. The evidence could be almost anything, a monument, or marker, perhaps a beacon of some sort, or even the remains of a base, encampment, or outpost, or artifacts, such as tools, associated with them. If the alien spaceship

Fig. 15.1. Someday evidence may be found, in the geologic record, of artifacts left by an extrasolar expedition.

(or, if unmanned, probe) crashed, or became so damaged that it could not take off, then it would have remained on Earth, possibly with its crew (assuming a second vessel was not available for rescue). Figure 15.1 depicts the remains of a hypothetical extrasolar ship within the geologic record.

Extrasolar beings could have landed on the Earth any time during its multibillion-year history except during its formative and consolidation phases. If artifacts, and possibly even the beings themselves, were left on Earth the possibility exists that they might have become incorporated into the geologic record. To date, nothing has been found in sedimentary rocks that, with one or two exceptions, is remotely suggestive of nonterrestrial intelligent life. Nor is there anything in the evolutionary history of fauna that would suggest a profound discontinuity. For example, an extrasolar expedition might have visited Earth during the early development of *Australopithicus*, *Homo habilis*, or *Homo erectus* and taught small colonies techniques they would not have discovered themselves for tens or hundreds of thousands of years. Again, there is no evidence that this ever happened, although certain features of the progression of man might be so interpreted by an agile and well-read science fiction writer.

From this discussion we conclude that the Earth was never visited by extrasolar beings and that life here has never been purposely tampered with. We rather must simply reemphasize that there is not a shred of evidence based on artifacts that our planet has been the target of an extrasolar landing expedition.

Evidence Based on the Catastrophic Results of a Landing Attempt

Extrasolar beings could have attempted a landing on Earth and failed, crashing to the surface and destroying themselves and their ship. Or, their spaceship could have become disabled as it was making a parabolic

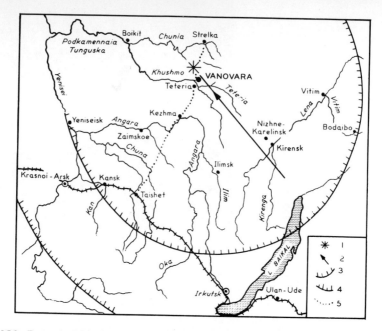

Fig. 15.2. Tungus region of central Siberia. Key chart at lower right shows (1) point of impact north of Vanovara, (2) probable direction of flight, (3) semicircle indicating limit of visual sightings of impacting object, (4) limit of noise produced, and (5) trail from rail station at Taishet to impact area. Diagram from USSR Academy of Sciences, courtesy *Sky and Telescope*.

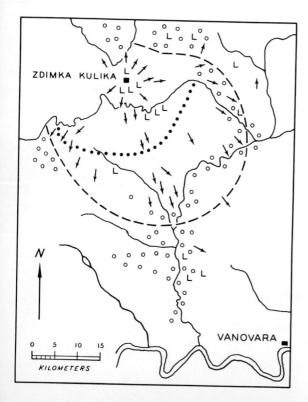

Fig. 15.3. E. L. Krinnov's sketch map. Fires occurred as far as the line of dots, and trees were felled out to dashed circle (arrows indicating direction of fall). The L symbols represent dead trees still standing and open circles mark existing old forest. Diagram from USSR Academy of Sciences, courtesy *Sky and Telescope*.

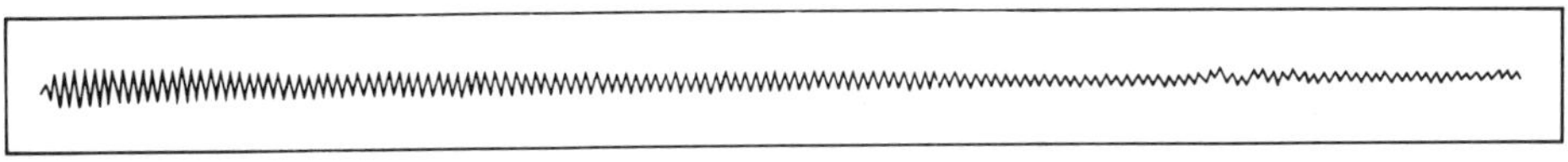

Empirical Evidence of Extrasolar Intelligence 293

Fig. 15.4. Seismograph recording of impact made at Irkutsk Magnetic and Meteorological Observatory, 550 miles south of Vanovara. Two minutes after explosion, waves began (left); then, 45 minutes later, the air wave arrived producing the three oscillations superimposed on record (right). Diagram from USSR Academy of Sciences, Courtesy *Sky and Telescope*.

approach, with intention not of landing but of simply orbiting, and subsequently destroyed itself. That such an event did happen in historical times, in fact in our own century, is believed by a few Soviet investigators. It all has to do with the so-called "Tungus Mystery" (Figs. 15.2—15.5).

Fig. 15.5. Three views of Tungus area (a) southern swamp, (b) felled trees 5 miles from center of impact, and (c) area of dead but still standing trees. Photos made by F. L. Krinov in 1929. Courtesy American Museum of Natural History and USSR Academy of Sciences.

THE TUNGUS MYSTERY

The following paragraphs are taken from a review of a chapter of a book by Parry prepared by Ordway.[1] This subject is treated at some length because, whatever the explanation of the phenomenon, something did happen and the results of what happened can still be seen today. The interpretation of the phenomenon has long been the subject of lively debate in the Soviet Union.

"Without doubt the most fascinating chapter of the book has to do with the so-called 'Tungus Mystery,' the unknown circumstances surrounding a tremendous explosion that occurred in the wastes of Siberia on the morning of 30 June 1908. According to Parry, Soviet scientists and science enthusiasts have, since 1946, continuously been asking the question: 'Was the giant messenger a meteorite from space or perhaps rather a spaceship from some other planet?' It has been proposed that during a landing attempt the spaceship exploded because of certain tragic circumstances that the crew was unable to foresee.

"Speculation, which has grown to become controversy, arose from the fact that while a huge area of devastation was discovered in the wilds of Siberia (and has been the target of many expeditions from Tsarist times), no evidence has been unearthed to prove it was caused by an impacting meteorite. However heated the arguments may be in the pages of Soviet journals, there are three points regarding this mystery on which everyone seems to agree: (1) the Tungus explosion is a mystery, awaiting a complete scientific explanation, (2) its force has no parallel in the known annals of our planet, (3) it is lamentable that the Tsarist Government of Nicholas II failed to explore the site of the explosion immediately after it occurred. It must be added that even the Soviet Government (until very recently) has not thrown significant research talent into the solution of the enigma. Only on the 50th anniversary of the explosion, in June 1958, was a well organized team of geologists, chemists, geochemists, astronomers, physicists, and other specialists sent by the Academy of Sciences into the Tungus region. In the summer of 1959, a second expedition explored the area, followed by nonacademy expeditions in the summer of 1960.

"According to top Soviet scientists, if the Tungus devastation had been caused by a meteorite, its size must have been tremendous, at least several thousands of tons and probably (in the view of Academician Vasily G. Fesenkov) many millions of tons. The force of the explosion was probably

---

[1] Parry, Albert, *Russia's Rockets and Missiles*. New York: Doubleday and Company, 1960. The review is by Frederick I. Ordway, III, and appeared in *Space Intelligence Notes*, 1, No. 2 (1 December 1960), 12–16, published by the Space Systems Information Branch, George C. Marshall Space Flight Center, National Aeronautics and Space Administration, Huntsville, Alabama.

equal to about one million tons of man made explosives. F. Zigel wrote that it was impossible to argue 'the fact that the explosion was equal to that of several tens of atomic bombs.'

"The impact of the explosion was so great that it produced air pressure waves throughout Asia and Europe, which were registered as far west as London. The director of the Kew Observatory noted that in addition there was an unusual airglow over England at midnight following the Siberian explosion. Professor Kirill P. Florensky, a geochemist, records that many seismological observatories in the world registered subterranean jolts emanating from central Siberia, and also states that during the three nights following the Tungus explosion there was hardly any darkness in Europe, North Africa, Central Asia, and Western Siberia (great luminous clouds were seen thousands of miles from the Tungus area; there was also a strong murkiness in the upper strata of the Earth's atmosphere).

"On the night of the great explosion many people 'saw the fiery body as it sped at a fantastic rate across the skies, leaving a luminescent trail behind it, from south toward north until it descended over the Taiga near Vanavara, a tiny trading post on the Podkamennaya Tunguska shore. A number of terrified on-lookers saw its shape as elongated: like a huge chimney, some said; resembling an enormous log, thought others.' The explosion came with an energy estimated conservatively at about 10^{21} ergs, a force equal at least to that produced by the greatest natural catastrophes known to man, such as volcanic eruptions, earthquakes and severe hurricanes.

"For distances of 20 to 30 miles from the site of the blast houses were shaken, glass broken, animals and people knocked down. 'At the farthest known point . . . at a place 50 miles south of the blast, a farmer named Semyonov was sitting that morning on the high porch of his house when he saw a great light of fire in the North. A hot wave came over him, and he feared his clothes would catch fire. The light was gone, and oppressive darkness came momentarily, then a tremendous explosion sounded, and this threw him from the porch to the ground. He was knocked down unconscious, and when he awoke he saw ruins where his house had stood a while ago.' A huge forest 700 to 800 square miles in area was totally devastated, and the sound of the explosion has been reported as deafening 400 miles away.

"Despite all this, 'there was no definite report of a crater, either at the time or any point since 1908.' Furthermore, no metal or stone pieces of a possible meteorite have ever been found since the explosion, either by local Siberians or by visiting scientists.

"The first official Soviet expedition to the Tungus area occurred in 1927, although its leader, Leonid Kulik had visited there in 1921 as a side trip during an Academy of Sciences meteoritic expedition to other parts of Siberia. At that time he had distributed 2,500 copies of a questionaire to witnesses of the great explosion. During the 1927 expedition, Kulik (a foremost mineralogist and meteoritic expert) discovered that the rotted trees killed by the explosion were established in a curious radial fanlike pattern, tops pointing outward, the roots all toward a common center. Inexplicably, at the so-called common center many surviving trees stood upright; no crater whatsoever could be discovered. In the swamp where the explosion occurred a number of small pits were found, the largest being filled with water. Kulik and his men drained some of the water-filled holes and drilled deeply into them, but found nothing but peat and more water; no vestige of meteoritic remains could be discovered.

"Kulik's group made four expeditions into the area, and even had to help build a 62-mile access road from a trading post to facilitate explorations. By 1938 aerial photographs had been taken of the central part of the stricken forest, and some mapping occurred. This ended all efforts prior to the war.

"During the postwar years expeditions were dispatched which resulted in the discovery of tiny globules of metallic iron as well as samples some five millimeters in size containing 7 per cent nickel and 0.7 per cent cobalt, plus vestiges of copper and germanium (along with the iron). This, to many, seemed to prove a meteoritic origin of the crater; however, in 1958, Alexander H. Kazantsev presented a new set of arguments in favor of the spaceship origin.

"One: No remnants of a meteorite, small or large, were ever found in the Tunguska area. The analysis of the soil samples done in 1957, far from proving any meteorite origin, completely vindicates the advocates of the spaceship theory. Iron containing nickel and cobalt? Why, that was a splinter off the nickel-and-cobalt steel shell of the spaceship! Traces of copper and germanium? 'Quite naturally,' agrees Kazantsev. 'Remember that the ship must have had electrical and technical instruments, also copper wires, and surely means of communication—semiconductors containing germanium.'

"Two: No crater was ever found in the Tunguska area. And yet, according to estimates by astronomers, the meteorite (had there been one) must have had a mass of a million tons and travelled at the speed of $18\frac{1}{2}$ to $37\frac{1}{4}$ miles per second. It should have created our planet's largest crater —had it been a meteorite.

"Three: The advocates of the meteorite theory never satisfactorily explained the fact that, at the very center of the Tunguska explosion, some taiga forest remained intact, but with broken branches and with burns at points of such breaks.

"Four: The strange rays piercing thick clouds observed on the several nights following the Tunguska explosion were not typical of meteorite falls.

"Five: Nor was the picture of the explosion, as described by closest eye-witnesses on that June morning of 1908, usual for a meteorite: the fiery pillar with black smoke, going up into the cloudless sky, there to become a black mushroom.

"Six: The mushroom cloud, from what we know since Hiroshima, surely should make us think of an atomic explosion. And that is what the Tunguska blast of 1908 may in fact have been: an explosion of an interplanetary ship motivated by atomic power.'

"In bolstering the argument for an atomic spaceship catastrophe, Kazantsev accounts for the absence of the crater on the basis that the spaceship exploded in the air. Here is how he describes the situation:

" 'The explosion wave rushed downward, and the trees directly below the point of the explosion remained standing, having lost only their crowns and branches. The wave burned the points of those breaks on the trees, and hit the permafrost, splitting it. Underground waters, responding to the tremendous pressure of the blow, gushed up as those fountains seen by natives after the explosion. But where the explosion wave struck at an angle, trees were felled in a fanlike pattern.

" 'At the moment of the explosion, temperature rose to tens of millions of degrees. Elements, even those not involved in the explosion directly, were vaporized; and, in part, carried into the upper strata of the atmosphere where, continuing their radioactive disintegration, they caused that luminescent air. In part, those elements fell to the ground as precipitation, with radioactive effects.'

"The Soviet aircraft designer A. Yu Manotskov brought in the science of aerodynamics and trajectories to support the spaceship theory. According to him, the body was approaching our planet at a speed not typical for meteorites. He feels it significant that as it neared the Earth 'it braked its speed to 0.7 kilometers per second . . .' In order for a meteorite to have done the damage recorded, approaching at that speed, it would have had to weigh not a million tons but several billions of tons. 'The size of such a meteorite . . . would have been more than one kilometer in diameter, and such a meteorite would have certainly covered the whole sky . . . and certainly, such a tremendous body could not have disappeared without a trace.'

"Some of the arguments for the spaceship theory border on the fantastic. One hypothesis has been promoted that an interplanetary rocket coming to this planet may have consisted of anti-matter, and when it reached our atmosphere reacted explosively with it.

"Whatever the truth may be, one can summarize this present state of affairs by the statement of I. T. Zotkin, astronomer and member of the Florensky 1958 expedition: 'all members of the expedition unanimously concluded that (the area of the 1908 explosion) is not a meteorite crater and holds not a single trace of a meteorite crash against the ground.'

"In the last few years the nuclear explosion theory (whether it was caused by a spaceship or not) has gained added credence, since for the first time the soil in the explosion area has been carefully tested for radioactivity. More and more Soviet scientists are coming to the conclusion that the explosion was atomic in nature but not necessarily man-made. Some go so far as to say that a collision with an asteroid occurred.

"Even more recently, a theory has emerged that a crater may one day be found. Kirill P. Stanyukovich, the well known Soviet astrophysicist, has shown that the center of the ballistic wave explosion and the place where the remnants of the meteorite finally struck the ground may not necessarily coincide. 'Such remnants having fallen at a much more diminished speed than the meteorite had traveled before its ballistic wave explosion, and a small crater possibly having been formed by this fall, this crater may well lie a long way from the place where that Tunguska forest was demolished, and where most of the previous searches had been made.' Despite this, expeditions have searched out to 500 miles from the center of the explosion and found not a trace of a crater or any meteorite ores. Modern scientists are even denying that the tiny amounts of meteor-type ores found (iron and the nickel-cobalt-germanium globules) were not meteoritic but 'secondary dirt-formation.'

"The last expedition reported by Parry was headed by Dr. Gennady Plekhanov on the staff of the Betatron Laboratory of the Tomsk Medical Institute. He and his researchers covered more territory than any explorers before them, taking more than 300 samples of soil scores of miles from the epicenter of the 1908 explosion. They brought out nearly 100 plants to be tested for radioactivity and nearly 80 samples of ashes for similar analysis. Dr. Plekhanov showed that 'in the center of the catastrophe radioactivity is $1\frac{1}{2}$ to 2 times higher than along the 30 or 40 kilometers away from the center'. He could offer no explanation for the radioactivity."

We now leave Parry's reports on the mystery and examine later material. The next investigator to consider the matter in considerable detail was Volotov,[2] who conducted his research on the Tungus phenomenon during the summer of 1959. He records that the region of complete destruction extends for a distance of 20 to 22 km from the epicenter of the explosion, which was equivalent to 10 million tons of TNT. Based on the presence of standing trees for a distance of 5 km from the epicenter Volotov indicates that the explosion of what he assumes to have been a meteorite occurred in the air at a height of not less than 5 km. He makes an evaluation of the velocity of the meteorite at the time of the explosion and concludes that it was moving at between

---

[2] Volotov, A. V., "New Data on the Tunguska Catastrophes of 1908," *Doklady Academi i Nauk SSSR* **126**, No. 1 (1961), 84–87. Translation by Office of Technical Services, Dept. of Commerce, Washington, D.C.

3 and 4 km/sec. He does not believe the explosion was due to the conversion of kinetic energy to thermal energy, as in reentry heating, but rather to the conversion of internal energy. He determined its light energy was on the order of 1.1 to 2.8 × 10^{23} ergs, leading him to state that the ratio of light energy to full energy is of the same order as in a nuclear explosion.

Two books soon appeared on the subject, one entitled *In the Wake of the Fiery Stone* by I. Evgen'ev and L. Kuznetsova and the other *Visitor from the Cosmos* by Kazantsev (whose arguments we have just covered); they were reviewed, respectively, by Professor Stanyukovich and V. V. Fedynskii, and Yu G. Perel.[3] The former book is praised as providing an excellent account, in popular form, "of the scientific data concerning the Tungus meteorite," which is precisely what the reviewers assume it to have been. The second book is dealt with very harshly, not only because it presents the opinion that the Tungus explosion was caused by a crashing Martian spaceship but because there is a "consistent and conscious deception of the reader, in the pursuit of one definite goal: to show that he alone, Kazantsev, has discovered the true nature of the complex phenomena, contrary to all the 'conjectures' of the representatives of official science."

The reviewer now summarizes the latest Soviet scientific opinion regarding Tungus, namely that it is almost definitely of natural orgin. He admits, however, that it is "impossible to say (and it is not maintained by anyone) that all the questions relating to the fall of the meteorite have been comprehensively answered. However, investigations in this direction have not been discontinued by any means, but are continuing . . . the investigations which have been conducted refute completely both the conjectures and the direct fabrications adduced by A. Kazantsev in the defense of his 'hypothesis.' " The Russian Tungus Mystery, like the US Flying Saucer Mystery, refuses to die out largely because there remain a few unknowns that scientists cannot completely explain, and partly because scientists themselves are at odds over interpretations. Two nonspaceship, nonmeteorite interpretations close our discussion of the subject.

Florensky, who led three well-organized Soviet expeditions into the site of the destruction in 1958, 1961, and 1962, emphatically rejects[4] the atomic spaceship explosion theory. Regarding the reported radioactivity in deep rings of fallen trees, he states that his expeditions examined carefully the question and "established that this radioactivity is

---

[3] *Soviet Astronomical Journal*, 36, No. 2 (1961), published by the USSR Academy of Sciences.

[4] Florensky, Kirill P., "Did a Comet Collide with the Earth in 1908?" *Sky and Telescope* 26, No. 5 (November 1963), 268–69; also *Komsomolskaya Pravda* in a 1961 interview.

fallout from modern atomic bombs, which has been absorbed into the wood. There is nothing specific in the nature of the radioactivity in the area of the Tungus fall."

If it was not due to a spaceship, he also doubts it was caused by a conventional meteorite, suggesting rather that it was a comet. He cites many facts leading him to believe a comet did indeed collide with the Earth back in 1908: "the unusually loose structure, which led to breakup in the atmosphere; the dust tail, pointing away from the sun, that caused unusual sunsets over nearly all Europe; the nature of the orbit; and lack of big fragments." The comet, he believes, was moving at about 5 km/sec when the frozen gases in its head were melted by friction as it plunged into the Earth's atmosphere, setting off a tremendous blast above the surface.

An alternative hypothesis supposes that a meteorite composed of antimatter collided with, and disintegrated in, the Earth's atmosphere, producing—in terms of the relatively small assumed size of the meteorite —an enormous quantity of energy. The most recent exponents of the theory, Cowan, Atluri, and Libby,[5] cautiously suggest that the antimatter hypothesis is plausible to explain the known facts and that it warrants thorough study and investigation. One of the clues that led to the hypothesis is the observed increase of carbon 14 in tree rings in many parts of the world corresponding to the time of the explosion in Siberia. It is theorized that, should a collision of antimatter with the atmosphere occur, a considerable increase of carbon 14 in the atmosphere would be registered. Since trees absorb carbon dioxide from the air they would also absorb its contained carbon 14—and this would show up in the annual rings.

Evidence Based on Intangibles

The fictional literature, particularly fantasy and the weird tale, are replete with examples of strange happenings, unusual appearances, and inexplicable occurrences, many of which are supposed to have been caused by beings from other worlds, or at least beings living on our own world but in other dimensions or times concerning which it may occasionally be possible for a human to have an awareness. Such awareness is generally associated with some dire event or unfortunate circumstance. The events described by fantasy authors, however sincere some may be, are considered intangible, and certainly are unprovable, evidence of an alien life existing beyond the comprehension of man.

Legends, myths, primitive religions, and folklore to varying extents

---

[5] Cowan, Clyde, C. R. Atluri, and W. F. Libby, "Possible Antimatter Content of the Tunguska Meteor of 1908," *Nature*, **206**, No. 4987 (29 May 1965), 861.

are, or border on, fantasy and to varying extents contain elements that could be partially explained by the appearance, during the history of mankind, of superior beings from other worlds. Virtually all religions are based on mysterious gods and messiahs who come to Earth, or make themselves known to man, under supernatural circumstances. Should any of them have been extrasolar representatives, a proposition almost too fantastic to be considered seriously, then some of the abilities attributed to them might be explained. Also, the careful study of legends and folklore conceivably could turn up evidence for the arrival on Earth of extrasolar beings during prehistoric or early historic times.

Another intangible that must be mentioned is language. If extrasolar beings visited the Earth during the time when *Homo* was struggling upward, they might have introduced words into whatever language or protolanguage our forebears employed. Philologers might be able to detect, in ancient or modern languages, sounds or constructions that do not appear to have sprung naturally from the mind and vocal organs of man. Of course, any such sounds or constructions that early man may have learned from an extrasolar expedition presumably would have become so modified as man accomodated the "words" to his physical abilities of pronunciation that they would almost certainly be unrecognized as alien today.

Not quite "intangible" are the alleged influences of extraterrestrial beings on tangible objects of Earth (Fig. 15.6), for example, on designs of money, on sculptures, on paintings, on architecture, and so forth. Doubtless, many of us who view what passes for art in this modern age feel prone to attribute it to an alien influence. But we would be hard put to prove it. There are those, however, who feel they have proof enough of the arrival on Earth hundreds of years ago of some sort of creatures from space.

Fig. 15.6. Hypothetical evidence of extrasolar influence on primitive terrestrial art.

Two examples, cited and refuted by Voronin,[6] will serve our purposes.

Before giving his examples Voronin writes that if "strictly proven traces of visits to the earth by intelligent beings from other worlds, supposedly in the distant past, are found, this will confirm the real existence of other extraterrestrial civilizations." However, he does not believe that any such traces have been found. The first claim he refutes has to do with early Etruscan gems and other cut stones which are alleged by some to have "images of 'creatures in cosmic helmets,' and a 'cosmic rocket.'" He uses several pages and seven figures to discharge the assertion as nonsense. "The recognition of 'cosmic ships' and 'astronauts' in the images of the gems could only be made by those who did not know the history of the development of antique gems, did not bother to learn it and heedlessly sought cheap sensation and, consequently, lead people unfamiliar with these questions into delusion." So much for the gems.

The second example takes us to Scotland. A writer named Ketman contributed an article to a French magazine which was subsequently published in 1960.[7] In it he "expressed the thought that certain ancient round castle towers in Scotland and Ireland are enigmatic" in that the walls of the towers seem to have been fused at a high temperature. ("les remparts de granit ont été vitrifiés, sur les bords, par de chaleurs extraordinaires"). The question arises, perhaps inevitably in some minds, that this is evidence of an atomic explosion. Voronin goes on to destroy this absurd conclusion, pointing out that the Scotch and Irish were building good homes and forts as far back as the 5th century B.C. Stone embankments were for a long time subjected to a "special firing on the surface" we call vitrifaction. After the stones were laid (without binder), the structure, in this case the tower, was "fired by burning peat at its walls," providing the desired binding which, incidentally, was waterproof.

Sagan also has examined the possibility of extrasolar contact with Earth during historical times.[8] He provides several interesting accounts of legends and myths, including incidents from the *Bible,* that are somewhat suggestive of such contacts. However, he holds little hope that research along these lines will prove fruitful, writing that a "completely convincing demonstration of past contact with an extraterrestrial civilization may never be provided on textual and iconographic grounds alone."

---

[6] Voronin, M. A., "O poiskakh sledov stivilizatsii v inykh mirakh" ("In quest of signs of civilization in other worlds"), *Priroda,* No. 11 (1962), 78–83. (Translated by the National Aeronautics and Space Administration, Rept. TT F-8590, Washington D.C., March 1964).

[7] Ketman, Georges, "Les cartes boulversantes de Piri Reis," (The Staggering Accounts of Piri Reis") *Science et Vie,* 9, No. 516 (1960); republished in the USSR in *Yuny tekhnik,* No. 12 (1960), p. 45.

[8] Sagan, Carl, "Direct Contact Among Galactic Civilizations by Relativistic Interstellar Spaceflight," *Planetary and Space Science,* 2 (1963), 485.

Other possible evidence of an extrasolar visit during geologic history is cited by Cade.[9] A mysterious object known as "Dr. Gurlt's Cube" was found in 1885 in a block of coal of Tertiary age, described as "almost a cube, two opposite faces being rounded, so that the dimensions were 67 mm × 47 mm (the last measurement being between the rounded faces). A deep incision ran all round the cube near its center. The weight of the object was 785 gm (26 oz), and in composition it resembled a hard nickel-carbon steel." The sulphur content was reported to have been too low to be associated with naturally occurring pyrites. Among several possible explanations is that it is of artificial origin. The cube reposes in the Salzburg Museum in Austria.

Evidence Based on Sightings of Alleged Extraterrestrial Objects

Ever since man reached the intellectual stage where he began to take notice of the mysteries of nature he gazed in wonder at the many inexplicable objects in the skies. For thousands of years he watched with amazement eclipses, meteors, comets and an occasional nova. Cloud formations, reflected light, auroral displays, lightning, dust storms, and other terrestrial atmospheric phenomena also stirred his imagination and often incited a deep seated fear, or at least awe, of the unknown. Since the beginnings of literature all manner of interpretations have been given, many dealing with gods, demons, angels, and mysterious messengers from other worlds.

SCIENCE FICTION AND FANTASY LITERATURE

In more modern times science fiction and fantasy writers have delighted in explaining unknown sights in the skies and heavens as being caused by visitors from other worlds, be they located in our Solar System, in other solar systems in the Milky Way galaxy, or in other galaxies. One of the most fantastic and haunting of these tales is Lovecraft's "The Colour Out of Space,"[10] involving a mysterious, meteorite-like object that landed on Earth. Despite painstaking examinations by scientists, no explanation could be given of its nature and origin. It was invulnerable to all reagents and solvents. Its "colour . . . was almost impossible to describe; and it was

---

[9] Cade, C. M., "Communicating with Life in Space," *Discovery*, 24, No. 5 (May 1963), 36–41.

[10] Lovecraft, H. P., "The Colour Out of Space," in his anthology *The Outsider and Others*. Sauk City, Wisconsin: Arkham House, 1939. (The short stories in this volume were collected by August Derleth and Donald Wandrei.)

only by analogy that they (the investigators) called it colour at all. Its texture was glossy, and upon tapping it appeared to promise both brittleness and hollowness." There was no way to adequately describe or analyze it, causing the conclusion to be reached that it "was nothing of this earth, but a piece of the great outside; and as such dowered with outside properties and obedient to outside laws."

In keeping with a good fantasy story, the inanimate visitor from beyond had horrible effects on the vegetation, animals and humans living in the neighborhood where it struck, creating madness in the minds of the latter. The mystery is never solved. "Something terrible came to the hills and valleys on that meteor and something terrible—though I know not in what proportion—still remains." But of this we can be sure, it was "no fruit of such worlds and suns as shine on the telescopes and photographic plates of our observatories. This was no breath from skies whose motions and dimensions our astronomers measure or deem too vast to measure. It was just a colour out of space—a frightful messenger from unformed realms of infinity beyond all Nature as we know it; from realms whose mere existence stuns the brain and numbs us with the black extra-cosmic gulfs it throws open before our frenzied eyes."

A whole book could be written on purported evidence of extraterrestial visitors or objects of extraterrestrial origin either artificially made or endowed with fearsome powers such as the "colour out of space." With the advent of rocket and prespace age thinking, the world went through probably the most intense period in history of sightings of vehicles of supposed extraterrestrial origin. Since the public had become fairly well convinced that intelligent life does not exist on other planets in the Solar System, the vehicles that were supposedly observed by thousands of sane persons were assumed by many to come from somewhere beyond.

UNIDENTIFIED AERIAL PHENOMENA

Unknown objects seen soaring across our skies are generally called *unidentified aerial phenomena, unidentified flying objects,* or, more popularly, *flying saucers* (Fig. 15.7). According to Menzel and Boyd[11] the "overture to the *Flying Saucers* opera took place in the summer of 1947, presenting the main themes that were to develop with fantastic variations during

---

[11] Menzel, Donald H. and Lyle G. Boyd, *The World of Flying Saucers: A Scientific Examination of a Major Myth of the Space Age.* Garden City, New York: Doubleday and Company, 1963. This book is by the Director of the Harvard College Observatory and by a professional editor and writer of scientific literature. The senior author has been closely associated with the flying saucer investigations and has had access to Air Force files, including those of the Aerial Phenomena Group of the Aerospace Technical Intelligence Center.

Fig. 15.7. Pair of alleged flying saucers based on descriptions provided by the Air Force Aerial Phenomena Group.

the fifteen-year-long drama that followed: mysterious apparitions in the sky, alleged interplanetary visitors, government investigators, growing public excitement, civilians who zealously encouraged the hysteria, and, as a climax, an elaborate hoax that produced material 'evidence' to prove the existence of spaceships."

The Menzel-Boyd work is without question the most authoritative to appear on the subject and should be consulted for details of dozens of case histories, including subsequent rational interpretations of the "flying saucer" sightings. Early in the book the authors emphasize that both Air Force and scientific investigators "have been able to account for almost every reported 'spaceship' as the result of failure to identify some natural phenomenon. Some were the product of delusion or deliberate hoaxes. A few remain technically 'unknown' because, although the probable expanation is obvious, too few facts are available to permit a positive identification. No such report suggests the possibility of interplanetary craft cruising in our skies."

OFFICIAL UNITED STATES AIR FORCE REPORT

We requested from the Air Force an official statement on what are now called *unidentified aerial phenomena* (UAP). This is reproduced in the following paragraphs.[12]

"The Air Force investigation and analysis of unidentified aerial phenomena sightings over the United States is known as Project Blue Book. This program is related directly to the Air Force's responsibility for its role in the defense of the United States. The air defense role is divided into

---

[12] Courtesy of Project Blue Book Information Office, Headquarters U.S. Air Force (SAF-OIPB), The Pentagon, Washington, D.C.

four phases: detection, identification, interception, and destruction (if national security is threatened). The UAP investigation effort falls under the second of these four phases.

"The Air Force interest in UAP is three-fold: first, as a possible threat to the security of the United States and its forces; second, to determine the technical or scientific characteristics of any such unidentified aerial phenomena; and third, to explain or identify all sightings.

"The Air Force is often asked just what is an unidentified aerial phenomenon? It is any aerial phenomenon which by performance, aerodynamic characteristics, or unusual features, does not conform to known aircraft or missiles, balloons, birds, kites, searchlights, aircraft navigation and anticollision beacons, astronomical bodies or phenomena, jet engine exhaust, condensation trails or known meteorological phenomena, and remains unidentified long enough to be reported.

"The initial investigation of an unidentified aerial phenomenon sighting is the responsibility of the nearest Air Force unit in the area of the sighting. If this organization cannot identify the object, the sighting and results of the preliminary investigation are reported to the Office of Aerial Phenomena for further investigation and evaluation. With this move, the responsibility for release of information regarding the sighting is transferred to U.S. Air Force Headquarters. The Office of Aerial Phenomena is part of the Air Force's research and development arm known as the Air Force Systems Command. Each case is objectively and scientifically analyzed without regard to the outcome. If necessary, as in many cases, part or all of the scientific disciplines available to the Air Force can be invoked to assist the office in arriving at a conclusion. This is also true of the many facilities available in the Air Force, such as materials laboratories where many of the parts of missiles and space vehicles are tested. Other government facilities outside the Air Force are also utilized from time to time.

"The personnel involved in the investigative and evaluation efforts of this program are a selected group of highly qualified scientists. They approach each case with no preconceived ideas and proceed to analyze each facet of each case on a scientific basis.

"The general conclusions are: To date, no unidentified aerial phenomenon has given any indication of threat to the National Security; there has been no evidence submitted to or discovered by the Air Force that unidentified sightings represented technological developments or principles beyond the range of our present day scientific knowledge; and finally, there has been nothing in the way of evidence or other data to indicate that these unidentified sightings were extraterrestrial vehicles under intelligent control.

"Several thousand balloons are released in the United States every day. There are several types of balloons: weather balloons (ozone-sondes and radiosondes), private man-carrying balloons used as sporting devices and the large research balloons which have diameters of 300 feet. The larger balloons when airborne at night carry running lights which contribute to an unusual appearance when observed. Reflection of the sun on balloons at dawn and sunset produce rather startling sights, particularly when the observer's position is no longer in direct sunlight (in the earth's shadow) when the balloon, because of its altitude, is exposed to the sun. Large balloons can move at speeds of over 200 mph when located within high altitude jet windstreams. Sometimes these balloons appear to be flattened on top or because of the sun's rays reflecting through the plastic material of the balloon from panel to panel, appear to be saucer shaped and to have lights mounted on the bag itself. Experienced pilots, although they may have observed high altitude balloons before, are sometimes startled by their strange appearance under varying atmospheric conditions.

"Ground and air observers are susceptible to autokinesis, a physiological phenomenon sometimes causing psychological reactions. For example, a pilot who stares at a fixed light in an otherwise dark environment will soon experience the illusion that the light has begun to move erratically. If he stares at the light long enough, he may become almost hypnotized by it, so that it takes up practically all his attention. Size-distance illusion results from staring at a point of light which approaches and recedes from the observer. In the absence of additional distance clues, accurate depth perception is extremely difficult. Instead of seeing the light advancing and receding, the observer has the illusion that it is expanding and contracting at a fixed distance. The cause for autokinetic phenomenon is not fully understood. It can be prevented or dispelled by continually shifting fixation from point to point. Ground observers seeing a point of light in the night sky can also experience this phenomenon.

"Airliners, conventional and jet, have been using an anti-collision beacon which, when viewed at slant range and under varying atmospheric conditions, has produced UAP reports. This light is a white strobe type light usually involving two or more lights operating alternately. The beacon is intense and flashes for only a fraction of a second. It has been seen as many as 50 miles away and at that distance only the light could be seen, thereby producing a rather startling sight.

"Many modern aircraft, particularly swept and delta wing types, are reported as UAPs under adverse weather and sighting conditions. When observed at high altitudes and usually at some distance, aircraft can have appearances ranging from a disc to rocket body shapes due to reflection of sunlight off the bright surfaces or when only jet exhausts are visible. Vapor or condensation trails will sometimes appear to glow fiery red or orange when reflecting sunlight. Afterburners (reheating of jet exhaust gases) are often reported as UAPs and can be seen for long distances.

"Astronomical sightings include bright stars, planets, comets, fireballs and bolide meteors and other celestial bodies. When observed through haze, light fog, moving clouds or other obscurations or unusual conditions, the planets Venus, Jupiter, Mars and Saturn have been reported as unidentified aerial phenomena. Stellar mirages are sometimes a source for an UAP report.

"There are several meteor showers occurring each year. These sometimes produce rather spectacular displays and are suspect when evaluating reports during these periods. Recently, widespread reports were received of flashing objects in the sky. Investigation revealed that these objects were associated with the earth's passage through meteor trails known as the Cygnids and the Quadrantids.

"Because of the number of man-made objects in orbit the separate category of Satellites was established. Keeping track of man-made objects in orbit about the earth is the job of the Space Detection and Tracking System (SPADATS). This sophisticated electronic system takes in complex space traffic data instantly from tracking stations within the far-flung air defense lines and supplies information on a rapid basis into the aerospace warning system. Inputs to this system are instantly fed from the Ballistic Missile Early Warning System (BMEWS) sites, airborne and other radar installations which maintain a constant surveillance of space activity. Additionally, other space surveillance activities include the use of ballistic tracking and other large telescopic cameras.

"The category labeled 'Other' includes reflections, searchlights, birds, kites, blimps, clouds (particularly lenticular formations), sun and moon dogs, spurious radar indications, hoaxes, firework displays, flares, ice crystals, and other meteorological phenomena.

"The 'Insufficient' category includes all sightings where essential details of information are lacking thereby prohibiting a valid conclusion. These will sometimes include lack of corroborating information from an additional witness or witnesses. For instance, a sighting in the city of New York reported by just one person with no other corroboration. Other missing details which are necessary in order to reach a conclusion include description or size, shape or color of an object, direction and elevation, exact time and location, etc.

"A sighting is considered 'Unidentified' when a report apparently contains all the pertinent data necessary to normally suggest at least one valid hypothesis on the cause or explanation of the sighting, but the description of the object and its motion cannot be correlated with any known object or phenomena.

"Totals in any one specific year, as they appear on the statistical sheets, will be at variance with previously published totals for that same year. This situation developed during the review of all cases in the specific year. It was discovered that as high as 30 separate reports were

Table 15.1
Sightings of Aerial Phenomena at first Described as Unidentified (for the year 1963)

Month	Explanation of the Phenomenon								Totals
	Astronomical[a]	Aircraft	Balloon	Insufficient	Other[b]	Satellite	Unidentified	Pending	
January	4	4	1	5	1	2	0	0	17
February	5	3	1	3	3	2	0	0	17
March	8	2	1	4	7	7	0	0	29
April	4	7	1	4	2	8	0	0	26
May	3	6	3	3	6	0	2	0	23
June	3	10	6	12	4	24	1	0	60
July	9	9	6	3	3	11	1	0	42
August	11	15	4	6	7	4	2	1	50
September	12	7	1	4	3	12	2	0	41
October	10	6	0	6	3	6	4	0	35
November	8	2	0	2	6	2	0	1	21
December	5	1	1	3	3	3	3	2	21
Totals	82	72	25	55	48	81	15	4	382

[a] see Table 15.2   [b] see Table 15.3

Table 15.2
Cases Explained by Astronomical Phenomena

Month	Explanation of the Phenomenon			Total
	Meteors	Stars & Planets	Other	
January	3	1	0	4
February	3	2	0	5
March	8	0	0	8
April	3	0	1[a]	4
May	3	0	0	3
June	1	1	1[a]	3
July	5	4	0	9
August	4	5	2[b,c]	11
September	7	4	1[d]	12
October	7	3	0	10
November	6	2	0	8
December	5	0	0	5
Totals	55	22	5	82

[a] Sun Dog.   [b] Moon.   [c] Parhelia.   [d] Aurora.

all concerned with the same sighting. These have now been consolidated and are carried as one sighting rather than the individual sightings as previously listed. There have been no sightings dropped from the records. It is consolidation only.

"The Air Force does not deny the possibility that some form of life exists on other planets in the universe; however, there has been no evidence submitted to or discovered by the Air Force that proves or tends to prove such a contention. The Air Force continues to extend an open invitation to any individual or group, which feels they possess evidence that would prove or tend to prove the existence of extraterrestrial vehicles operating within the earth's near space envelope, to submit their evidence for analysis . . ." Tables 15.1 to 15.3 were prepared from data supplied by the Project Blue Book Information Office.

From the past excerpt we can only conclude that the Earth is not being visited by detectable extrasolar objects. That there may be extrasolar probes or stations in the Solar System is something about which there is as yet no conclusive evidence.

Evidence Based on the Detection of Signals

If we could detect artificially-emitted radio or other signals from some nonhuman-created source beyond the Earth it would be evidence of the existence of intelligent beings. Or, if we could turn up indisputable evidence that such signals had been detected in the past we could come to the same conclusion. Unfortunately, the problem of visual and radio signals is akin to the problem of unidentified flying objects: both have been reported as emanating from nonterrestrial sources, but careful postinvestigation does not sustain the claims of many enthusiastic supporters.

For hundreds of years people have speculated how one might signal optically a hypothetical race on Mars and, in turn, how Martians might signal to us. With the advent of radio, interest was accelerated, but not so much in broadcasting to the Martians but rather in receiving signals from them . . . either from stations on the planet or from spaceships cruising the void. Sagan[13] and many other scientists have mused over whether, at our level of astronomical know-how, hypothetical Martian astronomers could detect life on Earth. He does not think any of our largest engineering works would be visible, but that the entire city of Los Angeles might be marginally visible on a clear night. And our largest nuclear explosions might be viewed fleetingly. But, if they had "radio reception equipment, and chose to scan Earth in narrow wave bands,

---

[13] Sagan, Carl, "The Quest for Life Beyond the Earth," *Harvard Alumni Bulletin*, **66**, No. 12 (4 April 1964), 508–13.

Table 15.3 Cases Explained by Miscellaneous Phenomena

Explanation of the Phenomenon

Month	Hoaxes and Hallucinations, Unreliable Reports, Psychological Causes	Missiles and Rockets	Reflections	Flares and Fireworks	Mirages and Inversions	Search and Ground Lights	Clouds and Contrails	Chaff	Birds	Physical Specimens	Radar Analysis	Photo Analysis	Satellite Decay	Other Causes	Totals
January	0	0	0	0	0	0	0	1	0	0	0	0	0	0	1
February	1	0	0	0	0	1	1	0	0	0	0	0	0	0	3
March	3	3	0	0	0	0	0	0	1	0	0	0	0	0	7
April	2	0	0	0	0	0	0	0	0	0	0	0	0	0	2
May	0	2	0	1	0	0	1	0	0	1[a]	1[d]	0	0	0	6
June	1	0	0	0	0	1	0	0	0	0	0	1[e]	1	0	4
July	0	2	0	0	0	0	0	0	0	0	0	1[f]	0	0	3
August	2	0	0	0	0	0	1	1	0	1[b]	0	0	1	2[g,h]	7
September	0	1	1	0	0	0	0	0	0	0	0	0	0	1[j]	3
October	1	1	0	0	0	0	0	0	0	1[c]	0	0	0	0	3
November	2	3	0	1	0	0	0	0	0	0	0	0	0	0	6
December	1	1	0	0	0	0	0	0	0	0	0	0	1	0	3
Totals	13	13	0	2	0	2	3	2	1	3	1	2	3	3	48

[a] Parachute. [b] Rock salt. [c] Rock. [d] ECM (Electronic Counter Measures). [e] Part of equipment in which photographer was riding. [f] Lens flare. [g] Grass fly eggs. [h] Objects attached to aircraft. [j] Hole in ground.

they would certainly be rewarded—if that is the word—by television transmission from Earth. There would be an intensity maximum when the North American continent faces Mars, and it would perhaps be possible to determine that this radio frequency emission was not entirely random noise."

Despite this theoretical exercise, few scientists look forward to finding intelligent life on Mars;[14] nor, for that matter, on any planet in the Solar System. Therefore, if alien artificial signals, presumably either electromagnetic or laser, are intercepted it would mean, *ipso facto,* that they are of extrasolar origin. They would not necessarily have to come from an extrasolar planetary system; rather, they might originate from interstellar spaceships. But in any case they would not be from our system.

Occasionally, unexplained signals are picked up, but most have logical interpretations. Back in June of 1961 British and Soviet astronomers, including Sir Bernard Lovell, Professor Alla Masevitch, and Dr. Jouli Khodarev, intercepted signals that Sir Bernard described as "extremely short bursts," the longest lasting about 50 sec. Subsequently, analysis indicated they may have come from the first Soviet Venus probe, but the Soviets reported they had lost communications with it three months earlier on 2 March 1962. Despite this, Jodrell Bank reported it had picked up signals on the same wavelength over two months later, on 17 May. On 9-16 June, Masevitch and Khodarev went to Jodrell Bank where they were unable to contact the probe or to intercept the strange signals.

This is just one of many examples, and is given for illustrative purposes only and not to suggest that either the British or Russian radio tracking astronomers even considered the signals of 12 June to have been caused by alien intelligence. But similar signals have been so interpreted by the same sort of people who have eagerly jumped on the flying saucer bandwagon. Further consideration of communications and interstellar signalling is given in Chapter 16.

Evidence Based on Discoveries that May be Made on Nonterrestrial Worlds in the Solar System

If the Solar System has ever been visited by extrasolar expeditions there are two major possibilities: (1) that no landings were attempted on

---

[14] There are some, however. For example, the Soviet astronomer Docent F. Zigel, writing in *Zhaniyesila* (Moscow), No. 2 (1961), p. 20, informs us that at present "there is scarcely to be found a single Soviet astronomer who considers our Earth to be the only inhabited planet." He then cites various expert opinions that there is organic life on Mars and from this builds up a case for intelligent beings, if not now existing (and he strongly suspects they are still around) at least formerly inhabiting the planet. Whatever the case, their works are there for us to explore, including the canals and the two satellites (which he believes to be artificial—see next section), and, incidentally, to have been created between 1862 and 1877.

Fig. 15.8. A future expedition to a moon of Jupiter may turn up evidence of a visit there by an extrasolar spaceship.

314 Empirical Evidence of Extrasolar Intelligence

any of the planets or satellites, and (2) that landings were made. If the latter, then there are numerous subpossibilities, in fact as many as there are worlds in the Solar System on which landings could be made. It is perfectly possible that the extrasolar beings did not chose to land on Earth —or, if they did, they left no discoverable remains—but did land on other worlds, say the Moon, Mars, or the moons of the outer planets (Fig. 15.8). These would be logical Solar System bodies, all having low escape velocities and no (to extrasolar explorers, possibly troublesome) atmospheres— except Mars and Titan with only tenuous atmospheres. If they did land on these worlds, we may find evidence of it from unmanned probes or at least as a result of manned exploration. Landings on the high temperature, close-to-the-sun planets Mercury and Venus (Fig. 15.9) are improbable.

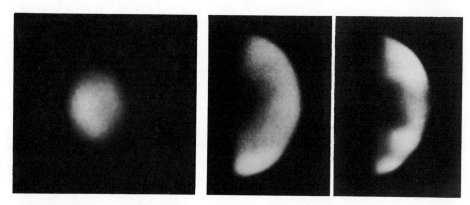

Fig. 15.9. Mercury and Venus, improbable targets of extrasolar expeditions.

The Moon, Mars, and the outer planet satellites would be relatively immune to the strong weathering processes that act on Earth and tend to destroy or modify whatever is on or near the surface. Certainly on the Moon there is not now, nor has there been for millenia (if there ever was), an atmospheric-hydrospheric envelope. We might expect, therefore, that any remains left by an extrasolar expedition on our satellite would still be in place, although perhaps damaged by cosmic infall if they were not buried a few feet or yards. In any event, we would not have to worry about searching for them in sedimentary formations hundreds or thousands of feet below the surface.

POSSIBLE MARTIAN EVIDENCE

In the case of Mars, if the activities of intelligent beings are discovered it would be quite an archeological task to determine if they were the result of native Martians or explorers from some alien star. We do not expect to find such evidence, but it may exist. It is well known that for many years the red planet was believed to be an abode of life. Scientific impetus for the belief was given by the American astronomer Percival Lowell in a series of writings and lectures during the early years of the 20th century. His best known work, entitled *Mars as an Abode of Life*,[15] made a persuasive argument for an inhabited planet. In six chapters he examined the genesis of a world, the evolution of life, and influence of solar radiation on Mars and its life, Mars and the future of Earth, the canals and oases, and proofs of life on Mars. The chapter on the canals and oases provides fascinating interpretations of their characteristics and purposes. Lowell concludes that the canals are used to send waters from the melting polar caps down to equatorial regions as the seasons change. He even figured out how fast the waters flowed. "It is possible to gauge the speed of the latitudinal sprouting of the vegetation, and therefore of the advent of water down the canals, by the difference in time between the successive darkenings of the canals of the several zones. Thus it appears that it takes the water fifty-two days to descend from latitude 72°N. to the equator, a distance of 2650 miles. This means a speed of 51 miles a day, or 2.1 miles per hour.[16] Figure 15.10 is a map of Mars constructed by Lowell in 1905, and Fig. 15.11 shows a double canal network.

When he comes to the beings that created the canals he cautions that while the advance of vegetable life as spring progresses can be observed, animal life cannot. "Not by its body, but by its mind, would it be known. Across the gulf of space it could be recognized only by the imprint it had

---

[15] Lowell, Percival, *Mars as an Abode of Life*. New York: MacMillan Co., 1910.
[16] Lowell, *Op. Cit.*, p. 181.

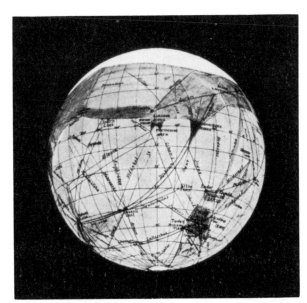

Fig. 15.10. Appearance of Mars in 1909. From Lowell, Percival, *Mars as an Abode of Life*. New York, 1910: Macmillan Co., p. 217.

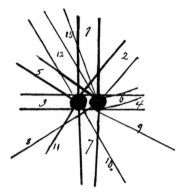

Fig. 15.11. Luci Ismenii, revealing the systematic method in which the Double Canals enter the Twin Oases 1. Euphrates, double. 2. Hiddekel, double. 3. Protonilus, double. 4. Deuteronilus, double. 5. Astaboras, double. 6. Djihoun, double. 7. Arnon, convergent double. 8. Aroeris. 9. Sados. 10. Pallacopas 11. Phthuth. 12. Naarmalcha, double. 13. Naarsares. From Lowell, Percival, Mars as an Abode of Life. New York, 1910: Macmillan Co., p. 190.

made on the face of Mars." And to Lowell the imprint was the elaborate system of canals (and oases) he patiently observed for many years.

We should not suppose that Lowell did not thoroughly examine natural explanations for the canals. But he could not get over their apparent regularity: ". . . this arrangement cannot be due to chance, the probabilities against the lines meeting one another in this orderly manner being millions to one."[17] And so ". . . it becomes apparent to any one capable of weighing evidence that these things which so palpably imply artificiality on their face cannot be natural products at all, but that the observer apparently stands confronted with the workings of an intelligence akin to and therefore appealing to his own. What he is gazing on typifies not the outcome of natural forces of an elemental kind, but the artificial product of a mind directing it to a purposed and definite end."

Although Lowell felt certain Mars was inhabited, he realized that the planet was a dying world, that the "process that brought it to its present pass must go on to the bitter end, until the last spark of Martian life goes out. The drying up of the planet is certain to proceed until its surface can support no life at all. Slowly but surely time will snuff it out. When the last ember is thus extinguished, the planet will roll a dead world through space, its evolutionary career forever ended."[18]

As astronomers gained increasing knowledge of the planet and of the requirements for life, they became ever more skeptical of Lowell's hypothesis and eventually it became all but abandoned except in the pages of science fiction. Many astronomers felt that no life whatever could exist on the planet; others that only simple plants might occur, but certainly no intelligent beings. The canals were relegated to the works of natural geologic or astronomic forces.

---

[17] Lowell, *Op. Cit.*, p. 195.
[18] Lowell, *Op. Cit.*, p. 216.

Fig. 15.12. Mars and its moons, Phobos and Deimos. Courtesy Gerard P. Kuiper

Today, we have little hope of finding more than plants and just possibly very simple forms of animal life on the planet; nor can we count on finding the remains of a former race of intelligent beings for, if it existed, it would be most unlikely that it allowed natural forces to "snuff" it out. Almost surely it would have mastered climate control, atomic energy, and space flight, and, if it could not accomodate itself to a dying world, it would have left for greener pastures, such as the Earth. So if we do find evidence of intelligence on the planet it is likely to have been caused by extrasolar expeditions.

Before leaving the planet Mars, mention must be made of the flurry of interest generated in 1959 by the Soviet astronomer Iosif Samuilovitch Shklovsky over the possibility that the two small moons of Mars (Fig. 15.12) are artificial rather than natural. Noting them to be extremely small (Phobos is from 10 to 15 miles across, Deimos 5 to 7, according to most estimates) and very close to the planet (5,800 and 14,600 miles from the center of Mars, respectively; Mars' radius is 2,108 miles), he emphasizes that the period of revolution of Phobos is absolutely unique in the Solar System—being shorter than the period of rotation of its primary. Moreover, the two moons revolve right on the equatorial plane in practically circular orbits (their eccentricities are 0.0210 and 0.0023, respectively). Furthermore, there seems to be a definite retardation of Phobos indicating that in about 15 million years it will decay into the atmosphere of Mars.

Shklovsky discounts friction between interplanetary matter and Phobos as a cause of its retardation in orbit (no such retardation is seen for Deimos), and quotes Jeffries' mathematical calculations to demonstrate that Martian tidal forces could not be responsible. His own mathematical investigations refute the possibility that a powerful magnetic field may be the cause, nor do they sustain the explanation that solar and planetary forces may be responsible. From this he arrived at the conclusion that there is no natural way to explain the origin and peculiarities of the tiny moon. In an interview with *Kosmosmolskaya Pravda* he asserted that it is "precisely the retarding effect of the upper, highly rarefied layers of the atmosphere that probably plays the decisive part in this case. But, for this retardation to be so considerable, in view of the extremely rarefied atmosphere of Mars at such an altitude, Phobos must have a very small mass and, hence, an average density of approximately one-thousandth the density of water." And, of course, a solid body cannot have such a density, leading Shklovsky to the conclusion that Phobos is hollow. "But can a natural cosmic body be hollow inside? No, and again no! Consequently, Phobos is of artificial origin. Consequently also, Phobos is an artificial satellite of Mars." And though the properties of Deimos are not so peculiar, it too is doubtless artificial. The Soviet astronomer is dismayed neither at their sizes nor their masses—equal to 100 or more million tons.

Having established that Mars is accompanied by two artificial satellites, Shklovsky is not so rash as to postulate intelligent life now existing on the planet. In fact, he thinks that it has not existed there for 2 or 3 billion years. Thus, the artificial satellites are incredibly old—if they are artificial. Of course, if they do turn out to be artificial it would still have to be demonstrated that they were built and orbited by Martians and not by beings of extrasolar origin.

Shklovsky's amazing hypothesis did not go long unchallenged. Among its articulate critics is Tombaugh[19] who finds little reason to interpret observations of the Martian moons in such a manner. First of all he points out that the orbiting of multimillion-ton moons would be a severe strain on any economy, particularly Martian since it could not be rich in natural resources. And he feels that evidence is decidedly against the emergence of intelligent life there, past or present. In particular, he argues convincingly against the planet ever possessing large bodies of water, showing that the "implications are drastic and far-reaching. No sedimentary rocks would have been laid down and consequently no conditions for the formation of coal and oil. That wonderful agent in geology, known as 'hydrothermal action,' which concentrates many valuable minerals into veins rich enough to make them worth mining, would be inoperative." Other telling arguments are given, but it is not our purpose to review them here; suffice it to say that his views are shared by virtually the entire astronomical and biological communities—no higher life now on Mars and no higher life in the past.

As for the peculiarities of the two satellites, there are, first of all, other small ones in the Solar System. The three outer moons of Jupiter are tiny. Moreover, Jupiter's fifth discovered satellite orbits at roughly the same time as its primary's rotation period. And others have nearly circular orbits. As for the acceleration tendencies of Phobos, which cause gradual orbital

---

[19] See, for example, Tombaugh, Clyde W., "Could the Satellites of Mars be Artificial," *Astronautics*, 4, No. 12 (December 1959), 29.

decay, Tombaugh quotes Fred Whipple as indicating a lifetime of 60, not 15, million years. Tombaugh believes many asteroids cross the orbit of Mars. "The number of such fragments should increase inversely as the cube of their diameters. If the hits were elastic, no change should ensue because as many would hit a satellite going inside the orbit as coming out. Because of friction and heat generated, there will be a slow loss of orbital kinetic energy and the satellite must gradually fall inward. Owing to the strong gravitational attraction of Mars on asteroids that brush by, they will be deviated toward Mars and hence funneled to greater concentration in the vicinity of Phobos' orbit than they would for Deimos. "Phobos would then no longer be required to have a low unnatural density." [from Tombaugh, *op. cit.*]

Evidence Based on Inference

Observations of certain stellar phenomena may be interpreted as having been caused by very advanced biological or mechanical communities. It can be assumed that as a civilization progresses it will develop the ability to do things on a scale that will inevitably lead to what might be called "cosmic engineering." The energy sources, and energy requirements, of highly developed extrasolar communities will doubtless be prodigious, yielding phenomena that may be detectable from Earth.

INFRARED RADIATORS

Two ideas of Dyson illustrate this concept. The first[20] illustrated by Fig. 15.13 involves the expenditure of enormous quantities of energy to disassemble a large planet, or planets, and rearrange its or their masses to create a spherical shell or a "loose collection or swarm of objects traveling in independent orbits around the star." The purpose of this exercise would be to provide comfortable living space, at the optimum distance from the host star, for a rapidly expanding population.

In developing his thoughts, Dyson considers "what would be the likely course of events if these (extrasolar) beings had originated in a solar system identical to ours. Taking our own Solar System as the model, we shall reach at least a possible picture of what may be expected to happen elsewhere. He does not say it will happen, only that it might. If a growth rate of 1 per cent a year is assumed, in only 3,000 years Earthly population and industry would grow by a factor of 10^{12}. Moreover, he calculates

---

[20] Dyson, Freeman J., "Search for Artificial Stellar Sources of Infrared Radiation," *Science*, **131**, No. 3114 (3 June 1960), p. 1667; see also letters concerning the article in *Science*, **132**, No. 3122 (22 July 1960), p. 250 and Dyson's reply in the same issue.

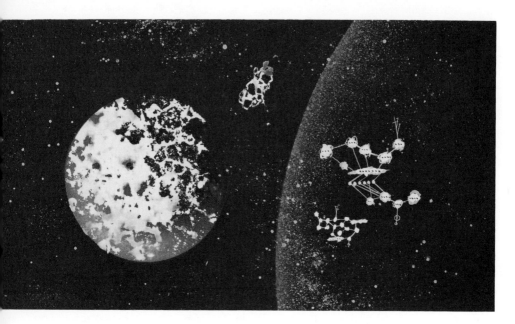

Fig. 15.13. Dismantling a planet and constructing what could be called space communities around parent star.

that some 10^{44} ergs of energy would be needed to disassemble and rearrange Jupiter and that if it were "distributed in a spherical shell revolving around the sun at twice the Earth's distance from it, would have a thickness such that the mass is 200 grams per square centimeter of surface area (2 to 3 meters thick, depending on the density)." He feels that such a shell could be made quite comfortable and "could contain all the machinery required for exploiting the solar radiation falling onto it from the inside."

Assuming that intelligent species permit their population to expand rapidly, Dyson expects that they "should be found occupying an artificial biosphere which completely surrounds its parent star." If this is so, their habitat would be "a dark object, having a size comparable with the Earth's orbit, and a surface temperature of 200° to 300°K. Such a dark object would be radiating as copiously as the star which is hidden inside, but the radiation would be in the far infrared, around 10 microns wavelength." He now proposes that a search be conducted for point sources of such radiation. While not so indicating in his article, he does state in his follow-up letter that "The discovery of an intense point source of infrared radiation would not by itself imply that extraterrestrial intelligence had been found." But it would be an indication that could be exploited by other detecting techniques, e.g., listening for signs of intelligent signals or sending out investigatory probes.

GRAVITY MACHINES

The second indication by inference of extrasolar intelligent communities developed by Dyson, already mentioned in Chapter 12 in a different context, has to do with "gravity machines."[21] He postulates that intelligent communities on a planet associated with a binary system of white dwarfs may derive vast amounts of energy not from their dim sunlight but from their prodigious stores of gravitational energy. He explains the operation of the scheme with the aid of Fig. 12.1.

Binary stars A and B revolve around each other. The beings using the gravitational machine are on body P, which may be a planet or a spaceship and which orbits the binary stars at a considerable distance (much greater than the separation of the stars). Along the dashed line object C, say a spaceship, is propelled whose trajectory is so computed that "it makes a close approach to B at a time when B is moving in a direction opposite to the direction of arrival of C. The mass C then swings around B and escapes with greatly increased velocity. The effect is almost as if the light mass C had made an elastic collision with the moving mass B. The mass C will arrive at a distant point Q with velocity somewhat greater than $2V$." At Q, the mass C could either be intercepted and its kinetic energy converted into useful form, or the device could become a means of propelling a spaceship to some predetermined location in space. The energy source of such a machine is the gravitational potential between the two stars. As the machine continues to operate it will pull energy out of the system, causing stars A and B to come close together. Eventually, they will be so close as to make it impossible for an object travelling along path C to pass between them.

Dyson calculates that white dwarfs with specified characteristics could accelerate "delicate and fragile objects to a velocity of 2000 km/sec at an acceleration of 10,000 g [gravity], without doing any damage to the objects and without expending any rocket propellant. The only internal forces acting on the accelerated objects would be tidal stresses produced by the gradients of the gravitational fields." He imagines that highly developed beings "might use white-dwarf binaries scattered around the galaxy as relay stations for heavy long-distance freight transportation."

Dyson concludes his essay by noting that the dynamics of stellar systems wherein gravitational radiation is significant have not been the objects

---

[21] Dyson, Freeman J., "Gravitational Machines," an essay prepared for and published by the Gravity Research Foundation, New Boston, New Hampshire, 1961.

of appreciable study. He observes that in any "search for evidences of technologically advanced societies in the universe, an investigation of anomalously intense sources of gravitational radiation ought to be included."

Despite all the possibilities described in this chapter, we have no evidence on Earth, on the Moon, on Mars, or anywhere else in the Solar System of any intelligence, native (e.g., Martian) or extrasolar, past or present. Since we have explored first hand only one world—the Earth—we cannot, of course, conclude that there have never been any intelligent extraterrestrial beings or expeditions in the Solar System. But with great confidence we can assert that up to now there is absolutely no *empirical* reason to believe that non *Homo* intelligent beings, biological or mechanical, have ever inhabited or visited the empire of the Sun. And there is no *direct* evidence as yet uncovered of the existence of extrasolar intelligence in the galaxy at large.

16
Communications with Extrasolar Intelligence

As we discovered in the preceding chapter, if we could detect artificially-produced signals emanating from beyond the Solar System we would have conclusive evidence of the existence of extrasolar intelligence, be it biological or mechanical. Signals might be picked up by accident, but more likely as the result of a planned search made for them. Such a search could be sporadic—as has already happened in Project Ozma (see below)—or sustained. The more sustained the search and the more complete the monitoring (in terms of stars covered), the greater the chances of success. In the relatively near future only a few possible sources of intelligent signals will be sought; hopefully, equipment will be directed to this purpose over periods of scores of years and not merely for a few weeks or months. These sources, logically, will be stars that astronomers feel are most apt to have planetary systems offering conditions conducive to the advent and development of higher manifestations of life.

"Man's will alone measures the distance between the possible and the impossible."

Mario Gonzales Ulloa

"The tumult of the time disconsolate To inarticulate murmurs dies away While the eternal ages watch and wait."

Henry Wadsworth Longfellow

Early Proposals for Interplanetary Communications

While it is relatively recent since man first thought of intercepting signals from extrasolar sources, for a long time he has mused over the possibilities of communicating with hypothetical beings on other worlds in the Solar System, notably the Moon and Mars. Rynin and Perelman[1] more than 35 years ago reviewed suggestions from the literature on how to signal across space, especially by optical means—radio was of only recent invention and at the time had not attracted the attention of would be signallers or searchers for signals. Among the schemes concocted for signalling from Earth to other worlds were huge bonfires built in geometrical patterns (illustrating, typically, a square or triangle), trees planted according to easily identifiable patterns, and canals built in deserts in such a manner as to indicate construction by intelligent creatures. Fire had advantages and disadvantages: advantages in that it would definitely indicate a signal if burning in a clearly artificial pattern, and disadvantages in that it would be expensive to maintain over long periods of time (a one-night fire would have little likelihood of being observed at the particular time it happened to be lit).

Perhaps not a signal in that it was to have been permanent, the famous mathematician Karl Friedrich Gauss's proposal (developed in the 1820s) for letting other worlds know there is intelligent life on Earth was to plant, in remote Siberia, a 10-mile wide pine forest in the form of a Pythagorean triangle (Fig. 16.1). Inside the triangle wheat was to have been sown to provide summer contrast with the dark green of the trees. In winter, the green of the trees would contrast strongly with the white snow, demonstrating seasonal changes.

Gauss hatched another scheme for across-space signalling, particularly to inhabitants suspected by many to live on the Moon. He suggested that large reflective mirrors be constructed and attempts be made to communicate at a time when the lunar beings would likely be observing the Earth and when twilight conditions would be such as to permit the mirrors to reflect light from the setting Sun and beam it to the Moon. As Perelman relates the plan, "it would be necessary to interrupt that light in regular time intervals; thus, it will be possible to send them (the lunar inhabitants) numbers that have the greatest significance in mathematics."

Some discussion is devoted to ideas of Nikola Tesla who is reported to have noticed "mysterious electrical signals while conducting tests at high altitudes" towards the latter part of the year 1900. Perelman reports that

---

[1] Rynin, Nikolai A., *Mezhplanetnye Soobshcheniya* Volume III: *Luchistaya Energiya v Fantaziyakh Romanistov i v Proektakh Uchenykh* (Interplanetary Communications Volume III: Radiant Energy in the Fantasies of the Novelists and in Projects of Scientists). Leningrad: Izdatel'stvo P.P. Soikin, 1930. And Perelman, Ya I., *Mezhplanetnye Puteshestuiya* (Interplanetary Voyages). Moscow and Leningrad: Government Press of RSFSR, 1929 (sixth edition). We gratefully acknowledge the efforts of Brounislav J. Soshinsky in translating material from these sources.

Fig. 16.1. Gauss's idea of planting 10-mile wide pine forest strips in the form of a Pythagorean triangle.

Tesla had informed a "British scientific journal" in 1901 "that there was a repetitive electrical fluctuation which he was unable to explain." But, he is said to have postulated that they were caused by currents emanating from the planets, leading him to speculate on means whereby artificially-induced energy could be sent across space to Mars in sufficient quantity to permit it to be captured by "an electrical receiver similar in nature to a telephone or telegraph . . . . I am certain," he is said to have written, "that with a properly constructed apparatus, it will be possible to transmit energy to other planets, as Mars and Venus, even at their greatest distance from the Earth. My method will give the practical answer to the question of transmission and reception of communications to and from planets."

An astronomer of note, Joseph Johann von Littrow, director of the Vienna Observatory, in 1840 advanced the plan of digging a circular ditch in the Sahara, some 20 miles in diameter, and filling it with water on top of which would be poured kerosene (Fig. 16.2). When lit, it was thought the huge circular fire would be visible to hypothetical inhabitants on the Moon, Mars, or other inner worlds. It was proposed that on subsequent nights similar ditches, having been dug so as to represent enormous squares and triangles, be ignited, and thus provide certain evidence of the habitability of Earth and of its desire to open up interplanetary communications.

Ananoff, in his "La Signalisation Interplanétaire"[2] ("Interplanetary

---

[2] Ananoff, Alexandre, *L'Astronautique*. Paris: Librairie Artheme Fayard, 1950, pp. 59–64.

Fig. 16.2. Von Littrow's scheme of digging circular ditch in Sahara, filling it with water, and setting fire to oil poured on top.

Signalling") briefly reviews a proposal developed by Charles Cros in his *Études sur les moyens de communication avec les planètes* (Studies of Methods of Communicating with the Planets) published by Gauthier-Villars in Paris in 1869. He felt that if a huge mirror were built of very low curvature it would be possible to focus sunlight on Mars and actually write simple numbers as the beam heated the desert sands (Fig. 16.3). Also mentioned is the project of one Schmoll who wanted to establish a network

Fig. 16.3. Plan conceived by Charles Cros for fusing sands of Mars by a low-curvature mirror on Earth.

of mirrors at Bordeaux, Cherbourg, Marseille, Stockholm, Amsterdam, Copenhagen, and on the shores of the Gulf of Bothnia to give the appearance from space of the Big Bear constellation. He reasoned that beings on Mars or other worlds could not help but notice the duplication of the well-known constellation and hence recognize it as an attempt at signalling —or at the very least comprehend that intelligent beings populated the Earth.

The astronomer William Henry Pickering (1858-1938) had a life-long interest in the planet Mars and in 1894 erected the observatory and telescope for Percival Lowell at Flagstaff, Arizona from which detailed observations of the planet were carried out. Among other accomplishments, in 1892, prior to the construction of the Lowell Observatory, Pickering discovered the first "oasis" on Mars, a roundish area at the juncture of several canals. So fascinated was he by the possibility that the red planet harbored intelligent life that he proposed a serious attempt be made to signal to Mars several months prior to the opposition of September 1909. Nothing came of his proposal except to draw attention to the impracticality of attempting to signal across millions of miles of space by optical means; and, conversely, to the possibility of electromagnetic methods of doing the same.

At about that time, Camille Flammarion, in his *La planète Mars et ses conditions d'habitabilité* (The Habitability of the Planet Mars) Tome II (Paris, 1909), looked into the problem and concluded that there are some opportunities for optically signalling the planet. Thinking of the possibility of our being able to intercept signals from Venus he wrote, "I have always had the impression that luminous signals projected by powerful electric light sources would be perceptible from here." And, he added, "it would be the same for nocturnal signals observed from the planet Mars." He later cautioned that the most insurmountable obstacle would be the difficulty of sending optical signals through the atmosphere without attenuation and distortion.

To Signal or to Listen?

As the 20th Century matured it became ever more evident that there were no nonterrestrial intelligent beings in the Solar System to whom to send, and from whom to receive, signals. Accordingly, until the last decade there was very little speculation concerning the possibility of communicating across space with extraterrestrial life, whatever might be its form and nature, wherever might be its home.

Partly as an outgrowth of deepening knowledge of the factors influencing the origin of life, and partly as a consequence of changing views on the origin and proliferation of planetary systems, science came to think of a universe in which habitable planets, if not abundant, were by no means exceedingly rare. It was only one step to the view that, if inhabited planets do revolve around other stars, it may be possible now or in the future to communicate with them. Not only would such communications have an enormous intellectual and philosophical impact on our civilization, but might well pave the way for an interchange of scientific and other information that could literally change the very nature of human activities and thinking processes.

At our stage of radio and optical maser (laser) technology it does not seem practical to announce to the universe at large, or even to selected nearby stars, that we are willing and able to communicate. For one thing the cost would be immense. Moreover, we could not expect any rapid answers to outgoing signals—the world would probably have to wait a great many years before it would have any assurance that its signals had been received and interpreted (transmission time to a star only 10 light years away would be 10 years, and one could not expect an answer for at least another 10 years). Moreover, even if an extrasolar society does receive the message, there is no assurance it would be willing to acknowledge it—at least not immediately. So we have to resign ourselves to signalling more or less continuously to many likely stars for periods measured in tens of years if we are to expect successfully to open up communications with extrasolar worlds.

We must also consider the desire of the planet Earth to reveal that it has progressed technologically to the point where it can communicate on an interstellar scale. There could well be apprehensions in many quarters as to whether or not it would be prudent to proclaim ourselves to the universe. Surely, an advanced extrasolar society would recognize from our manner of signalling that we have only recently emerged, scientifically speaking. If it were malevolent, such a revelation on our part might spell doom for terrestrial civilization. Yet this argument is hardly valid, for at any stage of our scientific and technological history we would be faced with the prospects of encountering a still more advanced community. At some time, perhaps not until the world is unified, the decision will be made to embark on an interstellar signalling program. Until then we must be content with planning on a serious, extensive, and selective listening program, hoping that worlds more advanced than ours will reveal themselves to us in a manner within our powers of comprehension.

Electromagnetic Communication

Virtually all speculation concerning interstellar communications has been concentrated on techniques involving the propagation of electromagnetic waves, particularly within the radio frequency limits from 10 to 10,000 Mc, or roughly between 30-meter and 1-cm wavelengths. Radio waves between 10 meters and 10 cm come through to the Earth's surface in the "radio window," shorter wavelengths being totally or partially absorbed, and longer wavelengths being reflected back into space by the ionosphere. Longer wavelengths are also ruled out because of cosmic radio interference. The problem of shorter wavelength absorption in planetary atmospheres would become unimportant if signalling were done from airless world to airless world or from transmitting and receiving stations in orbit above planetary atmospheres.

We have had two decades of what can be called space communications. In January, 1946 the US Army Signal Corps "bounced" radar signals off the Moon when it was at a distance of approximately 250,000 miles; this is shown in Fig. 16.4. The equipment employed for the historic experi-

Fig. 16.4. Radar signals being "bounced" off the Moon.

ment was a modified SCR-271 radar. The first radar contact with a planet occurred on 10 February 1958 when scientists at Massachusetts Institute of Technology's, Linclon Laboratory sent a signal to and received the echo from Venus at a time when the planet was about 28,000,000 miles distant. The Sun was probed the following year by Stanford University researchers, and in early 1963 contact with Mars[3] was made by the Goldstone Station

---

[3] The Martian contact was several thousand times more difficult than the Venusian experiments, requiring the use of a 100,000-watt transmitter and integration times of at least 15 hours. In comparison with Venus, Mars is a very poor reflector; moreover, due to its rotation, the red planet causes a spectrum spreading on the order of 400 cycles per second, compared with a few cycles for Venus.

of the California Institute of Technology. Since then radar contact with the inner Solar System has become fairly routine. As for the reception of radio signals from planetary bodies, Drs. Bernard F. Burke and Kenneth L. Franklin of the Department of Terrestrial Magnetism, Carnegie Institution discovered, in 1955, that Jupiter produced radio bursts between 10 and 21.4 meters, with a peak later determined to be at 15.8 meters. Centimeter and decimeter radiations were later discovered at 3.15 cm, 10.3 cm, 22 cm, and 68 cm. The following year, observations by the Naval Research Laboratory revealed natural radio waves coming from Venus, and in mid 1961 they were detected from Mercury by scientists at the Michigan Radio Astronomy Observatory.

From the beginning of the space age in October 1957 a wealth of experience has been gained in communicating with artificial objects placed in space in orbit around the Earth and in departure trajectories towards the Moon and planets. The US probe Mariner 2, launched on 27 August 1962, approached within 21,594 miles of the planet Venus on 14 December 1962. Its 109.5 day trip took it 180,200,000 miles along a heliocentric trajectory, during which time it was continuously tracked by stations located on the Earth's surface. When contact was lost on 4 January 1963 it had travelled 225,000,000 miles and was 54,300,000 miles from Earth (and 5,700,000 miles beyond Venus). Even more ambitious was the launching, on 28 November 1964, of Mariner 4 towards the planet Mars. After a 225-day journey that covered a total track of 325,000,000 miles, the probe photographed Mars on 14 July 1965—when the planet was more than 134,000,000 miles distant.

Since Karl G. Jansky first noted extraterrestrial radio "noise" coming from the Milky Way in 1931 the science of radio astronomy has grown. The first radio telescope was built in 1937 by Grote Reber, who discovered that radio frequency radiation of cosmic origin came not only from the center of our galaxy but also along the galactic plane with a secondary maximum in Cygnus.[4] Building on World War II technology, from 1945 on, the science of radio astronomy advanced at an extremely rapid pace and today large radio telescopes are located in all corners of the Earth.

---

[4] Jansky prepared a paper entitled "Electrical Disturbances Apparently of Extraterrestrial Origin," published in the *Proceedings of the Institute of Radio Engineers*, 21, No. 10 (October 1933), 1387–1398, which reviewed his observations from the summer of 1931 up to June 1933. In it he announced his conclusion that the detected electromagnetic waves came from outside the Solar System. Reber published papers in the same journal and, in the *Astrophysical Journal*, 100, No. 3 (November 1944), 279–287, he reviewed his survey made at a frequency of 160 Mc, which demonstrated that radio waves emanated strongly from the constellation of Sagittarius, with minor maxima in Cygnus, Cassiopeia, Canis Major, and Puppis. He found that the lowest minimum was in Perseus.

DETECTION OF INTERSTELLAR SIGNALS

Having possession of these great telescopes led a number of physicists and astronomers to propose that they be used to attempt to detect artificial electromagnetic radiations that may travel through the vast reaches of interstellar space. Among the earliest to make a practical, well thought out study of possible artificial signals of stellar origin were Cocconi and Morrison of Cornell University.[5] They start out with the assumption that there are other intelligent societies in the universe, possibly including some on planets circling stars relatively close to the Sun. Such societies would doubtless recognize the Sun as the possessor of a system of planets at least one of which might be capable of supporting life. They then conjecture that the nearby extrasolar community has long since established a channel of communication to us and is "patiently" awaiting our response, indicating that "a new society has entered the community of intelligence." The problem is to find out what the channel is.

They felt at the time of the preparation of their paper that only with electromagnetic waves would it be possible to effect communications between stars without dispersion in direction. They made the wise suggestion that the channel chosen by the extrasolar civilization would place a "minimum burden of frequency and angular discrimination on the detector," since the search is presumably being made to detect emerging societies and not highly advanced communities with which communications have presumably already been established. Radio frequencies chosen should be subject to minimum attenuation by the interstellar plasma and by atmospheres through which signals have to pass. Cocconi and Morrison selected the band from 1 Mc to 10,000 Mc as the most logical region for signals to be broadcast. Cosmic noise, the major deleterious factor in interstellar communications, increases at lower frequencies.

The next problem is to determine just what frequency within the practical range would be chosen by extrasolar societies. Most physicists and radio astronomers who have investigated the problem recognize that searching for signals (whose frequency is not known) over a wide range of possible frequencies would be exceedingly difficult. However, there is an astronomically significant frequency at 1,420.4056 Mc, the natural frequency radiated by neutral interstellar hydrogen (wavelength: 21 cm). We know that hydrogen atoms make up the vast majority of matter in the universe and that each generates a minute quantity of energy at the cited frequency. Other societies must realize the significance of this radio emission line and logically would transmit on a frequency close to 1,420 Mc.

---

[5] Cocconi, Giuseppe and Philip Morrison, "Searching for Intersellar Communications," *Nature*, **184**, No. 4690 (19 September 1959), 844–846.

Numerous advantages of signalling at this frequency are recognized. At the 21-cm hydrogen-line wavelength, signals could travel enormous distances along what in effect is a clear interstellar channel of communication provided by nature. And not only is open space quite transparent to 21-cm signals but they can pass with relative ease through planetary atmospheres.

Von Hoerner[6] has pointed out possible shortcomings of the 21-cm line, and suggests a modification. He notes that "the background of a signal would be much stronger within this line than beside it—so strong as to drown out a small signal—and the boundaries of the line are not well enough defined for us to place the signal exactly beside it." He therefore feels that we would do better to use a frequency twice that of the 21-cm line of neutral hydrogen. If this does not work out, "we would have to look for more sophisticated methods of producing contacting signals."

With a 1,000-ft radio telescope we could detect signals produced by an extrasolar society up to about 100 light years distant if they possessed transmitting equipment similar to the best we now have here on Earth; for example, the radar antenna at Millstone Hill, Mass. Obviously, we would expect a more advanced society to have far better equipment, but how much better is impossible to predict. It might well be capable of signalling to most parts of the Milky Way galaxy and perhaps to other nearby galaxies. The existing 85-ft radio telescope at Green Bank, West Virginia (Fig. 16.14) could pick up signals generated by Millstone Hill up to 10 light years away and Green Bank's companion 140-ft telescope could intercept them at least 15 light years away. The designed, partially constructed, and, in 1962, cancelled Navy 600-ft antenna at Sugar Grove, West Virginia could intercept signals 60 light years away, and 1,000-ft antennas like that at Arecibo, Puerto Rico, would be effective up to distances of 100 light years.

TRANSMISSION OF INTERSTELLAR SIGNALS

The planet Earth has so far shown no inclination to transmit signals to potential habitable-planet-possessing stars within, say, 10 to 20 light years. It could do so at considerable cost; but, prior to receiving substantial information that the chances are at least fairly good of an extrasolar society occurring within transmission range, our society has not seen fit to embark on a signalling program.

The ability of an antenna to send a powerful beam through space is largely determined by its size, which is measured in wavelengths, and on

---

[6] Von Hoerner, Sebastian, "The Search for Signals from Other Civilizations," *Science*, **134**, No. 3493 (8 December 1961), p. 1843.

the power of transmission. Similarly, on the diameter and sensitivity of the receiver will depend whether or not the signals are received and understood. If the Earth wants to detect a signal it should employ the largest possible receiving antenna; by the same token, if it wishes to probe to great distances in space it must build huge transmitting antennas. We can build larger transmitting antennas than we now have, but it is most unlikely we shall soon consider a sustained interstellar signalling program to help justify their construction.

If all societies take the attitude currently prevalent on Earth and occasionally listen but never transmit, interstellar communications could never come about. Therefore, someone has to transmit if someone else is to hear more than cosmic noise on highly sensitive inquiring receivers.

We do not suggest that the "someone" necessarily be us at our primitive stage of technology, although probably within the coming hundreds of years we shall be in a position to embark on a transmission program. Huang[7] has emphasized that even if alien societies are signalling to us "we still cannot receive the full information without being supported by a transmitting station which can reach the place where these particular people live." For one thing, we may suspect an artificial signal but are uncertain of its nature, especially if it is of short duration. Having a transmitter would allow us to beam a signal to the star suspected of having sent the message, thereby indicating that it had been received and we are ready to communicate. And, of course, if we definitely recognize the probing signal we would want to alert the star so that informational messages would be forthcoming.

It has proven feasible, even at our stage of development, to convince the authorities concerned with the allocation of radio telescope observing time that some efforts should be made to listen to possible artificial signals from space. After all, either a positive or negative report can be given. But, as Su-Shu Huang points out, "the person who is transmitting signals to other worlds can report neither success nor failure year after year . . . and has to explain his long silence to people who have financed the project." Truly, the incentives for transmitting are lacking to financiers and scientists alike because no results can be expected for at least 20 years (time for signals to reach, say, Tau Ceti and an answer to return—compare this with the expectation of receiving a signal at any time on a passively listening radio telescope).

Aside from the lack of incentives, a transmitting society is faced with the very real problem of deciding which targets should be probed. Much

---

[7] Huang, Su-Shu, "Problem of Transmission in Interstellar Communication," in *Interstellar Communication*, ed. A. G. W. Cameron. New York: W. A. Benjamin, Inc., 1963, pp 201–206.

depends on their analysis of the probabilities of an intelligent community existing within a given radius, measured in light years, from their solar system. This, in turn, will doubtless depend on where they are located in the galaxy. Societies in central sectors of high stellar concentration could reach many more stars at relatively close range than societies inhabiting the more sparsely populated outer regions. How many stars they could reach would depend on the power of their equipment and the duration of their probing signals. They might, for example, hit all attractive stars within 100 light years, sending 1-hr long messages once a year. Or, they might send fewer, and somewhat longer, messages to greater distances at specially chosen stars. Their incentives, quite naturally, would be for contacting closer stars because of understandable desires to achieve satisfying results as quickly as possible.

Laser Communication

As we saw earlier, before the widespread use of radio the only way envisioned to signal to, and be signaled from, other worlds was by optical devices. It was later recognized that conventional optical schemes were much too limited for long distance space communication and were gradually abandoned. Only very recently have optics come back into the picture, in the form of the optical maser or laser (whose name is derived from *light amplification by stimulated emission of radiation*). The latest lasers are made of gallium arsenide and are fired by the direct injection of electricity. Types able to emit short pulses of light do so at powers up to 50,000,000 watts. With relatively small antennas, very narrow beams can be directed across space. By using a filter as a frequency selector, it has been possible to produce light at a single frequency (optical masers usually produce energy at a number of different frequencies) and narrow its beam to 0.02 of a degree. Efficiencies of the latest gallium arsenide diode lasers are approaching 100 per cent.

Space experiments with optical masers have not yet been numerous but they have proved feasible. On 9 May 1962, Massachusetts Institute of Technology experimenters reflected a laser-produced beam off the lunar surface. With current techniques, it may now be possible to signal out to a light year, assuming that the radiated energy is of the order of 10^4 joules/pulse. For a laser beam to be detected by a receiver of practical size it must produce a sufficient number of photons and these must be susceptible to separation from the light emanating from the parent sun. The latter problem is particularly acute, leading Schwartz and Townes[8] to conclude

---

[8] Schwartz, R. N. and C. H. Townes, "Interstellar and Interplanetary Communication by Optical Masers," *Nature*, **190**, No. 4772 (15 April 1961), 205.

there is but little hope of "resolving the two spatially." They note that it would be difficult to distinguish between the Sun and Earth at 10 light years distance even if the latter produced light of the same intensity as the former. They then suggest that "one can perhaps more easily resort to high spectral resolution in order to discriminate between laser and stellar radiation." They feel that an optical maser system with the following characteristics should be detectable some 10 light years distant: 10-kw continuous power, $\sim$ 5,000-A (Angstrom) wavelength, $\sim$ 1-Mc bandwidth; reflector diameter 200 in.; and beam width 10^{-7} radian. Lasers appear to be marginally useful for Sun-like stars, but are quite attractive for very faint stars possessing habitable planets.

Oliver writes[9] that the mean objective diameter for a 60-db circuit for Earth-Moon communications by optical maser is 0.3 in., compared with a 3-in. objective (40-db signal-to-noise ratio) necessary for Earth-Mars contact and a 10-in. objective (30-db signal-to-noise ratio) for communicating across the entire Solar System. But when it comes to the stars we are faced with the requirement of "a tremendous increase in optical maser power output . . ." before communication becomes possible. For the time being, at least, lasers do not compete with radio systems.

Searching for Artificial Extrasolar Signals

If our society has not felt the desire to transmit signals to hypothetical extrasolar worlds, leading scientists within it have done much thinking about, and devoted considerable study to, means of searching for evidence of artificial signals emanating from stars in relative proximity to the Sun (Figs. 16.5 to 16.8). Some consideration also has been given to searching for signals from spaceships cruising in interstellar space (Fig. 16.9) and from extrasolar probes that may now be stationed in our own Solar System (Fig. 16.10). These three possibilities are examined in turn.

SIGNALS FROM EXTRASOLAR PLANETARY SYSTEMS

Signals from extrasolar societies, when they are first intercepted, will doubtless come from planets, moons, or asteroids in the alien solar system or from artificial satellites orbiting around them. Because of this, prime attention has been accorded the nearest stars to the Sun (Fig. 16.11) whose characteristics offer the greatest possibilities of possessing habitable planets.

---

[9] Oliver, B. M., "Some Potentialities of Optical Masers," *Proceedings of the Institute of Radio Engineers*, 50 (1962), 135.

Fig. 16.5. Extrasolar community transmits at random, hoping that a habitable planet intercepts, and understands, its signals.

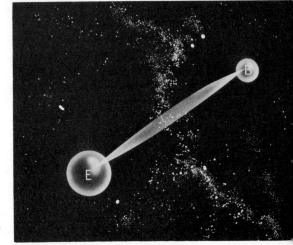

Fig. 16.6. Direct communications between an extrasolar planet and the Earth.

Fig. 16.7. If the Earth were located between two extrasolar planets it might detect communications between them.

Communications with Extrasolar Intelligence

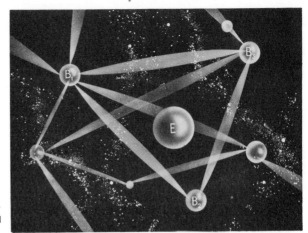

Fig. 16.8. Such an extensive extrasolar communications network could possibly be intercepted by Earth.

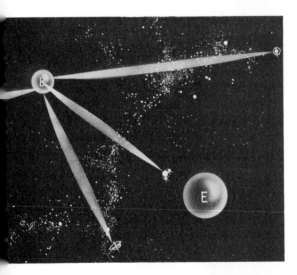

Fig. 16.9. Signals from extrasolar planet to cruising spaceships could be intercepted by Earth.

Fig. 16.10. An alien probe may be monitoring our Solar System, transmitting information back to its home planet.

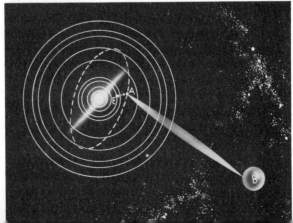

Fig. 16.11. Known stars within 16.5 light years of the Sun. Adapted from model of a globe constructed by Dr. S. L. Lippencott. Courtesy Dr. Lippencott and Sproul Observatory.

Fig. 16.12. Epsilon Eridani. Courtesy Mount Wilson and Palomar Observatories, Dr. I.S. Bowen, and Dr. Armin J. Deutsch. Copyright National Geographic Society — Palomar Observatory Sky Survey.

Fig. 16.13. Tau Ceti. Courtesy Mount Wilson and Palomar Observatories, Dr. I.S. Bowen and Dr. Armin J. Deutsch. Copyright National Geographic Society—Palomar Sky Survey.

These stars turn out to be Epsilon Eridani, Tau Ceti, and Epsilon Indi (see Table 2.4).

An actual search for artificial signals was undertaken during May, June, and July 1960 under Project Ozma at the National Radio Astronomy Observatory, Green Bank, West Virginia, using its 85-ft radio telescope. Under the direction of Frank D. Drake,[10] approximately 150 hr of observing time was expended during which 400 kc of bandwidth was explored in a futile attempt to intercept possible signals from 10.8-light-year distant Epsilon Eridani and 11.1-light-year distant Tau Ceti, photographs of which are shown in Figs. 16.12 and 16.13. The 85-ft steerable antenna is seen in Fig. 16.14, the radiometer in Fig. 16.15, and its block diagram in Fig. 16.16. The apparatus was sufficiently sensitive to have detected signals from a megawatt transmitter emitting through a 600-ft antenna located in a solar system that either of the two stars may posses *if* it was broadcasting during the time Drake and his associates were listening, *if* the signals were aimed at our Solar System, and *if* 1,420 Mc (21-cm line) was the selected frequency.

---

[10] See Drake, Frank D., "Project Ozma," *Physics Today*, **14**, No. 4 (April 1961), 40–46.

Fig. 16.14. 85-ft antenna, National Radio Astronomy Observatory.

Fig. 16.15. The radio telescope's radiometer. Courtesy Dr. Frank D. Drake.

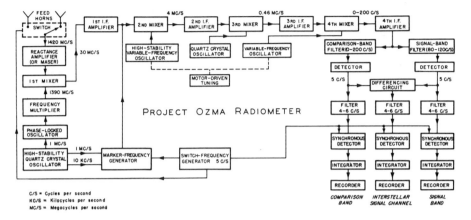

Fig. 16.16 Block diagram of the radiometer. Courtesy Dr. Frank D. Drake.

Drake[11] describes the Ozma block diagram of the receiver as follows. "This receiver uses two feed horns at the focus; the receiver switches between the horns, giving two beams, one of which is pointed on the target star, and the other in space along side the star. This allows us to distinguish true signals from the star from terrestrial interference which may creep in through side lobes. This switching also gives us the advantages of the Dicke radiometer. Since we use very narrow bandwidths, of the order of 100 cycles per second, we require very stable local oscillators in the superheterodyne receiver. In fact, our first local oscillator is phase locked to a highly stable quartz crystal oscillator. After the first conversion we use a standard high quality superheterodyne receiver, with four conversions, to bring the intermediate frequency to the very low frequency allowing us to filter out very narrow bandwidths. Two bands are used simultaneously, one wide, one narrow. Cosmic noise, which is broad band, will be present in both bands to the same extent, whereas a narrow-band signal will be in the narrow-band channel more strongly than in the wide-band channel. If we then difference the outputs from these two channels, cosmic noise will difference to zero whereas an intelligent signal will produce a net output. The device thus discriminates against cosmic noise. In order to achieve very high sensitivity, we employ a parametric amplifier at the front end of the receiver."

The results of Ozma disappointed few since few expected it would reveal the existence of extrasolar societies. For one thing there may be none on the stars selected, or if there are they may not have been

---

[11] Drake, *Op. Cit.*, p. 46.

transmitting during the 150 hours that man happened to be listening. Or they may have been transmitting towards other regions of space, or even towards us on some other frequency. To increase our chances of success a long-period search must be inaugurated, tuned in to given stars for at least a year or two at a time. If we could monitor each likely star out to a distance of 100 light years there should be a fair chance of turning up artificially-produced radio signals.

We must, of course, be prepared one day to intercept signals that may not be aimed expressly at us, but rather are being used to connect several extrasolar societies and/or their spaceships and probes. They could be transmitting to our general area of space but not to us in particular.

SIGNALS FROM EXTRASOLAR SPACESHIPS IN INTERSTELLAR SPACE

Transmissions may be intercepted from cruising spaceships operated by biological species or mechanical automata. The signals produced may be consciously aimed at us, they may be aimed at some other target and simply intercepted by us, or they may be regional probing signals emitted in the hopes of educing a return signal from an undiscovered intelligent community. Upon receiving the answering signal the spaceship could simply catalog it, or enter into communication with us, or physically visit the Solar System.

SIGNALS FROM EXTRASOLAR PROBES IN THE SOLAR SYSTEM

Radiating signals to possible planet-possessing stars situated within, say, 100 light years from an advanced community may seem to offer too little chance of success to make the endeavour worth while. Not only would a great many transmitters have to be devoted to signalling the many possible stars in a 100-light-year radius sphere, but they would have to be kept constantly in operation over long periods of time to justify the effort in the first place.

Faced with these thoughts, Bracewell[12] asks if an advanced society would not rather send probes into alien solar systems. He ventures the suggestion that such a society would intensely investigate nearby solar systems and that those belonging to suitable stars farther away might receive a "spray" of "modest probes." He postulates that each such vehicle "would be sent into a circular orbit about one of the thousand stars (which

---

[12] Bracewell, R. N., "Communications from Superior Galactic Communities," *Nature*, **186**, No. 4726 (28 May 1960), 670–671.

he suggests might be the object of the attentions of a typical extrasolar civilization), at a distance within the habitable zone of temperature. Armored against meteorites and radiation damage, and stellar powered, the probes could contain durable radio transmitters for the purpose of attracting the attention of technologies such as ours." Such a scheme would have double benefits. One, we would not have to worry about selecting the right star at which to aim our inquiring signals; and two, the visiting probes could produce a far stronger and clearer signal for us to analyze than could a transmitter located somewhere in another solar system. On this line of thinking, Bracewell concludes that we might "better devote our efforts to scrutinizing our solar system for signs of probes sent here by our more advanced neighbors."

Other conclusions of Bracewell are worthy of our attention. For one thing, he assumes that advanced communities are already tied into a communication network, suggesting that only the nearest such community will make contact with Earth. If Bracewell's arguments are valid, there is little likelihood of duplication of searching. Once the community has determined a technological civilization exists in the Solar System it would be expected to send to us a probe containing a "quite elaborate store of information and a complex computer, so that it could not only detect our presence, but could also converse with us." If such a probe is in the Solar System now it would probably transmit over a wavelength that we would be almost certain to use. Its transmission might be identical signals to those we produce, "having delays of seconds or minutes." If we find such repeat signals we would re-emit them, thereby informing the probe that we comprehend what is happening. When the probe realizes an intelligent community has arisen in the Solar System it may automatically put into operation instrumentation to analyze the civilization or it may await instructions from the extrasolar community from which it came.

Galactic Communication Networks

Within some 50 light years of the Sun there are approximately 100 stars whose spectral and other characteristics indicate they may possess planets, a few of which at least could harbor intelligent communities. The number increases as the radius of investigation from the Sun increases; for example, within a radius of 100 light years there could be up to 1,000 possible candidates. Those actually supporting intelligent communities are almost certain to be further advanced than ours—the chances that within 1,000 light years another community has only recently crossed the intelligence and emerging technological thresholds are very slim indeed. Such communities can be expected to be in radio contact, perhaps on a regional basis, perhaps also on a galactic basis.

Webb[13] feels that we should expect to find evidence of such a network rather than expend energies searching for signals supposedly beamed specifically towards our Sun in the hopes of determining if it possesses a planet on which intelligent life exists. He postulates that other communities will have communication networks for a variety of purposes not associated with searching out undetected societies. He sees "good reason to believe that the microwave spectrum will be used by any advanced society . . . [not only because] the microwave and millimeter wavelengths offer the greatest receiver sensitivity available throughout the entire spectrum [but because] geometric attenuation and antenna design tolerances are not excessive in this region."

A good deal of attention has been devoted to the feasibility of undertaking interstellar communications based on techniques now known on Earth or reasonably near-future extrapolations thereof. Trying to establish limits based on what *we* can do as a result of less than a century of progress is patently absurd and serves no useful purpose. As far as we now know the factors which determine the ultimate range of an electromagnetic communication system are three in number (quoted from Oliver):[14] "the energy radiated per symbol, the directive gain of the antennas, and the noise" (the symbols in the simplest form of modulation are "pulses or spaces, and the messages would be encoded into these in some way"). If the signal energy is approximately equal to the noise energy, in other words if the signal-to-noise ratio is about 1, detection will be barely possible. As the ratio increases so will the possibility of detecting whatever message is being transmitted.

Interstellar Flight

As we recall from Chapter 12, because of the vast distances involved, within the foreseeable future there is little probability of humans being able to undertake travel to even the nearest stars, which are just over 4 light years away. That some day man may achieve the ability to visit our stellar neighbors is possible; it is equally possible that some extrasolar biological societies may find interstellar travel feasible, conceivably even routine, and, of course, mechanical automata may move from star to star unconcerned over the elapsed time (though there would seem scant reason for them to do so except occasionally to monitor supply and trade operations and to migrate—possibly towards the center of the galaxy).

---

[13] Webb, J.A., "Detection of Intelligent Signals from Space," *Institute of Radio Engineers Seventh National Communications Symposium Record,* 1961, pp. 10–15.

[14] Oliver, Bernard M., "Radio Search for Distant Races," *International Science and Technology,* No. 10 (October 1962), pp. 55–60.

Rocket experts, physicists, and others have been writing on interstellar flight for many years, Robert Esnault-Pelterie being one of the first.[15] Without going into details on a variety of possible propulsion schemes, it does appear possible that man eventually will be able to construct interstellar vehicles, although the energy requirements will be prodigious. The limiting factor may not be energy but the human lifespan, leading Dyson[16] to say that interstellar travel "is essentially not a problem in physics or engineering but a problem in biology."

All investigations of the possibility of interstellar flight are faced with the seemingly unsurmountable time factor; even if light velocity could be attained by a spaceship—and we are quite certain it cannot—round trip travel times to the nearest stars would take over 8 years, and to stars offering fair probabilities for harboring habitable planets, some 20 years. With energy resources we can but dimly perceive on the far distant technological horizon, only fractions of the velocity of light are likely to be achieved, meaning that we shall either have to rule out interstellar travel with humans aboard or rely on some such concept as generation travel (one generation of carefully selected couples would embark on the voyage and a later generation would arrive). If generation travel proves undesirable, it may become possible to place crews and passengers in a combination of hypothermia and artificial hibernation, wherein bodily functions become suspended or nearly suspended until the destination is approached. Hypothetical extrasolar beings may have lifespans measured in thousands of years and would not be as concerned with the time element as man.

Some relief from the time problem comes from a consideration of the special theory of relativity having to do with time dilation. According to the theory, a clock moving through space, e.g., in a spaceship, will run slower than one ticking away in a fixed reference system, e.g., back on Earth. The two different times are shown by the relationship

$$T = T_0 \sqrt{1 - \left(\frac{v}{c}\right)^2}$$

---

[15] Esnault-Pelterie, Robert, *L'Astronautique*. Paris; Imprimerie A. Lahur, 1930. For a résumé of thinking and a bibliography of works on the subject up to 1958, see Adams, C.C., F.I. Ordway, III, H.E. Canney, Jr., and R.C. Wakeford, *Space Flight*. New York: McGraw-Hill Book Co., Inc., 1958 (Chap. 14 "Interstellar Flight"); and up to 1963, see Ordway, F.I., III, J.P. Gardner, M.R. Sharpe, Jr., and R.C. Wakeford, *Applied Astronautics*. Englewood Cliffs, N.J.: Prentice-Hall, Inc., 1963 (see Chap. 8 "Beyond the Solar System"). Stearns, E.V.N., *Navigation and Guidance in Space*. Englewood Cliffs, N.J.: Prentice-Hall, Inc., 1963 contains a comprehensive chapter on "Navigation in Interstellar Space."

[16] Dyson, Freeman J., in a letter to the editor published in *Scientific American*, **210**, No. 4 (April 1964).

where T is time as indicated by the clock in motion, T_0 the time according to the clock held in a fixed reference system, v the velocity of the clock in motion, and c the velocity of light in a vacuum. If a spaceship were able to attain 90 per cent of the velocity of light it seemingly would make the 4.3-light year round trip to Alpha Centauri in 9.56 Earth years (8.6 light years ÷ 0.90 c). Taking T_0 as 9.56, T comes out to 4.16, meaning that a time dilation reduction of time of about 5 years occurs on the spaceship.

In accordance with theories promulgated by Lorenz and Einstein, as velocity increases so does mass, with the result that a body reaching the speed of light becomes infinite in mass.

$$m = \frac{m_0}{\sqrt{1 - \left(\frac{v}{c}\right)^2}}$$

where m is the final mass, m_0 the rest mass, v the velocity of the spaceship, and c the velocity of light. Should v become equal to c, the mass of the spaceship would be infinite. Moreover, in accordance with the Fitzgerald contraction theory, the spaceship's length (L) would decrease as it moves forward at uniform velocity (based on a fixed reference point).

$$L = \sqrt{1 - \left(\frac{v}{2}\right)^2}$$

If the spaceship's rest length is L_0 and the apparently shorter length at v is L_x, we have

$$L_x = L_0 \sqrt{1 - \left(\frac{v}{2}\right)^2}$$

These equations show that time, from the point of view of the spaceship's crew and passengers, is shortened at near optic speeds, and that mass and length are also affected. All of this remains, of course, a theoretical exercise dependent on our some day being able to generate velocities approaching that of light.

Leslie R. Sheperd, Eugen Sänger, Ernst Stuhlinger, and others[17] have examined the mechanics of interstellar flight as well as nuclear, electric, photon and other propulsion systems. Purcell[18] makes some interesting calculations with what he calls the "ideal *nuclear fusion* propellant," burning hydrogen to helium with 100 per cent efficiency and creating an

---

[17] See sources cited earlier in this section and their bibliographies.

[18] Purcell, Edward, "Radioastronomy and Communication through Space," in *Interstellar Communication*, ed. A.G.W. Cameron. New York: W.A. Benjamin, Inc., 1963.

exhaust velocity of an eighth that of light. Even with this propellant, in order to attain 0.99 c "we need an initial mass which is a little over a billion times the final mass. To put up a ton we have to start off with a billion tons. . . ." Since this is the best fusion reaction in nature, we quickly see our limitations.

But, as Purcell writes, "This is no place for timidity, so let us take the ultimate step and switch to the perfect matter-antimatter propellant. Matter and antimatter annihilate; the resulting energy leaves our rocket with an exhaust velocity of c or thereabouts . . . to go up to 99 per cent the velocity of light only a ratio of 14 is needed between the initial mass and the final mass." But, energy must still be expended to decelerate for the extrasolar landing, then, on the homeward leg of the journey to the stars, accelerate again and decelerate again, making the ratio 10^4, or 40,000. If a 10-ton payload is envisioned, the takeoff weight would be 400,000 tons, half being matter, half antimatter.

Relativity and propulsion are far from the only problems involved in interstellar flight. For example, if it should be possible to accelerate a spaceship to near optic velocities over reasonable periods of time, the erosional effect of interstellar matter (see Chapter 12) on the vehicle might well be overwhelming; certainly, it would limit velocities just as the atmosphere limits velocities of bodies travelling through it—such as missiles, space carrier vehicles, and airplanes.

Fig. 16.17. Unmanned interstellar probe.

Fig. 16.18. Manned interstellar spaceship.

As far as we can now perceive, interstellar flight by manned vehicles is virtually an impossibility and will remain so for an indefinite time. Interstellar probes may become feasible, perhaps within a century, but they will never be competitive as information carriers once the Earth connects up with interstellar networks (assuming them to exist). Information is what we really want—it is what interstellar communications is all about; to obtain and transmit it on a continuous basis will be the job not of space vehicles but of receivers and transmitters. Figures 16.17 and 16.18 illustrate, respectively, an unmanned interstellar probe and a manned spaceship.

The question always comes up as to whether the velocity of light in a vacuum is a limiting factor in space travel. As far as we know today, it cannot be exceeded—if it could a substantial part of our physical knowledge would be affected, and many basic laws, theories and concepts would be overturned. Still, we cannot say categorically that light may *never* be exceeded by an artificial object, only that we cannot conceive it.

Sänger in 1956[19] pointed out a way by which the velocity of light for all practical purposes could be exceeded—i.e., from the point of view of the crew and passengers of an interstellar spaceship. Assuming photon

---

[19] Sänger, Eugen, "Die Erreichbarkeit der Fixsterne," in *Rendiconti del VII Congresso Internazionale Astronautico*. Rome: Stabilimento Fotomeccanico del Ministero Defesa-Aeronautica, 1956, pp 97–113.

Communications with Extrasolar Intelligence

propulsion some day to become feasible, relativistic mechanics would cause the time dilation effect to become operative as velocity increased, from which he concludes "that within the life-span of the crew and with limited mass-ratios of the rocket, every distance in space, up to the nebulae millions of light-years distant can be covered, so that, expressed in technical terms, and from the standpoint of the crew, the vehicle seems to be able to move with considerable super optic-velocity."

The Language of Communications

Many suggestions have been made as to what sort of messages to look for and what sort of messages to send out—initially and after contact. Simple pulses or prime numbers would be a logical start if we are going to signal before having been contacted ourselves. If we receive signals from an alien society, perhaps our first reaction would be to repeat exactly what we had received. Being confronted with echoes of its own broadcasts would surely demonstrate to the extrasolar society that its inquiring signals had been received and recognized.

An interesting approach is given by Oliver[20] and is reproduced in Figs. 16.19 and 16.20. He assumes the radio astronomers begin receiving from Epsilon Eridani a seemingly artificial series of pulses and spaces which is

---

[20] Oliver, *Op. Cit.*, p. 60.

Fig. 16.19. Hypothetical message from Epsilon Eridani (see text). Courtesy Dr. Bernard M. Oliver.

```
1000000000000000000000000000000000000000100001110000000000000100
0000000100010000000100010000000000000000000000000000000000000010
0010000100000010000000000001000000000000100010000010001001001000 0
0100010001000000011100000000000001000000000000100000000000000000
0000000000000000000000000000000000000000000000000001000000000010
0010000011000100000000000000000000000000000000000000000000011000
0110000110000110000110000100000000001001001001001001001001001010
0101010010010000110000110000110000110000110000000010000000000001
1111010000000000000000001000000000001000010000000000001011011
1001000000000000111110100000000000000000000000000000001000000000
0000001000100111000000000000101000000000000000101001000011001010
1110010100000000000000101001000010000000000100100000000000000000
0100100000100000000000011110000000000000011111000000111010100000 0
10101000000000001010000000100000000000010001000000000000010100010
0000000000000000010001001000100010011011001110110110100000100010
00101010100010001000000000000000100010001001001000100010000000
10000000000001110000011110000011100000001111101000001010100000 1
0100001000100000100000000010000010000111000010000010000011000
0000001000001000100010001000001000001000011000010000010001000100
01000010000011000000001100001101100011011000001100111
```

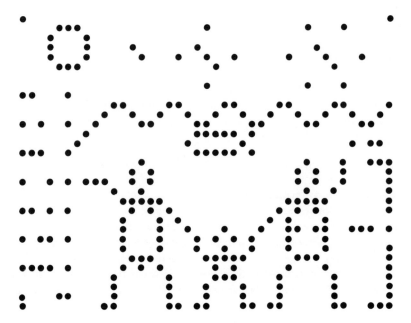

Fig. 16.20. Interpretation of message in Fig. 16.19. (see text). Courtesy Dr. Bernard M. Oliver.

quickly interpreted to be a message. Every 22 hr 53 min it is repeated, giving rise to the conclusion that it is coming from a planet revolving around that star whose day equals that period of time. He writes that the "pulses occur at separations which are integral multiples of a minimum separation." By inserting ones for the pulses and zeros in the blanks he arrives at the binary series in Fig. 16.19, which is made up of 1,271 entries, the product of 31 and 41. This fact "strongly suggests that we arrange the message in a 31 by 41 array," which he does, yielding Fig. 16.20 (leaving blanks for the zeroes and placing a dot for each pulse). The correct interpretation follows:

"Apparently we are in touch with a race of erect bipeds who reproduce sexually. There is even a suggestion that they might be mammals. The crude circle and column of dots at the left suggests their sun and planetary system. The figure is pointing to the fourth planet, evidently their home. The planets are numbered down the left hand edge in a binary code which increases in place value from left to right and starts with a decimal (or rather a binary) point to mark the beginning.

"The wavy line commencing at the third planet indicates that it is covered with water and the fish-like form shows there is marine life there. The bipeds know this, so they must have space travel. The diagrams at the top will be recognized as hydrogen, carbon, and oxygen atoms, so their

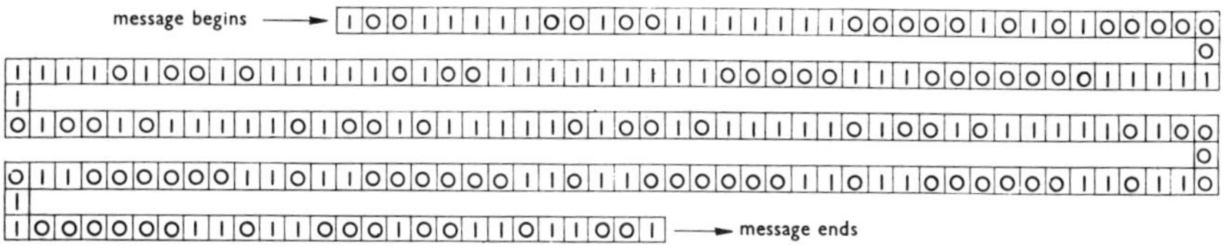

Communications with Extrasolar Intelligence 353

Fig. 16.21. Another hypothetical message (see text). Courtesy C. M. Cade and *Discovery*.

life is based on a carbohydrate chemistry. The binary number six above the raised arm of the right figure suggests six fingers and implies a base twelve number system. Finally, the dimension line at the lower right suggests that the figure is eleven somethings tall. Since the wavelength of 21 cm on which we received the message is the only length we both know, we conclude the beings are 231 cm, or 7 ft, in height."

A very similar scheme was developed by Cade[21] and is depicted by Figs. 16.21 and 16.22. The equal length pulses are arranged at "integral multiples of a minimum separation equal to one pulse length." As before, pulses get ones and blank intervals zeroes, yielding a sequence of 209 of both (the product of the prime numbers 19 and 11). The first array, seen in Fig. 16.22, has no meaning, but the alternate one shows an erect biped.

Alien signals purposely sent to elicit response from a listening society will probably continue for long periods of time carrying the same message. Therefore, once we suspect a message, confirmation would be forthcoming because of its repetitive nature. The cycle of repetition would be impossible to predict for much would depend on the concept of time inherent to the transmitting community. It would assuredly be wise were we to tape record

---

[21] Cade, C. M., "Communicating with Life in Space," *Discovery*, **24**, No. 5 (May 1963), 37–38.

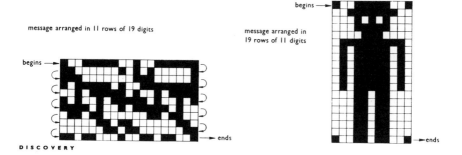

Fig. 16.22. Alternate interpretations of message in Fig. 16.21. Courtesy C. M. Cade and *Discovery*.

all signals and play them back at multiples and fractions of the speed at which we received them.

Many other schemes have been suggested, ranging from Hogben's[22] pioneering Astroglossa, a sort of mathematical Morse code, to the more recent Lincos developed in a 224-page book by Freudenthal.[23] The later describes "lingua cosmica," a linguistic vehicle the author proposes to transmit to extrasolar planetary systems by "radio signals of various duration and wave length." All information is proposed to be sent in accordance with abstract mathematical relationships rather than in the form of television pictures, a technique Freudenthal rejects because: "As long as we cannot tell the receiver, the principle, and the code of decomposition and arrangements, we cannot expect that he will understand television messages and transform them into images." He thinks of Lincos as if it is speech, not writing, and talks of "Lincos phonemes" rather than letters, of Lincos phonetica rather than spelling. He borrows the notations in his lingua cosmica from mathematics, logic, astronomy and Latin, and also uses Lincos words "written in entirely arbitrary fashion, such as *Ha, Hb, Hc*. . ." Since the language has to be taught to whomever receives it, only facts "which may be supposed to be known to the receiver" will at first be transmitted, enabling deciphering to be readily accomplished. Since mathematics is the most abstract and most (probably) universally known subject, the Lincos program, at its start, is mathematical in nature.

The Number of Communicating Civilizations in the Milky Way Galaxy

About as many estimates as there are scientists interested in the problem have been made concerning (1) how many suitable stars there are in the Milky Way capable of supporting attractive planets, (2) how many such planets there actually are, (3) on how many of them intelligent communities occur, (4) how long such communities may last, and (5) at any one time, how many communicating societies there may be in the galaxy.

Estimates range from relatively few to many, depending partially on guesses as to what happens to biological civilizations after they cross the technological threshold, how soon mechanical automata become dominant, and the desires and requirements very advanced communities may have to

---

[22] Hogben, Lancelot, "Astroglossa, or First Steps in Celestial Syntax," *Journal of the British Interplanetary Society*, 11, No. 6 (November 1952), 258–274.

[23] Freudenthal, Hans, *Lincos: Design of a Language for Cosmic Intercourse, Part I.* Amsterdam: North-Holland Publishing Co., 1960 (a volume in the series *Studies in Logic and the Foundations of Mathematics* ed. by L. E. J. Brouwer, E. W. Beth and A. Heyting).

communicate in the first place. One of the most optimistic figures in the literature on the numbers of stars with habitable planets is given by Dole[24] who concludes that "the mean distance between a star with a habitable planet and its closest neighbor with a habitable planet is about 24 light years." Others expect to find planets harboring intelligent beings considerably further apart, up to, and perhaps more than, 300 light years.[25] As we shall see in the next chapter, due to interstellar colonizing, inhabited planets may possibly be as close as a mere 10 light years.

Once an intelligent community has come into existence speculation arises as to how long it will last. We have no idea if most, many, or only a small percentage of civilizations that learn how to harness nuclear energy will destroy themselves. If they do not destroy themselves, do they inevitably continue to advance intellectually? Or are decay and degeneration the rule? No one knows. Even though a society may endure for hundreds of thousands, millions, or billions of years, it is impossible to predict whether or not it will desire to communicate, particularly with emerging civilizations. If these are very numerous, one more or one less could make little difference to an established community. The incentives for further acquisition of knowledge may become stifled, the urge to advance blunted, and the (to us) innate sense of curiosity may be dulled (perhaps in particular concerning beings just rising above the technological horizon).

Assuming a given society can be expected to endure for a million years, be it mechanical and/or biological, it would be interesting to know at what stage of development it is most likely to search for emerging communities. An analysis of our own case may shed some light on this consideration. We seem about ready to conduct long-term listening programs and, probably within 100 years, would be willing to undertake a more or less permanent transmitting program to selected stars within, say, 50 light years. Depending on the outcome of these endeavours, within 1,000 years from now at the most we should be able and desirous to inaugurate a massive search program, including the use of interstellar probes if necessary, to determine the occurrence of communities out to at least 1,000 light years. If we find no evidence of extrasolar life we may well become discouraged and give up. But, if we do make contact with an extrasolar communications network, it would be most unlikely that we would ever

---

[24] Dole, Stephen H., *Habitable Planets for Man*. New York: Blaisdell Publishing Co., 1964, p. 105.

[25] Cameron, A. G. W., "Communicating with Intelligent Life on Other Worlds," *Sky and Telescope*, **26**, No. 5 (November 1963), p. 260. He assumes 0.03 stars per cubic parsec, which comes out to 90 parsecs (or roughly 300 light years) between civilized solar systems.

Fig. 16.23. Communicating between the Milky Way and Andromeda galaxies.

withdraw from it. It would seem, then, that we may be contacted by a community ranging upwards from 1,000 years more advanced than we.

Intergalactic Communications

At a time when we are experiencing the infancy of interplanetary communications and are far from the realization of practical interstellar communications, it may seem absurd to even think about intergalactic communications. Yet the Russian astronomer Iosif Samuilovitch Shklovsky, some of whose ideas we have reviewed in Chapter 15, has done just that. In an interview[26] conducted at the Shternberg Astronomical Institute he advocated looking for signals produced by a supercivilization located

---

[26] Sullivan, Walter, "Search for Life in Nebula Urged," *New York Times* (3 Dec. 1962; dateline of interview 25 Nov. 1962).

around a star in the Andromeda galaxy. His reasoning is summarized as follows.

It is unlikely that there are many advanced civilizations in our own galaxy, perhaps only one within a radius of 700 light years from us. Those few that do exist would have to construct immensely powerful beacons to draw the attention of neighboring societies. The few and far between neighbors would be hard pressed to know where, amid the billions of stars in the Milky Way, to look for such signs of intelligence. But, if there are intelligent communities in Andromeda, a small target approximately 1 deg across, they would be revealed rather quickly by a powerful radio telescope.

Shklovsky speculates that an extremely advanced galactic community might harness 1 to 2 per cent of the energy of its parent star and convert it to radio energy for communications purposes, thus being able to span the 1,500,000 light years from Andromeda to the Milky Way (Fig. 16.23). Agreeing that the search should be at 1,420 Mc, he proposes scanning up to 2 Mc on either side of this basic frequency.

17
Summary and Conclusions

There appears to be a very good possibility that extrasolar intelligent societies are widespread in the universe and in our own galaxy, the Milky Way. If such societies are as prevalent as appears possible, then an extensive intragalactic communication network may exist; it is already technically feasible to attempt detection of intelligent signals from a distance of at least 10 light years and probably up to 100 light years. Since a vast amount of advanced scientific knowledge would be available if we could establish communications with superior extrasolar societies, it appears desirable to increase considerably our efforts towards detecting them.

A large portion of the research bearing on the subject of extrasolar intelligence is of independent general interest to other branches of the scientific community; therefore, very little research effort would have been wasted in the event of sustained failure to detect evidence of extrasolar intelligence.

Present research in artificial intelligence implies the most profound importance of this subject to very advanced technological societies. Superintelligent synthetic automata, originated by biologi-

"The eternal silence of these infinite spaces terrifies me."
Blaise Pascal

"The Universe is not hostile, nor yet is it friendly. It is simply indifferent."
John Haynes Holmes

"Order is Heav'n's first law"
Alexander Pope

"Fortune favors the prepared mind."
Louis Pasteur

"Observe how system into system runs, What other planets circle other suns."
Alexander Pope

cal societies, may be widespread in the universe, both as executives for biological societies and as independent entities.

In this final chapter we summarize facts, hypotheses, and speculations from many fields of science having a bearing on the probable prevalence of extrasolar intelligence. Assumed probability factors for each major field of consideration are explained, and then these factors are combined to yield the estimated number of stars in the Milky Way that have developed intelligent interstellar communicating societies in the past. Then we discuss the uncertainty of the longevity of these societies, which would greatly affect estimates of their present prevalence. This is followed by a brief review of the prospects and problems of communication with extrasolar intelligence.

Future research in many fields of science, engineering, and social science, all pertinent to the subject of extrasolar intelligence, is mentioned. These research areas include many of primary independent interest to scientists in their respective fields, and some of interest to those scientists mainly concerned with extrasolar intelligence.

The studies of the probabilities affecting the existence of extrasolar intelligence have many interesting ramifications. The mere realization of the possible, or probable, existence of extrasolar societies must have some impact on human thinking and philosophy.[1] The social impact will increase as an awareness of the implications of extrasolar intelligence filters slowly down to lower and lower educational levels, reaching an ever larger fraction of our society. But the bare realization of the probabilities concerning the existence of extrasolar intelligence is a sufficient initial goal for most scientists. If studies indicate a high probability that intelligent communities exist within our communication range, then it is inevitable that scientists and engineers will take an active part in establishing their nature. The problem of communications with the probable extrasolar societies would quickly be solved and Earth would stand to receive an undreamed of flood of advanced knowledge concerning all scientific and social disciplines.

Since the data are limited upon which the probabilities concerning the

---

[1] In 1960 the Brookings Institution prepared for the National Aeronautics and Space Administration a study ("Proposed Studies on the Implications of Peaceful Space Activities for Human Affairs") on the long-term social effects of space activities. Among many other topics, it dealt with the possible reaction of terrestrial social institutions should contact be made with an extrasolar society. If such a society were discovered, an all-out effort would probably be made by Earth to communicate with it, leading possibly to changes that could mean the collapse of civilization. The report noted that "societies sure of their own place have disintegrated when confronted by a superior society, and others have survived even though changed. Clearly, the better we can come to understand the factors involved in responding to such crises the better prepared we may be."

existence of extrasolar intelligence are based, and because their potential impact is very great, much additional research is desirable in order to increase the quantity of such data and, thereby, the confidence level of our statistics. An important aspect of the study of extrasolar intelligence is the elucidation of those scientific fields which are ripe for research emphasis.

The Probable Prevalence of Extrasolar Intelligence

Of the roughly 100 billion detectable galaxies in the known universe our own galaxy, the Milky Way, does not seem to be unique in any respect. The Milky Way is a typical spiral galaxy and spiral galaxies fall into one of several classes of common galaxies (See Chapter 2). A study of the probable existence of extrasolar intelligence in the Milky Way would apparently be applicable to most other galaxies, so the estimated number of intelligent societies in the Milky Way could be multiplied by 100 billion to obtain an approximate estimate of the number of intelligent societies in the known universe (for greater accuracy, the number of stars in each individual galaxy would have to be determined).

There exist from 150 to 200 billion stars in the Milky Way, some of which are suitable and some of which are unsuitable for the development of intelligent life. Stars that do not remain for a sufficiently long period of time on the main sequence would be unsuitable for the development of higher forms of life, since their energy output appears to be unstable. Developing life would probably be destroyed by the overheating and expansion of these stars.

The development of life with a human level of intelligence took about 4.5 billion years (counting from the creation of the Earth) and it will be assumed that this is the average time for this degree of evolution on an ideal planet. When a very small sample exists the confidence level is low, but the more atypical this sample is assumed to be the more improbable it becomes. Until the detailed structure of DNA and its genetic effects are understood, a precise estimate of the length of time required for any particular level of evolution must be somewhat in doubt. This evolutionary time, 4.5 billion years, is a critical assumption since the average lifetime of solar type stars is about 10 billion years. If the average time required for chemical and biological evolution approached or exceeded this average stellar lifetime, then the number of intelligent societies would be sharply reduced.

Many astronomers consider stars of spectral type O, B, A, and early

F unsuitable for the development of life, and there has been some debate about the suitability of the faint red stars. Huang feels the probability is low of a planet being located in the very restricted life zones of these stars, making them unlikely candidates for the development of life.[2] However, Cameron has argued that the number of planets falling in the life zone would be approximately constant down to the latest spectral classes.[3] Since there is some scientific uncertainty here, and since cosmological or stellar dynamic variables might easily remove planets from a small life zone, the faint red stars will be excluded from the set of suitable stars in this calculation.

Huang, in a detailed study of the dynamics of stellar systems, showed that binary and multiple star systems are not generally suitable for the development of life.[4] Planetary orbits of multiple stars would usually be unstable and would frequently move into regions of intense heat or cold. The already noted problem of stars having stable energy output is multiplied in multiple star systems where any one unstable star will make the entire system unsuitable for the development of life. The only binary systems which might be suitable would be special cases where two close stable stars would have fairly distant planets, or two very distant stable stars would have planets revolving close about one of the stars. From this paragraph we conclude that a majority of the binary and multiple star systems of the Milky Way are unlikely abodes of life.

In making assumptions similar to these, Cameron has concluded that the number of independent stars in the Milky Way, suitable for the development of higher forms of life (N_s) must be about 6 billion.[5]

$$N_S = 6 \cdot 10^9 \text{ stars}$$

The development of higher forms of life requires a stable platform such as a planet lying for a very long period in the life zone of a star. What fraction of the suitable stars can be expected to have planetary systems? The origin of the Solar System has been a major problem for hundreds of years and no fully and generally acceptable hypothesis appears likely in the near future. For many years accidental theories of the origin of planetary systems were in vogue, involving the collision or near collision of

---

[2] Huang, Su-Shu, "Occurrence of Life in the Universe," *American Scientist*, 47, No. 3 (Autumn—September 1959), 397–402.

[3] Cameron, A. G. W., "Stellar Life Zones," in *Interstellar Communication*, ed. A. G. W. Cameron. New York: W. A. Benjamin, Inc., 1963.

[4] Huang, *Op. Cit.*, and "Life Supporting Regions in the Vicinity of Binary System," *Publications of the Astronomical Society of the Pacific*, 72, No. 425 (April 1960), 106–114.

[5] Cameron, A. G. W., "Future Research on Interstellar Communication," in *Interstellar Communication*.

stars. Since collisions are generally regarded as rare events, the frequency of occurrence of planetary systems was assumed to be very low.

Harlow Shapley has pointed out that this is not necessarily true since the cosmological theory of the expanding universe implies that the astronomical bodies were much more densely packed in the distant past, and collisions must have been much more frequent.[6] Be that as it may, natural planetary origin theories are now receiving general acceptance, implying that planets are formed as a regular by-product of stellar formation processes. Our own Solar System may be taken as one bit of empirical evidence for the existence of planetary systems, and the occurrence of binary and multiple stars may also be taken to imply the existence of planetary systems (Chapter 3). Recent studies of periodicities in the motions of some nearby stars have indicated the existence of dark companions. Also, the sharp reduction in rotation rate for stars later than F_5 probably indicates that the excess angular momentum has been transferred to unseen planetary worlds. Considering the foregoing implications and empirical evidence it will be assumed that the fraction (f_s) of the 6 billion suitable stars having planetary systems is one:

$$f_s = 1$$

If all 6 billion suitable independent stars are assumed to have planetary systems, we can ask what fraction of these planetary systems (f_P) can be expected to have at least one planet capable of supporting the development of higher forms of life. This factor is determined by the average number of planets in a planetary system, the average extent of the life zone of the star, the size of the planets falling within the life zone, and the chemical milieu of the planets in the life zone. Since the Solar System has nine planets, it may again be assumed that this is the average number of planets in a planetary system. In the Solar System there are apparently two planets (Earth and Mars) lying within the life zone, but it will require instrumented probes to verify conditions on Mars (which lies at the very outside edge). In a detailed study of this problem Cameron has concluded that the average number of planets lying in the life zone of any type of star would be 1.4.[7]

The mass of the planet is another significant factor. Too small a mass would cause the loss of protective atmosphere, and too large a mass would result in a very dense atmosphere completely blocking radiation from the solid or liquid surface; moreover, high pressure at lower levels would tend to dissociate the atoms of many molecules. Too large a planetary mass would also hinder the evolution of large mobile animal life because of high

---

[6] Shapley, Harlow, *Of Stars and Men*. Boston: Beacon Press, 1958.

[7] Cameron, A. G. W. "Stellar Life Zones," in *Interstellar Communication*.

gravitational force. However, Bode's Law indicates that there is something systematic about the distribution of the masses of the planets of the Solar System, and this may lead to an enhanced expectancy of a planet of appropriate size existing in the life zone of a star. The chemical milieu of the planet should include an initially reducing atmosphere made up of such elementary molecules as hydrogen, ammonia, water vapor, and methane, but this is assumed to be the chemical environment of all primitive planets. It is also necessary that a planet contain a sufficient number of heavy elements permitting it to have a distinct solid surface, which may become partially or completely covered by liquids. Considering all these factors, we conclude that the fraction (f_P) of the 6 billion planetary systems which can be expected to have at least one planet that is suitable for the development of higher forms of life is about 0.5.

$$f_P = 0.5$$

Therefore, the total number of planetary systems in the Milky Way having at least one planet suitable for the evolution of higher forms of life is about 3 billion.

If 3 billion independent stars in the Milky Way have at least one planet which is well suited for the development of higher forms of life, by being very well situated in the life zone of the star, by having the right mass, by initially having a reducing atmosphere, and by having a substantial solid core, what fraction of these stars (f_L) will actually give rise to life? Chemical evolution from primitive elementary molecules is an inevitable result of the application to them of various forms of energy. Virtually any kind of energy (such as heat, ultraviolet light, electricity, or cosmic rays) when applied to a mixture of primitive atmospheric molecules will produce a mixture of amino acids which are the building blocks of proteins, the main constituent of living organisms.

Rapid progress is now being made showing how the basic building blocks of DNA can develop naturally from primitive molecules. DNA is a self-replicating molecule that directs the formation of all living organisms. The random polymerization of very large numbers of DNA molecules must inevitably yield many which cause the formation of assemblages of protein molecules that in some ways facilitate the survival and reproduction of those particular DNA molecules. This phenomenon of autocatalysis is the beginning of biological evolution. For these reasons, most biologists feel that chemical evolution followed by the development of biological life is an inevitable development on any ideal planet. Therefore, all the 3 billion stars having planets ideally suited to the development of life inevitably will develop proliferate life on their planets.

$$f_L = 1$$

What fraction (f_I) of the 3 billion stars in the Milky Way, having planets containing proliferate living organisms, also have evolved a species possessing a level of intelligence at least equal to that of humans? Intelligence is no more and no less than the information processing capability of an animal nervous system, and human intelligence is just a single point on a continuous spectrum of types and degrees of intelligence.[8] Every animal, no matter how primitive, has intelligence of some level, and intelligence is definitely a factor in natural selection. Intelligence provides a means for an animal to perceive and analyze the environment and also to act upon the environment. Although many other factors play prominent roles in natural selection, intelligence will always be a significant selective factor at all its levels. Therefore, the evolution of a human (or a space communication) level of intelligence is inevitable on a planet where many varied species flourish over sufficiently long periods of time. Stated another way, the evolution of a human level of intelligence is merely a probability function of time on a suitable planet.

Again, our only feasible alternative is to assume that the time required to evolve a human level of intelligence on Earth (4.5 billion years) is average for all planets having proliferate biological species. Therefore, the fraction (f_I) of the 3 billion stars in the Milky Way having proliferate life, which have developed at least one intelligent species must be assumed to be 1 or at least very close to 1.

$$f_I = 1$$

Some scientists have speculated that many intelligent species would not be technically inclined or that they might lose their interest in science and hence communication with other intelligent species. All endeavors of intelligent beings may be conveniently divided into four classifications: (1) physical science, (2) social science, (3) engineering science, and (4) operational science (the human operation of engineering devices). When interpreted broadly, these four branches of science include every facet of valid human knowledge; to varying extents, they all contribute to advancing technology which in turn promotes growth in the sciences. The pursuit of science and technology, then, is the only means to better understanding and control of the whole physical world; the idea of an intelligent society or automaton completely losing its interest in science and technology is inherently absurd.

All biological societies that have evolved human level or higher intelligence will almost inevitably become technologically capable of some

---

[8] MacGowan, Roger A., "On the Possibilities of the Existence of Extraterrestrial Intelligence," in *Advances in Space Science and Technology, Volume 4*, ed. Frederick I. Ordway III. New York: Academic Press Inc., 1962, pp. 64–80 and 105; MacGowan, Roger A., *Intelligence: A Definition and Military Implications*, Army Missile Command Technical Report No. COMP-TR-1-64, Redstone Arsenal, Ala.: Army Missile Command Computation Center, 16 January 1964.

degree of interstellar communication. The only thing which could conceivably prevent or restrict interstellar communication by intelligent biological societies or automata would be fear of interstellar competition or warfare, but it has been hypothesized previously that the very advanced level of technology required for interstellar transportation also implies a proportional capability for complete physical integration of brains or information processing systems. Therefore, competition and warfare are probably unknown in interstellar society. The fraction (f_c) of the 3 billion stars in the Milky Way having at least one planet on which an intelligent (human level or greater) society has evolved, and which has engaged in interstellar communication, is assumed to be 1.

$$f_c = 1$$

We conclude that 3 billion stars in the Milky Way have evolved intelligent communicating societies, but we need to know how many still exist. In order to plan for communicating with extrasolar intelligence, it is desirable to know the present number and distribution of intelligent societies.

R.N. Bracewell, a pioneer researcher on extrasolar intelligence, was one of the first to recognize the critical significance of the longevity of intelligent technical civilizations.[9] Of the several factors contributing to reduce the lifetime of technical societies, the one given greatest consideration among scientists is self-annihilation. We regard the likelihood of a major atomic war as very small, and even if it should occur, it would not completely destroy our technological civilization. If the human annihilation hypothesis of the atomic-war pessimists is accepted, then we would need to estimate the length of time required for a new intelligent species to evolve. This might require up to 50 million years. The length of time that an intelligent society would have existed in the space communicating phase would only have been about 50 years, since any societies lasting longer than this, while having the technological capability for self-destruction, must have found lasting solutions to the problem of social competition. Therefore, the assumption of general social self-annihilation implies that stars evolving intelligent societies would have a society in the interstellar communication phase only about 50 years out of every 50 million years, or one-millionth of the time.

The age of our galaxy, the Milky Way, is at least 15 billion and perhaps as many as 25 billion years. Most of its stars appear to have been formed in the first few billion years. It has already been assumed that the average time required to evolve an intelligent species on an ideal planet is

---

[9] Bracewell, R. N., "Communications from Superior Galactic Communities," Nature, **186**, No. 4726 (28 May 1960), 670–671.

about 4.5 billion years, and the average solar type star can be expected to survive for about 10 billion years. Therefore, it will be assumed that the average length of time since the ideal stars and planets first evolved intelligent species is several billion years.[10] If the self-annihilation hypothesis is accepted, it must be concluded that the fraction of the ideal stars (f_{cT}) in a space communication phase at any particular time during the last several billion years is about 10^{-6}

$$f_{cT} = 10^{-6}$$

This pessimistic assumption would indicate that only about 3,000 communicating intelligent societies would be likely to exist at the present time in our galaxy, and the average distance between intelligent societies would be on the order of 1,000 light years (see Fig. 12.2). In this event, the time required for an electromagnetic signal to travel to an adjacent star and a reply to return is about 2,000 years, a time period that greatly exceeds the assumed lifetime of the intelligent communicating societies. Therefore, no interstellar communications or communications network could exist, if the self-annihilation hypothesis is accepted.

J.P.T. Pearman, another noted scientist, has suggested that the distribution of lifetimes of intelligent societies may be bimodal, since some emerging intelligent societies may survive the early critical stage of development, and then they could survive for astronomical periods of time.[11] He guesses 1,000 years and 100 million years for the two modes. Another factor that could contribute to a bimodal distribution of lifetimes would be the feedback effect of the existence of an interstellar communication network, which could provide social survival information to those emerging societies fortunate enough to detect the signals in time.

As mentioned earlier, we do not accept the pessimistic hypothesis of self-annihilation. It seems infinitely more probable that most emerging technical societies will survive for long periods, improve their technological control, and develop intelligent automata.[12] The intelligent automata should be capable of beginning interstellar emigration in a time on the order of hundreds of years, leaving the biological society behind. Eventually, after a time on the order of thousands of years, some of the members of the biological society could conceivably emigrate to ideal planetary systems and colonize them. (During this period many intelligent automata may have been developed and may have emigrated.) This possible slow spreading

---

[10] Cameron, A. G. W., "Future Research on Interstellar Communication," in *Interstellar Communication*.

[11] Pearman, J. P. T., "Extraterrestrial Intelligent Life and Interstellar Communication: An Informal Discussion," in *Interstellar Communication*.

[12] MacGowan, Roger A. "On the Possibilities of the Existence of Extraterrestrial Intelligence," in *Advances in Space Science and Technology, Volume 4*, p. 80–103.

and colonization of all ideal planets by biological societies, and more rapid emigration (perhaps toward the center of the galaxy) of intelligent automata should, after a period of millions of years, result in all 3 billion stars, having ideal planets, being inhabited by intelligence.

$$f_{cT} = 1$$

It is also possible that many biologically unsuitable stars toward the center of the galaxy are inhabited by intelligent automata. This implies the likelihood of a very extensive interstellar communication network and even some local direct contacts by intelligent biological neighbors. The average distance between neighboring intelligent communities may be as few as 10 light years.

Another possibility is that reproduction control would be sufficient to reduce the size of the biological society (after many generations) to the point that the entire society might undertake interstellar emigration. This would probably imply a preparation time in the interstellar communication phase on the order of 1,000 years or more, and then there would not be a sudden cessation of communications since the emigration would necessarily be a very slow star-to-star, stepwise process. With this assumption, communications could come from a biological society in one general astronomical region for a period possibly on the order of 10,000 years, whereas re-evolution of another intelligent society in that general area might require 1,000,000 years or more. Then, the fraction of the time that communications might be received from an ideal planet or its general area would be on the order of 10^{-2}, or the fraction of the ideal planets communicating (or having communications from its general area) at any particular time would be about 10^{-2}.

$$f_{cT} = 10^{-2}$$

This indicates a distance between communicating societies on the order of tens of light years and, therefore, a very extensive interstellar communication network.

In summary, the distribution of intelligent societies in our galaxy is determined almost entirely by one most uncertain factor—the longevity of an intelligent communicating biological society or automaton on a particular planet or in its general astronomical vicinity. The number of suitable stars is quite large and not subject to great doubt, and the fraction of these suitable stars having planetary systems is assumed by many astronomers to be very close to 1 (although empirical evidence for this is still somewhat limited; and there is, as yet, no complete and widely accepted precise theory of planetary origin).

The fraction of solar systems having at least one ideal planet (for the development of life) we guess to be 0.5, and many astronomers would accept a value of similar magnitude; however, the limited empirical evidence for this is the large number of bodies in our own Solar System and their systematic distribution. The probability of life developing spontaneously on an ideal planet would be placed at or close to 1 by many biochemists, although admittedly every step of chemical evolution and the origin of life is not completely understood.

The evolution of intelligent species is a function of time and some random variables, but intelligence, in the broad sense, is so important a factor in natural selection that we estimate this probability to be 1. Some biologists would dispute this, but they are probably not giving sufficient weight to the intelligence factor (at all levels) in natural selection. It seems obvious that all intelligent societies will develop accelerating technologies, and all would attempt, sooner or later, interstellar communication and contact.

The longevity of an intelligent communicating society in a given area could conceivably vary anywhere from a few years to several billions of years; and, therefore, the fraction of the ideal planets containing intelligent communicating societies or automata at any particular time could vary from close to 0 to greater than 1 (because of colonization and emigrating automata). The reasons for this uncertainty are the lack of thought concerning (1) the social implications of construction of artificial intelligent automata, and (2) other very long-range effects of technology on biological societies (for example, biochemical genetic engineering and integration of synthetic components into biological organisms).

The following equation gives the estimated total number (N) of stars in the Milky Way which have developed intelligent interstellar communicating societies at some time in the past:

$$N = N_S \cdot f_S \cdot f_P \cdot f_L \cdot f_I \cdot f_C = 3 \cdot 10^9$$

where

$6 \cdot 10^9 = N_S = $ The number of independent stars in the Milky Way, suitable for the development of higher forms of life.

$1. = f_S = $ The fraction of the suitable stars having planetary systems.

$0.5 = f_P = $ The fraction of the planetary systems that can be expected to have at least one planet suitable for the development of higher forms of life.

$1. = f_L = $ The fraction of these planetary systems having an ideal planet upon which proliferate biological life will actually have developed.

$1. = f_I = $ The fraction of these planetary systems having a planet containing proliferate biological species that have evolved a species with a human level of intelligence.

$1. = f_C = $ The fraction of these planetary systems having a planet which has evolved an intelligent species that has engaged in interstellar communication.

fraction of the money the United States budgets the National Aeronautics and Space Administration (NASA) would be adequate for expanded efforts to detect extrasolar intelligent signals. Most of the equipment required for this research effort would also be useful for other general astronomical research, so little or nothing would have been wasted even if we fail to detect such signals. Considering this, research efforts for detecting extrasolar intelligent signals should be considerably expanded and accelerated.

Future Research Concerned with Extrasolar Intelligence

The extremely limited data supporting the fundamental components of the equation of probability of existence of extrasolar intelligence leave much to be desired. Research in many of the areas described could provide additional data and yield an increased confidence in the probability estimates. Of course, this new information could also radically change the currently accepted probability estimates. Wherever a decrease in the degree of speculation of our estimates can be achieved readily and economically, new research should be pursued aggressively.

Research promising to yield significant data on which to base estimates concerning the existence of extrasolar intelligence may be conveniently divided into two classes: (1) research of general interest to the scientific community and not exclusively oriented toward extrasolar intelligence, and (2) research specifically oriented toward extrasolar intelligence, which does not hold particular independent interest for the scientific community at large. There is no perfectly sharp line of demarcation between these two classes, but the distinction is clear enough to be of practical use. Fortunately, the great majority of the research required for expanded knowledge concerning extrasolar intelligence is of completely independent interest to the general scientific community. Therefore, there is no danger of wasting money and research talent on a project that might be completely unproductive. If future research projects having implications for extrasolar intelligence should greatly reduce the probable number of extrasolar intelligent societies, then most of these individual research projects would still be continued or completed since they are of broad concern to scientists and vital to the normal growth of scientific knowledge.

Another significant point concerning these required research projects is their relative independence of one another. A single, integrated project having a central organization directing all activities is not really necessary. The activities of many of the world's leading research scientists could be expanded and accelerated by merely increasing their research budgets slightly. They could continue their already outstanding work in their institutions and would not be required to change location.

Adequate research coordination and communication could be achieved through periodic government-sponsored, multidisciplinary seminars and the publication of the proceedings. One very limited seminar on extrasolar intelligence, sponsored by the Space Science Board of the National Academy of Sciences, was held in November 1961 at the National Radio Astronomy Observatory at Green Bank, West Virginia. Otto Struve served as chairman; other attendees were D.W. Atchley, Jr., M. Calvin, G. Cocconi, F.D. Drake, S.S. Huang, J.C. Lilly, P.M. Morrison, B.M. Oliver, C. Sagan, and J.P.T. Pearman. The Soviets held a similar meeting of minds between 20 and 23 May 1964 at the Byurakan Astrophysical Observatory in the Armenian SSR. The participants included I.S. Shklovsky, N.S. Kardashev, Yu. N. Pariyskiy, V.I. Slysh, S.E. Khaykin, V.S. Troitskiy, V.A. Kotel'nikov, N.A. Smirnova, N.L. Kaydanovskiy, and Ya. B. Zel'dovich; V.A. Ambartsumyan served as conference director.

ASTRONOMICAL RESEARCH

Many prominent astronomers have already made major pioneering contributions to the study of extrasolar intelligent life. The apparent preeminence of astronomers undoubtedly results from the fact that new theories and information pertaining to planetary system origin and frequency are a necessary prerequisite to sensible speculation concerning extrasolar intelligence.

Many areas of classical astronomical interest also contain implications for extrasolar intelligence research. As stellar characteristics and evolution are of general interest, so stellar formation and the origin of planetary systems are of special interest. Indirect empirical evidence indicating the possible presence of other planetary systems, in support of some current natural planetary origin theories, is also of great concern. Astrometric studies of small perturbations of nearby stars indicating the presence of unseen companions of varying sizes offer good prospects for further research. Research into the variation in stellar rotation rates with spectral class also fall into this category. Another astronomical field of continuing importance is Solar System research and planetary dynamics.

All of this research is of general interest to astronomers and it will certainly be pursued without any special subsidies. However, some extra encouragement in the form of small government appropriations to a few key researchers would greatly accelerate projects of special import to those investigating the possible existence of extrasolar societies.

BIOCHEMICAL RESEARCH

Biochemical research is unquestionably the most productive research frontier of our current era. The discovery of the structure of DNA by

Watson and Crick promises to lead directly to a very detailed understanding of all fundamental aspects of biochemical genetics. Knowledge of the effects of variations in DNA, the role of RNA, and the production of proteins is increasing at a remarkable rate. Much of this comes under the heading of general biochemical research, but some prominent biochemists are particularly concerned with chemical evolution under assumed primitive planetary conditions and also the origin of biological life.

General biochemical research seems to be very well supported and the present rate of progress is rapid, but research specifically aimed at laboratory simulation of chemical evolution seems rather limited. It is desirable to expand the activities of present researchers and to encourage other researchers to enter this field, which cannot help but be of great interest to all biochemists.

BIOLOGICAL RESEARCH

Biological research, encompassing especially the fields of anthropology, paleontology, and evolution, also is highly pertinent to studies of extrasolar intelligence. Certain critical questions are still considered controversial by some prominent biologists, meaning that much additional research is necessary. A number of biologists feel that intelligence, and certain gross morphological characteristics which are assumed to be essential accompaniments of high intelligence, are so important as factors in natural selection that a few distinct morphological categories for intelligent biological life can be predicted; furthermore, the evolution of highly intelligent life is assumed to be inevitable if given a sufficiently long period of evolution on a habitable planet.

Some biologists maintain that the uniqueness of the evolutionary steps which have led to *Homo sapiens* implies that the evolution of an intelligent species must be a very rare event, even considering the entire universe. They feel that the formation of the major classes of animal life was due to extraordinary events and circumstances. Others assert that the whole chain of evolutionary steps, which led to *Homo sapiens* on Earth, would have to be duplicated almost exactly on other planets beyond the Solar System.

Further elaboration of these considerations, either in the form of additional research or extensive scientific discussion and analysis, is obviously desirable. We want to know, for example, if there are only a very small number, a moderate number, or a very large number of possible evolutionary paths leading to some species having a human level of intelligence. And, are there only a few, a moderate number, or a very large number of genetically possible species having a human level of intelligence?

Here it is being assumed that we are simply concerned with ideal planetary environments with respect to temperature range, mass, density, atmospheric composition, etc., and that these ideal planets contain prolific biological life. This question can only be answered with mathematical precision after more biochemical research provides a precise mathematical-statistical understanding of genetics, but further study of anthropological and paleontological evidence may be able to give some indication of the final answer.

PSYCHOLOGICAL RESEARCH

A mathematical understanding of intelligence and related psychological phenomena is not only of interest in connection with biological evolution, but also encompasses the most profound implications for the development of synthetic intelligent automata. Virtually undreamed of concentrations of synthetic intelligence may be distributed throughout our galaxy and the universe.

At the present time psychologists, mathematicians, electronic engineers, and neurologists are making significant strides in research in artificial and biological intelligence, but most of the scientific community has not been made aware of the progress that has been made. This lack of awareness of a general improved understanding of the fundamentals of intelligence and related psychological phenomena can be corrected by an increased diffusion of research results in the scientific literature.

Research on artificial and biological intelligence, recognized by many leading universities to be of profound importance, is being provided substantial support. Military organizations have also begun to recognize the need for the solution of a variety of problems involving the parallel processing of two dimensional information patterns, problems invariably leading to research on neural networks—an integral part of artificial and biological intelligence research.

Advanced psychological research leading to a precise mathematical understanding of intelligence, drives, emotions, etc., is fairly well supported. The implications for extrasolar intelligence are not a primary consideration in support of such research, and there appears to be no reason at present to attempt to justify an acceleration of it on the basis of these implications. (However, there are many other justifications of the utmost significance for accelerating psychological research.)

INTERSTELLAR COMMUNICATION

Unquestionably, the most pertinent field for studies of extrasolar intelligence is interstellar communication. Research is being done and hardware is being developed leading to improved means of both interplanetary and interstellar communication. Laser, radiotelescope, radio transmitter and antenna, and radiometer research are all progressing relatively rapidly.

Some of this research is being supported by military budgets, some by the government space program, some by universities, and some by astronomical observatories. Unfortunately, those aspects of communication research specifically pointed toward the detection of intelligent extrasolar signals are largely unsupported. The expensive construction of large fixed and movable radio telescopes is proceeding briskly, and broad, basic research in lasers seems to be producing satisfactory results. Space communications research to facilitate communication with lunar and planetary probes is extremely urgent and has been royally supported by the space program. Military research support for satellite communication systems, and more especially for laser development, has also been quite substantial.

But what is really needed at the present time is modest but steady support for a long range project with the explicit goal of detecting extrasolar intelligent signals. A modest amount of time is required on some of the larger radio telescopes such as the one at Arecibo in Puerto Rico, and a competent team should be working continuously on the development of radiometers specifically designed to detect intelligent signals.

One or two digital computer specialists should be working in close cooperation with the radiometer team developing computer programs for the detailed statistical analysis of recorded radiometer data. Uncertainty as to frequency, phase, pulse width, etc. would require an astronomical number of calculations in many passes through the recorded data. Thus, time on a high speed digital computer would also be required. As scientific projects go, this would be a very small effort, both in cost and manpower, and the potential gain in scientific knowledge per unit of cost would undoubtedly be greater than the same ratio for any other conceivable research project.

Project Ozma, under the capable and visionary leadership of Dr. Frank D. Drake, was the first, albeit brief, effort of this nature. It is almost inconceivable that leading nations like the United States and the Soviet Union would spend billions of dollars per year on a major space program and not devote a few tens of thousands of dollars a year to maintain a small Ozma-type project to detect extrasolar intelligent communications. There can be no question but that the possible scientific gains warrant the outlay of sufficient research funds to maintain continuously a small but productive research team to carry out modest search efforts.

SOCIAL EVOLUTION RESEARCH

It has been shown that the most critical and uncertain factor determining the present density of extrasolar intelligent societies is their longevity. Related to this problem are the social interaction of intelligent biological societies, and the development and interaction of synthetic intelligent automata.

The major nations of the world at present spend but modest sums for research on social evolution, while earmarking prodigious amounts in efforts to optimize the functioning of their existing social organizations. This shortsighted policy results to a large extent from the conservatism of vested interests—economic, political, military, religious, and even scientific. For centuries physical scientists have honored an unwritten nonaggression pact with religious and political entities, with the result that scientific techniques and methods are rarely applied to the more basic social and philosophical problems of mankind; and even when they are, the results are frequently not widely disseminated in the general scientific literature. With one hand, some organizations dominated by physical scientists have ridiculed the methods and achievements of social scientists; and with the other, these same organizations have strangled progressive efforts by the social scientists.

Social science is an extremely complex subject and it is a function of technology—which is evolving at a fantastic rate. Such long range developments as genetic engineering, population control, and synthetic intelligent automata promise to exercise startling effects on social evolution.

Not only estimates of the prevalence of extrasolar intelligent societies, but also the immediate future well-being of our present human society, hinge on greatly improved understanding of social evolution. This can only be achieved by greater cooperation and mutual support by physical and social scientists all over the world. The result may be the accelerated advancement of science for the welfare of human society here on Earth—and for the determination of man's ultimate relationship with the universe around him.

Bibliography[1]

"Information Processing" (UNESCO). London: Butterworth's Scientific Publications, 1959.

"Signalisation Interplanétaire," *La Nature*, No. 2944 (1 January 1935), 1.

Trudy Sektora Astrobotaniki (Transactions of the Astrobotany Dept. of the Academy of Sciences, Vol. 5). Translations of five articles dated December 1960 by Technical Information Center, Wright Patterson Air Force Base, Ohio (contributions on astrobiology by Kozlova, Suvorov, Stanko, Suslov and Parshina).

"Vsesoyuznoye Soveshchaniye no Probleme Vnezemnykhtsivilizatsiy" (Trans. All-Union Conference on the Problem of Extraterrestrial Civilizations), Byurakan, 20–23 May 1965. Yerevan, lzd-vo An Armyanskoy SSR, 152 p.

Abell, G. O., "Membership of Clusters of Galaxies," in *Problems of Extra-Galactic Research*, ed. G. C. McVittie. New York: Macmillan Co., 1962.

Abelson, P. H. (ed.), *Researches in Geochemistry*, New York: John Wiley and Sons, Inc., 1960.

———, "Amino-acids Formed in 'Primitive Atmospheres'," *Science*, **124**, No. 3228 (9 November 1956), 935 (abstract).

Ahrens, Louis H., "Oldest Rocks Exposed," *Geological Society of America Special Papers*, **62** (1955), 155.

Alfvén, Hannes, *On the Origin of the Solar System*. Oxford: Clarendon Press, 1954.

———, "On the Origin of the Solar System," *New Scientist*, **7**, No. 182 (12 May 1960), 1188.

———, "Non-solar Planets and the Origin of the Solar System," *Nature*, **152**, No. 3868 (18 December 1943), 721.

Allen, Tom, *The Quest: A Report on Extraterrestrial Life*. Philadelphia: Chilton Books, 1965.

---

[1] Journal and book literature. For sources in the report literature, see "Extraterrestrial Life—Part 1," National Aeronautics and Space Administration, Rept. SP-7015, Sept. 1964, Washington, D.C.

Alter, G. and J. Ruprecht, *The System of Open Star Clusters and Our Galaxy.* New York: Academic Press, 1963.

Anders, Edward, "The Moon as a Collector of Biological Material," *Science,* 133, No. 3459 (14 April 1961), 1115.

———, "Origin, Age and Composition of Meteorites," *Space Science Reviews,* 3 (December 1964), 583.

——— and Frank W. Fitch, "Search for Organized Elements in Carbonaceous Chondrites," *Science,* 138, No. 3548 (28 Dec 1962,) 1392.

Anderson, P., *Is There Life on other Worlds?* New York: Crowell-Collier Co., 1963.

Arnold, James R., and S. A. Tyler, "Fossil Organisms from Precambrian Sediments," *Annals of the New York Academy of Sciences,* 108, Art. 2 (29 June 1963), 451.

Anfinsen, Christian B., "Steps Towards Breaking the Genetic Code," *New Scientist,* 11, No. 247 (10 August 1961), 324.

Arrhenius, Svante, *Worlds in the Making.* New York: Harper and Brothers, 1908.

Ascher, Robert and Marcia Ascher, "Interstellar Communication and Human Evolution," *Nature,* 193, No. 8419 (10 March 1962), 940.

———, "Recognizing the Emergence of Man," *Science,* 147, No. 3655 (15 January 1965), 243.

Asimov, Isaac, *Fact and Fancy.* Garden City, N.Y.: Doubleday & Co., Inc., 1962 (see Chap. 11).

———, "Hello, CTA-21—Is Anyone There?" *New York Times Magazine,* 29 November 1964.

Augusta, Josef and Zdeněk Burian, *Prehistoric Man.* London: Paul Hamlyn, 1960.

Baade, Walter, *Evolution of Stars and Galaxies.* Cambridge, Mass.: Harvard University Press, 1963.

Baldwin, Ralph B., *The Face of the Moon.* Chicago: University of Chicago Press, 1949 (see Chaps. 4 & 5 on Terrestrial and Fossil Terrestrial Meteorite Craters).

Barghoorn, Elso S. and Stanley A. Tyler, "Microorganisms from the Gunflint Chert," *Science,* 147, No. 3658 (5 February 1965), 563.

Basa, A. B., and E. J. Hawrylewicz, *Life in Extraterrestrial Environments.* Chicago: Armour Research Foundation Report No. 3194–4 (NASA Contract NAS r-22), 19.

Beer, Gavin de, *Atlas of Evolution,* London: Nelson, 1965.

Bernal, J. D., "Evolution of Life in the Universe," *Journal of the British Interplanetary Society,* 12, No. 3 (May 1953), 114.

———, *The Physical Basis of Life.* London: Routeledge and Kegan Paul, 1951.

Bickerton, A. W., "Partial Impact: A Possible Explanation of the Origin of the Solar System, Comets, and other Phenomena of the Universe," *Transactions, New Zealand Institute,* 11 (1 August 1878), 125.

Birkeland, Kr., "Sur l'origine des planètes et de leurs satellites, *Comptes Rendus des Séances de l'Académie des Sciences,* **155** (4 November 1912), 892.

Blum, Harold F., *Time's Arrow and Evolution.* Princeton, N.J.: Princeton University Press, 1951.

———, "Perspectives in Evolution," *American Scientist,* **43**, No. 4 (October 1955), 595.

Boehm, George A. W., "Are We Being Hailed from Interstellar Space?" *Fortune,* **63**, No. 3 (March 1961), 144.

Bonner, William, *The Mystery of the Expanding Universe.* New York: Macmillan Co., 1964.

Borek, Ernest, *The Code of Life.* New York: Columbia University Press, 1965.

Boton, E. A., "An Instrumented Search for Extraterrestrial Life," *Space Science Reviews,* **3** (December 1964), 715.

Boycott, Brian B., "Learning in the Octopus," *Scientific American,* **212**, No. 3 (March 1965), 42.

Bracewell, R. N., "Communications from Superior Galactic Communities," *Nature,* **186**, No. 4726 (28 May 1960), 670.

———, "Radio Signals From Other Planets," *Proceedings of the IRE,* **50**, No. 2 (February 1962), 214.

———, "Life in the Galaxy," in *A Journey through Space and the Atom,* ed. S. T. Butler and H. Messel. Sydney (Australia): Shakespeare Head Press Pty. Ltd., 1962.

Brancazio, Peter J. and A. G. W. Cameron (eds), *The Origin and Evolution of Atmospheres and Oceans.* New York: John Wiley & Sons, 1964.

Briggs, M. H., "The Chemical Origins of Life," *Spaceflight,* **11**, No. 3 (July 1959), 69.

———, "Evidence of Extraterrestrial Origin for some Organic Constituents of Meteorites," *Nature,* **197**, No. 4874 (30 March 1963), 1290.

———, "The Chemistry of the Lunar Surface," *Journal of the British Interplanetary Society,* **18**, No. 11 (July-August 1962), 386.

———, "The Detection of Planets at Interstellar Distances," *Journal of the British Interplanetary Society,* **17**, No. 2 (March-April 1959), 59.

———, "Terrestrial and Extraterrestrial Life," *Spaceflight,* **11**, No. 4 (October 1959), 120.

———, "Superior Galactic Communities," Spaceflight, **3**, No. 3 (May 1961), 109.

———, and Gregg Mamikunian, "Chemistry of the Solar System" in *Advances in Space Science and Technology,* Vol. 9, ed. Frederick I. Ordway III. New York: Academic Press, 1966.

———, "The Distribution of Life in the Solar System: An evaluation of the Present Evidence," *Journal of the British Interplanetary Society,* **18**, Nos. 11-12 (September-December 1962), 431.

———, and Gregg·Mamikunian, "Organic Constituents of the Carbonaceous Chondrites," *Space Science Reviews,* **1** (1962-63), 647.

Brown, Harrison, *The Challenge of Man's Future.* New York: Viking Press, 1954.

———, "Planetary Systems Associated with Main Sequence Stars," *Science,* **145**, No. 3637 (11 Sept. 1964), 1177.

Berrill, N. J., *Worlds Without End.* New York: Macmillan Co., 1964.

Budden, K. G. and G. G. Yates, "A Search For Radio Echoes of Long Delay," *Journal of Atmospheric and Terrestrial Physics,* **2**, No. 5 (1952) 272–281.

Bussard, R. W., "Galactic Matter and Interstellar Flight," *Astronautica Acta,* **6** (1960) 179.

Cade, C. M., "Communicating with Life in Space," *Discovery,* **24**, No. 5 (May 1963), 36.

———, "Are We Alone in Space?" *Discovery,* **24**, No. 4 (April 1963), 27.

Calvin, Melvin, "Origin of Life on Earth and Elsewhere," *Proceedings of Lunar and Planetary Exploration Colloquium,* **1**, No. 6 (25 April 1959), 8. Also *Annals of Internal Medicine,* **54**, No. 5 (May 1961), 954, and *Armed Forces Chemical Journal* (March–April 1961), 26.

———, "Chemical Evolution and the Origin of Life," *American Scientist,* **44**, No. 4 (July 1956), 248.

———, "Communication: From Molecules to Mars," *The AIBS Bulletin,* **12**, No. 5 (October 1962), 29.

———, *Chemical Evolution.* Eugene: University of Oregon Press, 1961.

———, and Susan K. Vaughn, "Extraterrestrial Life: Some Organic Constituents of Meteorites and Their Significance for Possible Extraterrestrial Biological Evolution," in *Space Research,* ed. H. K. Kallman Bijl. New York: Interscience Publishers, Inc., 1960.

Cameron, Alastair G. W. (ed.), *Interstellar Communication.* New York: W.A. Benjamin, Inc., 1963.

———, "The Formation of the Sun and Planets," *Icarus,* **1**, No. 1 (May 1962), 13.

———, "Communicating with Intelligent Life on Other Worlds," *Sky and Telescope,* **26**, No. 5. (November 1963), 258.

———, "Origin of the Solar System," *Astronautics and Aeronautics,* **2**, No. 8 (August 1964), 16.

Cameron, Roy, E., "The Role of Soil Science in Space Exploration," *Space Science Reviews,* **2**, No. 2 (August 1963), 297.

Carles, Jules, *The Origins of Life.* New York: Walker & Co., 1963.

Cassidy, William A., Luisa M. Villar, Theodore E. Bunch, Truman P. Kohman, and Daniel J. Milton, "Meteorites and Craters of Campo del Cielo, Argentina," *Science,* **149**, No. 3688 (3 September 1965), 1055.

Chamberlin, Thomas Chrowder, "Fundamental Problems in Geology," *Year*

Book No. 3, pp. 195–258. Washington D.C.: Carnegie Institute of Washington,

———, *The Origin of the Earth*. Chicago: University of Chicago Press, 1916.

———, *An Attempt to Test the Nebular Hypothesis by the Relations of Masses and Momenta*. Chicago: University of Chicago Press, 1900; *Journal of Geology*, **8** (1900), 58.

———, "On a Possible Function of Disruptive Approach in the Formation of Meteorites, Comets and Nebulae," *Astrophysical Journal*, **14**, No. 1 (July 1901), 17.

———, and Rollin D. Salisbury, *College Textbook of Geology*. New York: H. Holt & Co., 1927.

Cherry, Colin (ed), *Information Theory*. London: Butterworth's Scientific Publications, 1960.

Clark, W. E., *The Antecedents of Man*. New York: Harper and Row, 1963.

Clark, Legros W. E., *Fossil Evidence for Human Evolution*. Chicago: University of Chicago Press, 1954.

Clarke, Arthur C., *Profiles of the Future*. New York: Harper & Row, 1962 (see Chap. 18 "The Obsolescence of Man").

Clayton, Donald D., "Chronology of the Galaxy," *Science*, **143**, No. 3612 (20 March 1964), 1281.

Cleator, P. E., "Extra-Terrestrial Life," *Journal of the British Interplanetary Society*, **2**, No. 1 (May 1935), 3.

Cloud, Preston E., Jr., "Significance of the Gunflint (Precambrian) Microflora," *Science*, **148**, No. 3666 (2 April 1965), 27.

Cocconi, G. and P. Morrison, "Searching for Interstellar Communications," *Nature*, **184**, No. 4690 (19 September 1959), 844.

Cohen, Gaston, "La météorite d'Orgueil a-t-elle apporté les traces d'une vie extra-terrestre?" *Science Progrèss et La Nature*, No. 3345 (January 1964), 18.

Coon, Carlton S., *The Origin of Races*. New York: Alfred A. Knopf, 1963.

Cowan, Clyde, C. R. Atluri, and W. F. Libby, "Possible Antimatter Content of the Tunguska Meteor of 1908," *Nature*, **206**, No. 4987 (29 May 1965), 861.

Crick, F. H. C., "On the Genetic Code," *Science*, **139**, No. 3554 (8 February 1963), 461.

———, "Towards the Genetic Code," *Discovery*, **23**, No. 3 (March, 1962), 8.

———, "The Structure of Hereditary Material," *Scientific American*, **191**, No. 4 (October 1954), 54.

Dauvillier, *L'Origine Photochimique de la Vie*. Paris: Masson et Cie, 1958.

DeWitt, Bryce S., "Gravity," in *Advances in Space Science and Technology*, Vol. 6, ed. Frederick I. Ordway III. New York: Academic Press, 1964.

Deych, A. N., "61 Lebyedya, Kak Troinaya Systema," *Priroda*, **33**, No. 5 (1944), 99.

Dole, Stephen H., *Habitable Planets for Man*. New York: Blaisdell Publishing Co., 1964.

———, and Isaac Azimov, *Planets for Man*. London: Metheun & Co., 1965.

Drake, Frank, D., "How Can We Detect Radio Transmissions from Distant Planetary Systems," *Sky and Telescope*, **19**, No. 3 (January 1960), 140.

———, *Intelligent Life in Space*. New York: Macmillan Co., 1962.

———, "Project Ozma," *Physics Today*, **14**, No. 4 (April 1961), 40.

Dyson, Freeman, J., "Search for Artificial Stellar Sources of Infrared Radiation," *Science*, **131**, No. 3414 (3 June 1960), 1667.

———, *Gravitational Machines*, Gravity Research Foundation essay, New Boston, N.H., 1961.

———, J. Maddox, P. Anderson, and E. A. Sloane, "Artificial Biosphere (letters)," *Science*, **132**, No. 3421 (22 July 1960), 250.

Ehrlich, Paul R. and Richard W. Holm, *The Process of Evolution*. New York: McGraw-Hill Book Co., Inc., 1963.

Ehrensvärd, Gösta, *Life: Origin and Development*. Chicago: University of Chicago Press, 1962.

———, *Man on Another World*. Chicago: University of Chicago Press, 1965.

Engel, A. E. J., "Geologic Evolution of North America," *Science*, **140**, No. 3563 (12 April 1963), 143–152.

Ernst, B. R. and T. J. DeVries, *Atlas of the Universe*. London: Thomas Nelson and Sons, 1961.

Fesenkov, V. G., "Mars and Organic Life," *Priroda*, No. 2 (1963), 22.

Firsoff, V. A., *Life on Earth and Beyond*. London: Hutchison & Co., 1963.

Florensky, Kirill P., "Did a Comet Collide with the Earth in 1908?" *Sky and Telescope*, **26**, No. 5 (November 1963), 268.

Florkin, M. (ed), *Aspects of the Origin of Life*. London: Pergamon Press, 1960.

——— (ed), *Life Sciences and Space Research, Vol. 2*. New York: Interscience Publishers, 1964.

Fox, Sidney W., "How Did Life Begin?" *Science*, **132** (22 July 1960), 200.

——— (ed), *The Origins of Prebiological Systems*. New York: Academic Press, Inc., 1965.

Fowler, William A., and Fred Hoyle, *Nucleosynthesis in Massive Stars and Supernovae*. Chicago: University of Chicago Press, 1965.

Freudenthal, Hans, *Lincos, Design of a Language for Cosmic Intercourse*. Amsterdam: North-Holland Publishing Co., 1960.

Friend, J. L. (session chairman), "Organic Matter and Life in Meteorites," *Proceedings of Lunar and Planetary Exploration Colloquium*, **2**, No. 4 (15 November 1961), 49.

Gadomski, Jan, *The Stellar Echospheres Within a Radius of 17 Light Years Around the Sun,* translation No. T115. Santa Monica, Calif.: Rand Corporation, 11 June 1959 (original version: "Die Sternenökosphären im Radius von 17 Lichtjahren um die Sonne," in *VIIIth International Astronautical Congress Proceedings,* ed. F. Hecht. Vienna: Springer-Verlag, 1958, p. 127).

———, "Fünf Arten von ökosphärischen Planeten," in *IXth International Astronautical Congress Proceedings, Vol. II.* Vienna: Springer-Verlag, 1959.

Gaffron, H. "Photosynthesis and the Origin of Life," in *Rhythmic and Synthetic Processes in Growth.* Princeton: Princeton University Press, 1957.

Gambling, W. A. "Possibilities of Optical Communications," *Engineering,* **198** (18 December 1964), 776.

Gamow, George, *One, Two, Three . . . Infinity.* New York: Viking Press, 1947.

———, *A Planet Called Earth.* New York: Viking Press, 1963.

———, *The Creation of the Universe.* New York: Viking Press, 1961.

———, and J. A. Hynek, "Recent Progress in Astrophysics: A New Theory by C. F. von Weizsäcker of the Origin of the Planetary System," *Astrophysical Journal,* **101**, No. 2 (March 1945), 249.

Gatland, Kenneth W. and Derek D. Demster, *The Inhabited Universe.* London: Allan Wingate, 1957.

Gindilis, L. M., "Possibilities of Contacting Extraterrestrial Civilizations," *Zemlya i vselennaya,* No. 1 (1965), 18.

Golay, M. J. E., "Note on the Probable Character of Intelligent Radio Signals from other Planetary Systems,'" *Proceedings of the Institute of Radio Engineers,* **49**, No. 1, Part I (May 1961), 959.

Goldschmidt, V. M., "Geochemical Aspects of the Origin of Complex Organic Molecules on the Earth, as Precursors to Organic Life," *New Biology,* **12** (April 1952), 97.

Golomb, Solomon W., "Extraterrestrial Linguistics," *Astronautics,* **6**, No. 5 (May 1961), 46.

Good, I. J., "Life Outside the Earth," *The Listener* (3 June 1965); expanded as "The Human Preserve," *Spaceflight,* **7**, No. 5 (September 1965), 167.

Gratton,, L. (ed), *Star Evolution.* New York: Academic Press, 1964.

Gulick, Addison, "Phosphorus as a Factor in the Origin of Life," *American Scientist,* **43**, No. 2 (Summer 1955), 479.

Halacy, D. S., Jr., *Cyborg: Evolution of the Superman.* New York: Harper and Row, 1965.

Haldane, J. B. S., "The Origins of Life," in *The New Biology,* No. 16 (April 1954), 12 (also includes J. D. Bernal's "The Origin of Life," N. W. Pirie's "On Making and Recognizing Life," and J. W. S. Pringle's "The Evolution of Living Matter").

———, "Genesis of Life," in *The Earth and Its Atmosphere,* ed. D. R. Bates. New York: Basic Books, Inc., 1959.

Hawkins, J. K., "Self-Organizing Systems—A Review and Commentary," *Proceedings of the IRE (Special Issue on Computers),* **49,** No. 1 (January 1961), 31.

Hawrylewicz, Ervin, Betty Gowdy and Richard Ehrlich, "Microorganisms Under a Simulated Martian Environment," *Nature,* **193,** No. 4814 (3 February 1962), 497.

Hermann, Joachim, *Leben Anderen Sternen.* Gutersloh, Germany: C. Bertelsmann Verlag, 1963.

Hogben, Lancelot, "Astraglossa, or First Steps in Celestial Syntax," *Journal of the British Interplanetary Society,* **11,** No. 6 (November 1952), 258.

Holmberg, E., "Invisible Companions of Parallax Stars Revealed by Means of Modern Trigonometric Parallax Observations," *Meddelanden Fran Lunds Astronomiska Observatorium,* **11,** No. 92 (1938), 23.

Holmes, A., "The Oldest Dated Minerals of the Rhodesian Shield," *Nature,* **173,** No. 4405 (3 April 1954), 612.

Horowitz, N. H., "Astrobiology Session," in *Proceedings of Lunar and Planetary Exploration Colloquium,* **2,** No. 1 (September 1959), 2.

———, "Is There Life on Other Planets?" *Engineering and Science,* **24,** No. 6 (March 1961), 11.

———, "On the Evolution of Biochemical Synthesis," *Proceedings of the National Academy of Science,* **31,** No. 6 (15 June 1945), 153.

Howell, F. Clark, *Early Man.* New York: Time, Inc., 1965.

Howells, William W., "The Evolution of 'Humans' on Other Planets," *Discovery,* **22,** No. 6 (June 1961), 237. (From his book *Mankind in the Making.* London: Secker & Warburg, 1961)

———, *Back of History: The Story of Our Own Origins.* New York: Doubleday and Co., 1963.

Hoyle, Fred, *Frontiers of Astronomy.* New York: Harper and Brothers, 1955.

———, "On the Origin of the Solar System," *Proceedings of the Cambridge Philosophical Society,* **40,** Part 3 (October 1944), 250.

———, *Of Men and Galaxies.* Seattle: University of Washington Press, 1965.

Huang, Su-Shu, "Some Astronomical Aspects of Life in the Universe," *Sky and Telescope,* **21,** No. 6 (June 1961), 312.

———, "Occurrence of Life in the Universe," *American Scientist,* **47,** No. 3 (September 1959), 397.

———, "Life Outside the Solar System," *Scientific American,* **202,** No. 4 (April 1960), 55.

———, "The Limiting Sizes of Habitable Planets," *Publications of the Astronomical Society of the Pacific,* **72** (December 1960), 489.

Hulley, J., "Dynamics of Life in the Universe," *Space Journal*, **1**, No. 5 (March–May 1959), 33.

Hunter, A., "Non-solar Planets," *Nature*, **152**, No. 3846 (17 July 1943), 66.

Hutchinson, G. E., "A Note on Two Aspects of the Geochemistry of Carbon," *American Journal of Science*, **247**, No. 1 (January 1949), 27.

Huxley, Julian (ed), *Evolution as a Process*. New York: Collier Books, 1963.

———, *Evolution—the Modern Synthesis*. New York: Harper & Bros., 1942.

Ingram, Vernon, *The Biosynthesis of Macromolecules*. New York: W. A. Benjamin, Inc., 1964.

Jackson, C. D. and R. E. Hohmann, "*An Historic Report on Life in Space: Tesla, Marconi, Todd,*" American Rocket Society Paper 2730–62. New York, 1962.

Jaspers, Karl, *Future of Mankind*. Chicago: University of Chicago Press, 1961.

Jastrow, Robert, and A. G. W. Cameron, *Origin of the Solar System*. New York: Academic Press, 1963.

Jeans, James H., *Astronomy and Cosmogony*. New York: Dover Publications, Inc., 1961 (reprint of 1929 edition).

———, "Origin of the Solar System," *Nature*, **149**, No. 3790 (20 June 1942), 695.

———, "Non-Solar Planetary Systems," *Nature*, **152**, No. 3868 (18 December 1943), 721.

———, "The Motion of Tidally-distorted Masses, with Special Reference to Theories of Cosmogony," *Memoires, Royal Astronomical Society*, **62** (1917–1923), 1.

Jeffreys, Harold, "On the Early History of the Solar System," *Monthly Notices of the Royal Astronomical Society*, **78**, No. 6 (April 1918), 424.

———, "Collision and the Origin of Rotation in The Solar System," *Monthly Notices of the Royal Astronomical Society*, **89**, No. 7 (May 1929), 636.

———, "Early History of The Solar System on the Collision Theory," *Monthly Notices of the Royal Astronomical Society*, **89**, No. 9 (Supplement 1929), 731.

———, "On Certain Possible Distributions of Meteoric Bodies in the Solar System." *Monthly Notices of the Royal Astronomical Society*, **77**, No. 2 (December 1916), 84.

———, "Origin of the Solar System," *Nature*, **153**, No. 3874 (29 January 1944), 140.

Jones, H. Spencer, *Life on Other Worlds*. New York: Macmillan Co., 1940.

———, *Is There Life on Other Worlds?* Smithsonian Report 3556 Washington, D.C.: Smithsonian Institution, 1940.

Kapp, Reginald O., *Towards a Unified Cosmology*. New York: Basic Books, 1960.

Keilin, D., "The Problem of Anabiosis on Latent Life: History and Current Concept," *Proceedings of the Royal Society of London*, **150 B** (17 March 1959), 149.

Keosian, J., "On the Origin of Life," *Science,* **131,** No. 3399 (19 February 1960), 479.

Kerkut, G. A., *Implications of Evolution.* London: Pergamon Press, 1960.

Kind S. S., "Speculations on Extraterrestrial Life," *Spaceflight,* **1,** No. 8 (July 1959), 288.

Komarov, V., "Light-years Away," *Krasnaya Zvezda* (17 July 1965), 6.

Kornberg, A., "Biologic Synthesis of Deoxyribonucleic Acid," *Science,* **131,** No. 3412 (20 May 1960), 1503.

Kraus, Bertram S., *The Basis of Human Evolution.* New York: Harper and Row, 1964.

Kuiper, Gerard P., "On the Origin of the Solar System," in *Astrophysics: A Topical Symposium,* ed. J. A. Hynek. New York: McGraw-Hill Book Co., Inc., 1951.

———, "On the Evolution of the Protoplanets," *Proceedings of the National Academy of Science,* **37,** No. 7 (15 July 1951), 383.

———, "Planetary Atmospheres and their Origin," in *Atmospheres of the Earth and Planets,* ed. Gerard P. Kuiper. Chicago: University of Chicago Press, 1952.

———, "The Law of Planetary and Satellite Distances," *Astrophysical Journal,* **109,** No. 2 (March 1949), 308.

———, and Barbara M. Middlehurst, *Stars and Stellar Systems.* Chicago: University of Chicago Press, 1963.

Kuroda, P. K. and W. H. Crouch, Jr., "On The Chronology of the Formation of the Solar System," *Journal of Geophysical Research,* **67,** No. 12 (November 1962), 4863.

Lang, T. G. and H. A. P. Smith, "Communications Between Dolphins...," *Science,* **150,** No. 3705 (31 December 1965), 1839.

Lapp, Ralph E., "How to Talk to People, if any, on Other Planets," *Harper's* (March 1961), 58. (See Also Chap. 9 and Chap. 10 in his *Man and Space.* Boston: Harper and Brothers, 1961.).

Layzer, David, "The Formation of Stars and Galaxies: Unified Hypotheses," in *Annual Review of Astronomy and Astrophysics,* Vol. 2, ed Leo Goldberg. Palo Alto, Calif.: Annual Reviews, Inc., 1964.

Leakey, L. S. B., *Olduvai Gorge 1951–1961, Vol. 1: A Preliminary Report on the Geology and Fauna.* New York: Cambridge University Press 1965.

Leakey, L. S. B., P. V. Tobias, and J. R. Napier, "A New Species of the Genus Homo from Olduvai Gorge," *Nature,* **202,** No. 4927 (4 April 1964), 7.

Lederberg, Joshua, "Exobiology: Experimental Approaches to Life Beyond the Earth," in *Space Research,* ed. H. K. Kallman Bijl. New York: Interscience Publishers, Inc., 1960; also in *Science,* **132,** No. 3424 (12 August 1960), 393.

———, "The Search for Life Beyond the Earth," *New Scientist,* **7,** No. 170 (18 February 1960), 386.

———, and D. B. Cowie, "Moondust," *Science,* **127,** No. 3313 (27 June 1958), 1473.

———, and Carl Sagan, "Microenvironments for Life on Mars," *Proceedings of the National Academy of Sciences*, **48**, No. 9 (15 September 1962), 1473.

Levin, B. J., "Origin of Meteorites," *Nature*, **204**, No. 4962 (5 December 1964), 946.

Lhote, Henri, "Discovering a Stone Age Museum," *Horizon* (May 1959).

Lovell, Bernard, "Search for Voices from Other Worlds," *The New York Times Magazine* (24 December 1961), 10.

———, *The Exploration of Outer Space*. New York: Harper and Row, 1962 (see section on possibility of life on other worlds).

Lyttleton, R. A., *Man's View of the Universe*. Boston: Little, Brown and Co., 1961 (see esp. Chap. 6. "The Origin of the Solar System" and Chap. 8. "The Origin of Comets").

———, "An Accretion Hypothesis for the Origin of the Solar System," *Monthly Notices of the Royal Astronomical Society*, **122**, No. 5 (June 1961), 399.

MacDonald, Gordon J. F., "Earth and Moon: Past and Future," *Science*, **145**, No. 3635 (28 August 1964), 881.

MacGowan, Roger A., "On the Possibilities of the Existence of Extraterrestrial Intelligence," in *Advances in Space Science and Technology*, Vol. 4., ed, Frederick I. Ordway, III. New York: Academic Press, 1962.

Macvey, John W., *Journey to Alpha Centauri*. New York: Macmillan Co., 1965.

Mamikunian, Gregg and Michael H. Briggs, *Current Aspects of Exobiology*. Pasadena: California Institute of Technology—Jet Propulsion Laboratory, 1965 (distributed through Pergamon Press, New York). This work includes an excellent bibliography of over 1,500 titles.

Mannel, Cliff, "Interstellar Communications," *Challenge*, **3**, No. 1 (Spring 1964), 24.

Margaria, R., "On the Possible Existence of Intelligent Living Beings on Other Planets," in *XIIth International Astronautical Congress: Proceedings-Volume 2*, ed. Robert M. L. Baker, Jr. and Maud W. Makemson. New York: Academic Press, Inc., and Vienna: Springer Verlag, 1963.

Marx, Gyorgy, *Messages from Outer Space,* ASTIA Report No. AD 270 779, (Translation from *Tizikai Szemele,* Vol. 10, No. 11 (1960), 335). Washington, D.C.: Defense Documentation Center.

McCrea, W. H., "The Origin of the Solar System," *Proceedings of the Royal Society of London*, **256**, No. 1285 (21 June 1960), 245.

McVittie, G. C. (ed), *Problems of Extra-Galactic Research*. New York: Macmillan Co., 1962.

———, *Fact and Theory in Cosmology*. New York: Macmillan Co., 1961.

Meinschein, Warren G., "Hydrocarbons in Terrestrial Samples and the Orgueil Meteorite," *Space Science Reviews*, **2**, No. 5 (November 1963), 653.

Menzel, Donald H. and Lyle G. Boyd, *The World of Flying Saucers*. New York: Doubleday & Co., Inc., 1963.

Mercer, D.M.A., "Messages to the Stars," *Courier* (January 1966), 4.

Mercier, A., *Conférence astronomique sur la planète Mars . . . Projet d'études sur les moyens pratiques d'execution de signaux lumineux de la terre à Mars.* Orléans: M. Marron, 1902.

Middlehurst, Barbara M. and Gerard P. Kuiper, *The Moon, Meteorites and Comets.* Chicago: University of Chicago Press, 1963 (Volume 4 of *The Solar System*).

Miller, Stanley L., "A Production of Amino Acids under Possible Primitive Earth Conditions," *Science,* **117,** No. 3046 (15 May 1953), 528.

———, "Extraterrestrial Life," in *Lectures in Aerospace Medicine,* Brooks Air Force Base, Texas: USAF School of Aerospace Medicine, 1962, 277.

———, "Production of Some Organic Compounds under Possible Primitive Earth Conditions," *Journal of the American Chemical Society,* **77,** No. 9 (12 May 1955), 2351.

———, "The Formation of Organic Compounds on the Primitive Earth," *Annals of the New York Academy of Sciences,* **69,** Art. 2 (30 August 1957), 260.

———, and Harold C. Urey, "Organic Compound Synthesis of the Primitive Earth," *Science,* **130,** No. 3370 (31 July 1959), 245.

Minsky, M. L., "Steps Toward Artificial Intelligence," *Proceedings of the IRE (Special Issue on Computers),* **49,** No. 1 (January 1961), 8.

———, *Heuristic Aspects of the Artificial Intelligence Problem,* G. P. Report No. 34–55, Cambridge, Mass.: Lincoln Laboratories, 1956.

Montagu, Ashley, *Culture and the Evolution of Man.* New York: Oxford University Press, 1962.

Moore, Patrick, *The Planet Venus.* London: Faber and Faber Ltd., 1956. (See Chap. 15, "Life on Venus.")

———, *Guide to Mars.* London: Frederick Muller Ltd., 1956. (See Chap. 10, "Life on Mars.")

———, "Life on the Moon," *Irish Astronomical Journal,* **3,** No. 5 (March 1955), 133.

———, and Francis Jackson, *Life in the Universe.* New York: W.W. Norton and Co., Inc., 1962.

———, *Life on Mars.* London: Routledge & Kegan Paul, 1965.

Moore, R. C., *Introduction to Historical Geology.* New York: McGraw-Hill Book Co., Inc., 1958.

Moulton, F. R., "On the Evolution of the Solar System," *Astrophysical Journal,* **22,** No. 3 (October 1905), 165–181.

———, *Astronomy.* New York: Macmillan Co., 1931.

———, "An Attempt to Test the Nebular Hypothesis by the Relations of Masses and Momenta", *Journal of Geology,* **8,** No. 1 (January-February 1900), 58–73.

———, "An Attempt to Test the Nebular Hypothesis by an Appeal to the Laws of Dynamics", *Astrophysical Journal,* **11,** No. 2 (March 1900), 103.

Mowbray, Lionel, "Communications avec le Cosmos et langage universel," *Science Progrès La Nature*, No. 3334 (February 1963), 79.

Muller, H. J., "Genetic Nucleic Acid: Key Material in the Origin of Life," *Perspectives in Biology and Medicine*, **5**, No. 1 (Autumn 1961).

———, "Life Forms to be Expected Elsewhere than on Earth," *The American Biology Teacher*, **23**, No. 6 (October 1961); *Spaceflight*, **5**, No. 3 (May 1963), 74.

Murthy, V. Rama and Harold C. Urey, "The Time of Formation of the Solar System," *Astrophysical Journal*, **135**, No. 2 (March 1962), 626.

Muses, C. A. (ed.), *Aspects of the Theory of Artificial Intelligence*. New York: Plenum Press, 1962.

Mutschall, Vladimir, "Soviet Long-range Space Exploration Program," *Aerospace Technology Division Rept. 65–94* (30 December 1965). Washington, D.C.: Library of Congress, pp. 21–28 ("Extraterrestrial Contacts").

Nagy, Bartholomew, Warren G. Meinschein and Douglas J. Hennessy, "Mass Spectroscopic Analysis of the Orgueil Meteorite: Evidence for Biogenic Hydrocarbons," *Annals of the New York Academy of Sciences*, **93**, Art. 2, (5 June 1961), 25.

———, G. Claus and Douglas J. Hennessy, "Organic Particles Embedded in Minerals in the Orgueil and Ivuna Carbonaceous Chondrites," *Nature*, **193**, No. 4821 (24 March 1962), 1129.

Nelson, The Earl of, *Life and the Universe*. New York: Staples Press, 1953.

———, *There is Life on Mars*. London: Werner Laurie, 1955.

Oliver, Bernard M., "First Picture from Another Planet," *Institute of Radio Engineers, Student Quarterly and E. E. Digest* (September 1962).

———, "Some Potentialities of Optical Masers," *Proceedings of the Institute of Radio Engineers*, **50** (1962), 135.

———, "Radio Search for Distant Races," *International Science and Technology*, No. 10 (October 1962), 55.

Oparin, Alexander I., *The Origin of Life on Earth*. New York: Academic Press, Inc., 1957. (third edition; first edition carried title *The Origin of Life*, and was published by the Macmillan Co., New York, 1938 and reprinted by Dover Publications, New York, 1953; the second edition came out in 1941 without substantial alteration . . . it is this edition that has been revised and enlarged to result in the present volume).

———, "The Origin of Life in Space," *Space Science Reviews*, **3**, No. 1 (July 1964), 5.

———, *Life: Its Nature, Origin and Development*. New York: Academic Press, 1962.

———, "The Problem of the Origin of Life," *The Modern Quarterly*, **6**, No. 2 (1951), 135.

———, "Life in the Universe," *Priroda*, No. 2 (1963), 14.

———, *The Chemical Origin of Life.* Springfield, Illinois: Charles C. Thomas, 1964.

———, "Extraterrestrial Life," *Izvestiya Akademii nauk SSSR, Seriya Biologicheskaya,* **28**, No. 1 (1963), 3 (Translation by the Joint Publications Research Service) Report OTS 63-21498, 4 April 1963, Washington, D.C.

———, A. G. Pasynskii, A. E. Braunshtein and T. E. Pavlovskaya, *The Origin of Life on Earth.* New York: Pergamon Press, 1959 (Proceedings of First International Symposium on subject, sponsored by USSR Academy of Sciences; English, French, German edition edited for the International Union of Biochemistry by F. Clark and R. L. M. Synge).

———, and V. Fesenkov, *Life in the Universe.* New York: Twayne Publishers, 1962.

———, *The Universe.* Moscow: Foreign Languages Publishing House, 1957.

Ordway, F. I., III, J. P. Gardner and M. R. Sharpe, Jr., *Basic Astronautics.* Englewood Cliffs, N.J.: Prentice-Hall, Inc., 1962 (see Chap. 6, "Astrobiology").

———, "Some Implications of Extrasolar Intelligence," in *Planetology and Space Mission Planning Monograph,* New York Academy of Sciences, 1966.

———, *Life in Other Solar Systems.* New York: E. P. Dutton & Co., Inc., 1965.

Oró, J., D. W. Noorer, A. Zlatkis, S. A. Kikström, and E. S. Barghoorn, "Hydrocarbons of Biological Origin in Sediments about Two Billion Years Old," *Science,* **148**, No. 3660 (2 April 1965), 77.

Ovenden, Michael W., *Life in the Universe.* Garden City, New York: Doubleday & Co., 1962.

Pagal, Bernard, "The Evolution of Stars," *New Scientist,* **14**, No. 285 (3 May 1962), 216.

Page, Thornton, *Stars and Galaxies: Birth, Ageing and Death in the Universe.* Englewood Cliffs, N. J.: Prentice-Hall, Inc., 1962.

Paschke, R., R. W. H. Chang and D. Young, "Probable Role of Gamma Irradiation in Origin of Life," *Science,* **125**, No. 3253 (3 May 1957), 881.

Payne-Gaposchkin, Cecilia, *Stars in the Making.* London: Eyre & Spottiswoode, 1953.

Pereira, F. A., "Introducao à Astrobiologia." São Paulo: Sociedade Interplanetaria Brasileira, 1958.

Pfeiffer, John, "DNA—A Complex Molecule is Deemed the Master Substance of Life," *Natural History,* **69**, No. 10 (December 1960), 9.

———, *The Thinking Machine.* Philadelphia: J. B. Lippincott Co., 1962.

———, *From Galaxies to Man.* New York: Random House, 1959.

Pikelner, S., *Soviet Science of Interstellar Space.* New York: Philosophical Library, 1963.

Pirie, N. W. (ed), *Biology of Space Travel.* London: Institute of Biology, 1961.

———, "On Making and Recognizing Life," *New Biology*, **16** (April 1954), 41.

Poole, J. H. J., "The Evolution of the Atmosphere," *Royal Dublin Society, Scientific Proceedings*, **25**, No. 15 (1 May 951), 201.

Ponnamperuma, Cyril and Ruth Mack, "Nucleotide Synthesis under Primitive Earth Conditions," *Science*, **148**, No. 3674 (28 May 1965), 1221.

———, "Life in the Universe," *Astronautics & Aeronautics*, **3**, No. 10 (October 1965), 66.

Posin, Dan Q., *Life Beyond Our Planet*. New York: McGraw-Hill Book Co., Inc., 1962.

Pringle, J. W. S., "The Evolution of Living Matter," *New Biology*, **16** (April 1954), 54.

Rankama, K. and T. G. Sahama, *Geochemistry*. Chicago: University of Chicago Press, 1950.

Reyn, N. F., and N. N. Paruyskiy, "Katastroficheskiye gipotezy proiskhodzheniya solnechnoy sistemy," *Uspekhi Astronomicheskikh Nauk*, **2** (1941), 137.

Richardson, R. S. *Exploring Mars*. New York: McGraw-Hill Book Co., Inc., 1954. (See Chap. 8 "Life on Mars.")

Robinson, Ivor, Alfred Schild and E. L. Shucking (eds.), *Quasi-stellar Sources and Gravitational Collapse*. Chicago: University of Chicago Press, 1965.

Romer, Alfred S. *Vertebrate Paleontology*. Chicago: University of Chicago Press, 1945.

———, *Man and the Vertebrates*. London: Penguin Books, 1954.

Rosenberg, Paul, "Communication with Extraterrestrial Intelligence," *Aerospace Engineering*, **21**, No. 8 (August 1962), 68.

Rosenblatt, Frank, *Principles of Neurodynamics*. Washington, D.C.: Spartan Books, 1962.

Rubey, William W., "Development of the Hydrosphere and Atmosphere, with Special Reference to Probable Composition of the Early Atmosphere." *Geological Society of America Special Paper*, **62** (1955), 631.

Rush, J. H., *The Dawn of Life*. Garden City, New York: Hanover House, 1957.

Ruskol, E. L., "On the Origin of the Moon," *Soviet Astronomy*, **7**, No. 2 (September-October 1963), 221 (From *Astromicheskii Zhurnal*, **40**, No. 2 (March-April 1963), 288).

Russell, H. N., *The Solar System and Its Origin*. New York: Macmillan Co., 1935.

Rutten, M. G., *The Geological Aspects of the Origin of Life on Earth*. Amsterdam: Elsevier Publishing Co., 1962.

———, "Origin of Life on Earth, its Evolution and Actualism," *Evolution*, **11** (March 1957), 56.

Ruzic, Neil P., "The Case for Life Beyond the Earth," *Industrial Research*, **7**, No. 1 (January 1965).

Sagan, Carl, "Biological Contamination of the Moon," *Proceedings of the National Academy of Science,* **46,** No. 4 (15 June 1960), 396.

———, "Direct Contact among Galactic Civilizations by Relativistic Interstellar Spaceflight," *Planetary and Space Science,* **2,** No. 5 (May 1963), 485.

———, "Indigenous Organic Matter on the Moon," *Proceedings of the National Academy of Science,* **46,** No. 4 (15 June 1960), 393.

———, "On the Origin and Planetary Distribution of Life," *Radiation Research,* **15,** No. 2 (August 1961), 174.

———, "Interstellar Panspermia," in *Biochemical Activities of Terrestrial Microorganisms in Simulated Planetary Environments,* Project 109 Report: Space Sciences Laboratories, University of California.

———, "Is the Early Evolution of Life Related to the Development of the Earth's Core?" *Nature,* **206,** No. 4983 (1 May 1965), 448.

Salisbury, Frank B, "Exobiology," *Proceedings of the First Annual Rocky Mountain Bioengineering Symposium.* Colorado Springs, Colo.: US Air Force Academy, 1964.

Sandage, Allan R., "Exploding Galaxies," *Scientific American,* **211,** No. 5 (November 1964), 38.

Savage, Jay M., *Evolution.* New York: Holt, Rinehart and Winston, 1963.

Scheffler, Israel, "Inductive Inference: A New Approach," *Science,* **127,** No. 3291 (24 January 1958), 177.

Schmidt. O. J. and B. J. Levin, "The Origin of the Solar System," *New Scientist,* **13** (8 February 1962), 323.

Schopf, J. W., E. S. Barghoorn, M. D. Maser, and R. D. Gordon, "Electron Microscopy of Fossil Bacteria Two Billion Years Old," *Science,* **149,** No. 3690.

Schrodinger, Erwin, *What is Life?* New York: Macmillan Co., 1946.

Schwartz, R. N., and C. H. Townes, "Interstellar and Interplanetary Communication by Optical Masers," *Nature,* **190,** No. 4772 (15 April 1961), 205.

Seboek, Thomas A., "Animal Communication," *Science,* **147,** No. 3661 (26 February 1965), 1006.

Sen, H. K. "Non-solar Planetary Systems," *Nature,* **152,** No. 3864 (20 November 1943), 601.

Shannon, Claude E., and J. MacCarthy (eds.), *Automata Studies.* Princeton: Princeton University Press, 1956.

Shapley, Harlow, *Of Stars and Men.* Boston: Beacon Press, 1958 (new, revised, and enlarged edition, 1964).

———, (ed.), *Source Book In Astronomy 1900–1950.* Cambridge Mass: Harvard University Press, 1960 (see esp. Section X on "Stellar Evolution").

———, *The View From a Distant Star.* New York: Basic Books, 1963.

——— (ed.), *Climatic Change.* Cambridge, Mass. Harvard University Press, 1953.

———, "Extraterrestrial Life," *Astronautics*, **5**, No. 4 (April 1960), 32.

Shklovsky, I. S., "Is Communication Possible with Intelligent Beings on Other Planets?" *Priroda,* **7** (1960), 21. Translated and issued as Report M-MS-IS-61-2 by the George C. Marshall Space Flight Center, Huntsville, Ala.

———, *Intelligent Life in the Universe* (Carl Sagan, ed), San Francisco: Holder-Day, 1965.

———, *Kosmicheskoye Radioizlucheniye* (Cosmic Radio Waves). Moscow, 1965.

Siegel, S. M., L. Halpern, G. Davis and C. Giumarro, "The General and Comparative Biology of Experimental Atmospheres and Other Stress Conditions," *Aerospace Medicine*, **34**, No. 11 (November 1965), 1034.

———, "Effects of Reduced Oxygen Tension on Vascular Plants," *Physiologia Plantarum*, **14** (1961), 554.

———, and L. A. Rosen, "Effects of Reduced Oxygen Tension on Germination and Seedling Growth," *Physiologia Plantarum,* **15** (1962), 437.

———, Constance Giumarro and Richard Latterell, "Behavior of Plants under Extraterrestrial Conditions: Seed Germination in Atmospheres Containing Nitrogen Oxides," *Proceedings of the National Academy of Sciences*, **52**, No. 1 (July 1964), 11.

———, L. A. Rosen and G. Renwick, "Further Studies on the Composition of Seedlings Grown at Sub-Atmospheric Oxygen Levels," *Physiologia Plantarum,* **16** (1963), 549.

———, L. A. Halpern, C. Giumarro, G. Renwick and G. Davis, "Martian Biology: The Experimentalist's Approach," *Nature*, **197**, No. 4865 (26 January 1963), 329.

Simpson, George Gaylord, *The Meaning of Evolution*. New Haven: Yale University Press 1949.

———, *The Major Features of Evolution*. New York: Columbia University Press, 1953.

———, "The Nonprevalence of Humanoids," *Science,* **143**, No. 3608 (21 February 1964), 769.

———, *This View of Life*. New York: Harcourt, Brace & World, Inc., 1964.

Sinton, W. M., "Spectroscopic Evidence for Vegetation on Mars," *Astrophysical Journal*, **126**, No. 2 (September 1957), 231.

———, "Spectroscopic Evidence of Vegetation on Mars," *Publications of the Astronomical Society of the Pacific,* **70** (February 1958), 50.

———. "Further Evidence of Vegetation on Mars," *Science,* **130**, No. 3384 (6 November 1959), 1234.

Slater, A. E., "The Probability of Intelligent Life Evolving on a Planet," in *VIIIth International Astronautical Congress Proceedings*, ed. F. Hecht. Vienna: Springer-Verlag, 1958, 395.

———, "Life in the Universe," *Spaceflight*, **5**, No. 6 (November 1963), 198.

Sonneborn, T. M. (ed.), *The Control of Human Heredity and Evolution.* New York: Macmillan Co., 1965.

Spencer, Dwain F., *Fusion Propulsion System Requirements for an Interstellar Probe.* Report TR-32-397. Pasadena, Calif. Jet Propulsion Laboratory, 15 May 1963.

Spitzer, Lyman, Jr., "The Dissipation of Planetary Filaments," *Astrophysical Journal,* **90,** No. 5 (December 1939), 675.

Stanyukovich, Kirill P., On an Interstellar Flight," *Krasnaya zvezda* (23 May 1965), 4.

Stapledon, O., "Interplanetary Man?" *Journal of the British Interplanetary Society,* **7,** No. 6 (November 1948), 213.

Stormer, Carl, "Short Wave Echoes and the Aurora Borealis," *Nature,* **122,** No. 3079 (3 November 1928), 681.

Strand, K. Aa., "61 Cygni as a Triple System," *Publications of the Astronomical Society of the Pacific,* **55,** No. 322 (February 1943), 29.

———, "The Orbital Motion of 61 Cygni," *Proceedings of the American Philosophical Society,* **86,** No. 3 (July 1943), 364.

Strughold, Hubertus, "Advances in Astrobiology," in *Proceedings of Lunar and Planetary Exploration Colloquium,* **1,** No. 6 (25 April 1959), 1.

———, "Planetary Ecology (Astrobiology)," *Lectures in Aerospace Medicine.* Brooks Air Force Base, Texas: USAF Aerospace Medical Center, 1960.

———, "An Introduction to Astrobiology," *Astronautics,* **5,** No. 12 (December 1960), 20.

———, "Life on Mars in View of Physiological Principles," in *Epitome of Space Medicine.* Randolph Air Force Base, Texas: USAF School of Aviation Medicine, 1951.

———, "The Possibilities of an Inhabitable Extraterrestrial Environment Reachable from the Earth," *Journal of Aviation Medicine,* **28,** No. 5 (October 1957), 507.

———, *The Red and Green Planet.* Albuquerque, New Mexico: University of New Mexico Press, 1953.

Struve, Otto, "McCrea's Theory of the Solar System's Origin," *Sky and Telescope,* **19,** No. 3 (January 1960), 154.

———, "On the Origin of the Solar System," *Sky and Telescope,* **15,** No. 8 (June 1956), 349.

———, *The Universe.* Cambridge, Mass: M. I. T. Press, 1962.

Stuart, Piggott (ed), *The Dawn of Civilization.* New York: McGraw-Hill Book Co., Inc., 1961.

Studier, Martin H., Ryoichi Hayatsu and Edward Anders, "Organic Compounds in Carbonaceous Chondrites," *Science,* **149,** No. 3691 (24 September 1965), 1455.

Stuhlinger, E., "Life on Other Stars," *Space Journal,* Part 1:**1**, No. 2 (Spring 1958), 10: Part 2:**1**, No. 3 (Summer 1958), 21.

Sullivan, Walter, "Contact with Worlds in Space Explored by Leading Scientists," *New York Times* (4 February 1962) (report on National Academy of Sciences—sponsored conference on interstellar communication held at Green Bank, West Virginia, November 1961.)

———, "Space Signals Said to Hint Life Afar," *New York Times* (26 October 1964)

———, *We Are Not Alone.* New York: McGraw-Hill Book Co., Inc., 1964.

Sweitzer, D. I., *Biological and Artificial Intelligence,* Literature Search No. 254, Pasadena, California: Jet Propulsion Laboratory—California Institute of Technology, 1960—and supplement, published in 1961.

Tasch, Paul, "Life Forms in Meteorites and the Problem of Terrestrial Contamination—A Study in Methodology. *Annals of the New York Academy of Sciences,* **105**, (9 September 1964), 927.

Tax, Sol (ed.), *Evolution after Darwin* (three volumes). Chicago: University of Chicago Press, 1960.

Ter Haar, D., "Stellar Rotation and Age," *Astrophysical Journal,* **110**, No. 3 (November 1949), 321.

———, "Further Studies on the Origin of the Solar System," *Astrophysical Journal,* **111**, No. 1 (January 1950), 179.

Tesla, Nikola, "Talking with the Planets," *Current Literature* (March 1901), 359.

Tilton, G.R., and S.R. Hart, "Geochronology," *Science,* **140**, No. 3565 (26 April 1963), 357–366.

Tikhov, G.A., "Is Life Possible on Other Planets?" *Journal of the British Astronomical Association,* **65**, No. 5 (April 1955), 193.

———, "What is Astrobotany?" *Spaceflight,* **11**, No. 3 (July 1959), 74.

Tobias, Phillip V., "Early Man in East Africa," *Science,* **149**, No. 3679 (2 July 1965), 22.

Tombaugh, Clyde W., "Could the Satellites of Mars be Artificial?" *Astronautics,* **4**, No. 12 (December 1959), 39.

Troitskaya, O.V., "About the Possibilities of Plant Life on Mars," *Astronomicheskii Zhurnal,* **29**, No. 1 (1952). (Translated by J. L. Zygielbaum: JPL Translation No. 8, Pasadena, California: Jet Propulsion Laboratory, 16 February 1960.)

Tyler, Stanley A. and Elso S. Barghoorn, "Occurrence of Structurally Preserved Plants in the Pre-Cambrian Rocks on the Canadian Shield," *Science,* **119**, No. 3096 (30 April 1954), 606.

Urey, Harold C., *The Planets: Their Origin and Development.* New Haven, Connecticut: Yale University Press, 1952.

———, "The Origin and Evolution of the Solar System," in *Space Science,* ed. Donald C. Legalley. New York: John Wiley and Sons, Inc., 1963, p. 123.

———, *Some Cosmochemical Problems.* University Park, Pa.: Pennsylvania State University, 1963.

———, "On the Early Chemical History of the Earth and the Origin of Life," *Proceedings of the National Academy of Science,* **38,** No. 4 (15 April 1952), 351.

Vallee, Jacques, *Anatomy of a Phenomenon.* Chicago: Henry Regnery Co., 1965.

Van de Kamp, Peter, "Stars Nearer than Five Parsecs," *Sky and Telescope,* **14,** No. 12 (October 1955), 498. [Also *Publications of the Astronomical Society of the Pacific,* **65** (April 1953), 73.]

———, "Barnard's Star as an Astrometric Binary," *Sky and Telescope,* **26,** No. 1 (July 1963), 8.

———, "Astrometric Study of Barnard's Star From Plates Taken with the 24-inch Sproul Reflector," *Astronomical Journal,* **68,** No. 5 (September 1963), 515.

Van der Pol, Balth, "Short Wave Echoes and the Aurora Borealis," *Nature,* **122,** No. 3084 (8 December 1928), 878–879.

Vaucouleurs, Gérard de, *Physics of the Planet Mars.* London: Faber and Faber Ltd., 1954. (See Part IV, chap. II, "The Dark Areas of Mars.")

Von Hoerner, S., "The Search for Signals from Other Civilizations," *Science,* **134,** No. 3493 (8 December 1961), 1839.

———, "The General Limits of Space Travel," *Science,* **137,** No. 3523 (6 July 1962), 18.

Von Neumann, J., *The Computer and the Brain.* New Haven: Yale University Press, 1958.

Von Weizsäcker, Carl Friedrich, *History of Nature.* Chicago: University of Chicago Press, 1959.

———, "Über die Entstehung des Planetensystems," *Zeitschrift Für Astrophysik.,* **22,** No. 5 (December 1943), 319.

Voronin, M. A., "On Searching for Traces of Civilization in Other Worlds," *Priroda,* No. 11 (1963), 78.

Wald, George, "The Origin of Life," *Scientific American,* **191,** No. 2 (August 1954), 44.

———, "The Origins of Life," *Proceedings of the National Academy of Science,* **52,** No. 2 (15 August 1964), 595.

Walter, W. Gray, *The Living Brain,* New York: W. E. Norton Co., 1953.

Watson, J. D., and F. H. C. Crick, "A Structure for Deoxyribose Nucleic Acid," *Nature,* **171,** No. 4356 (25 April 1953), 737.

Webb, Alan Wells, *Mars, the New Frontier—Lowell's Hypothesis.* San Francisco: Fearon Publishers, 1956.

Webb, J. A., "Detection of Intelligent Signals from Space," Institute of Radio Engineers Seventh National Communication Record: Communications— Bridge or Barrier, 1961, p. 10.

Wendt, Herbert, *In Search of Adam.* New York: Collier Books, 1963.

Whipple, F. L., *Earth, Moon, and Planets.* Cambridge: Harvard University Press, 1958.

———, "The History of the Solar System," *Proceedings of the National Academy of Science,* **52**, No. 2 (August 1964), 565.

Wiener, Norbert, "Some Moral and Technical Consequences of Automation," *Science,* **131**, No. 3410 (6 May 1960), 1355.

———, *The Human Use of Human Beings.* Garden City, New York: Doubleday, 1954.

———, *Cybernetics.* Cambridge: Technology Press, M. I. T., and New York: John Wiley and Sons, Inc., 1948.

Wilson, A. G., "Introduction to Problems Common to the Field of Astronomy and Biology: A Symposium," *Publications of the Astronomical Society of the Pacific,* **70**, No. 42 (February 1958), 41.

Wilson, A. T., "Synthesis of Macromolecules under Possible Primeval Earth Conditions," *Nature,* **188**, No. 4755 (17 December 1960), 1007.

Wolstenholme, Gordon (ed), *Man and his Future.* Boston: Little, Brown and Co., 1964.

Woltjer, L. (ed.), *The Distribution and Motion of Interstellar Matter in Galaxies.* New York: W. A. Benjamin, Inc., 1963.

Wynne, E. Staten, "Sterilization of Space Vehicles: The Problem of Mutual Contamination," in *Lectures in Aerospace Medicine.* Brooks Air Force Base, Texas: USAF Aerospace Medical Center, 1961.

Young, Louis B. (ed.), *Exploring the Universe.* New York: McGraw-Hill Book Co., Inc., 1963 (see especially Part 9 "Is There Other Life in the Universe?"; contains 10 selections by leading experts.)

Young, Richard S., "Basic Reasearch in Astrobiology," in *Advances in the Astronautical Sciences—Vol. 6.* New York: Macmillan Co., 1961.

Yovits, Marshall C., George T. Jacobi and Gordon D. Goldstein (eds.), *Self-Organizing Systems.* Washington, D. C.: Spartan Books, 1962.

Zeuner, F., *Dating the Past.* London: Methuen and Co., 1952.

Index

Absolute dating, 114–15
Accidental origin theories, 50–55
Adjustment:
  instinctive, 170
  psychosomatic, 175
Air Force reports on UAP, 305–10
Alphabetic data processing, 200–4
Amino acids, *see* Proteins
Andromeda, galaxies in, 12, 14, 18, 19, 20*n*.
Animal intelligence, 166, 179–83, 185–89
  of extrasolar animals, 240–42, 244–45
Archeozoic Era, 120–21, 123
Artifacts as evidence of extrasolar intelligence, 290–91
Asteroids, 34, 41–43, 61, 298
Atmospheric composition: (*see also* Environment)
  of inner planets, 36, 38–39, 40–41, 78–79, 115, 117–18, 276, 282, 314
  of outer planets, 43, 45–46, 47
Autocatalysis, 103–5, 107–8, 364
Automata, intelligent:
  development of, 176, 182–83, 220–35
    social implications of development, 233–35, 264–72
    thinking automata, 224–33
  extrasolar, 238, 243–45, 248, 259, 264–72
  instrumented probes and, 253–54

Balloons, 281–82, 306–8

Cambrian Period, 118, 119, 123, 126, 150
Carboniferous Period, 123, 128–29
Catalysis, 103–5
Cenozoic Era, 133–37 (*see also* Pleistocene Epoch)
Central processors, 209–10
Clouds, 9–11, 19, 308, 311 (*see also* Dust clouds)

Communication, 226, 229, 263 (*see also* Language; Signals and Extrasolar Intelligence)
  with extrasolar intelligence, 235, 256, 269, 324–57, 369–72
    electromagnetic, 330–36
    galactic networks, 258, 259, 345–46
    intergalactic, 356–57
    interplanetary, 325–29
    interstellar, 243, 249, 365–66, 375–76
    language, 351–54
    laser, 330, 336–37
    Milky Way, 354–56
  extrasolar intelligence capable of, 249–50
Computers, 161–62, 164–67, 172, 176, 190–219, 345
  automata as, 227–29, 231, 232
  future, 205–11, 219, 222
  neural networks as, 211–19, 223
  systems programs for, 199
Conditioning, psychological, 173–74, 175, 176
Control operations, 175–76
Cosmobiota, 90, 91, 156, 278–80, 292
Cretaceous Period, 130, 131, 132–33

Data processing, 162, 166, 195–96
  alphabetic, 200–4
Dating, techniques of, 114–15, 121
Deduction, 170, 172, 174, 175, 176, 216
Deimos, 38, 41, 238, 317
Deoxyribonucleic acid (*see* DNA)
Descartes' vortex theory, 58
Devonian Period, 123, 127–28, 150
Dinosaurs, 130–33
Direct dating, 114–15, 121
DNA (deoxyribonucleic acid), 90, 94–96, 98–100, 102, 106–7, 239, 361, 364, 373

400 Index

DNA (cont.)
  components of, 94–95, 98
  detection of extraterrestrial, 282–83
Drives, 169, 176, 197, 199, 240
Dust clouds, 63–64, 70–72

Earth, 15, 28, 36–39, 61
  environment of, conducive to development of life, 74–87
  evolution on:
    biological, 111–52
    chemical, 76, 89–108, 115–16
    social, 262–64
  origin of, 52, 62, 64
  UAP sightings on, 290–300, 303–10
Eddy hypothesis, 59–63
  modified, 62–63
Electromagnetic communication, 226, 229
  with extrasolar intelligence, 330–36
Emotion, 175–76, 197, 199
  drive, 176
  instinctive, 176
  thinking, 175–76
Environment, 240, 242 (see also Atmospheric conditions; Planetary environments)
  conditions of, appearance of life and, 115–19
  control of, 225
  controlled by extrasolar intelligence, 254–59, 268
  intelligent interaction with, 168, 170, 174–75, 180, 189
Evolution:
  of automata, 232–33
  biological, 108, 110–58, 180–83, 189, 224–25
    chemical to biological, 76, 89–108, 115–16, 364
    on extrasolar worlds, 152–58, 241, 374
    prehistoric, 116–51
  social
    on earth, 262–64
    research, 376–77

Fiction, 290, 300–4
  science, 238, 239, 303–4
  unexplained phenomena in, 300–3
Flying saucers, 303–10
  catastrophic landing, attempt of, 290–300
Fossils, 114–15, 119–22, 127, 129, 157, 276, 278, 280
  human, 138, 139–41, 144–46

Galactic communication, 345–57
  intergalactic, 256, 356–57
  in Milky Way, 354–56, 369–70
  networks for, 258, 259, 345–46
Galactic systems, 18–22, 271
Galaxies, 7, 8–22, 29 (see also Galactic communication; Galactic systems; Milky Way)
Game playing by computers, 199–200, 201, 205
Gravity machines, 257–58, 321–22

"Gulliver" device, 284, 285, 293

Hercules constellation, 9, 20, 34

Ice Ages, 129, 133–34, 146, 255
Induction, 170, 173, 175, 176, 216, 230
Inference about extrasolar intelligence, 319–22
Infrared radiators, 319–21
Input operations, 168–71, 172, 173, 175, 187, 189, 213
  of automata, 225–26, 229
  computer, 191–93, 196–97, 201, 211–12, 214, 216
Input-output devices, 206–7, 208
Instinct, 170, 175, 176
Integrated computer networks, 210–11
Intelligence, 160–377 (see also Animal intelligence; Automata; Computers; Thinking)
  artificial, 92, 152, 162, 182, 220–35
    research, 162–65, 196, 224
    social effects and implications, 233–35, 264–69
  biological, 162–89
  defined, 162
  evolution of, 134, 136–47, 150–51, 182–83
  extrasolar, 236–72, 288–377
    capabilities, 246–59
    characteristics, 236–45
    communication, 235, 249–50, 256, 258–59, 269, 324–57, 369–72
    empirical evidence, 288–322
    future research, 372–77
    interactions, 270–72
    probable prevalence, 361–70
    social effects, 260–72
  research on computer, 206, 209, 216–18, 219, 222–23
Introspection, 169–70, 173

J-band detector, 285
Jupiter, 42, 43–45, 61, 318
  possible life on, 78, 81, 332
Jurassic Period, 130–32

Kant's nebular hypothesis, 58, 59

Language, 203, 204, 301 (see also Communication)
  animal, 185, 187–88
  of communication with extrasolar intelligence, 351–54
Laplacian nebular hypothesis, 59
Lasers, 330, 336–37, 375
Life, 74–108 (see also Evolution; Intelligence)
  building blocks of, 93–102
  cellular and multicellular, 105–8
  defined, 91–93
  development of, 74–108
    origin and development, 88–108
    planetary environments and, 74–87
  extraterrestrial, 239–40, 275–86
    extrasolar, 152–58

Index   401

Life (cont.)
  on other world of solar system, 280–86
  search for, on earth, 275–80
  prehistoric, 113–51
  terrestrial, resistant to nonterrestrial conditions, 80–81, 82, 84

Mars, 36–38, 40–41, 61, 86, 278, 282, 292, 314
  possible life on, 280–81, 363
    communication, 310, 312, 326–29, 331–32
    environment conducive to life, 76, 78, 79, 80–81, 82, 83
    intelligent beings, 314–19
    scientific exploration, 238, 286
Masers (lasers), 330, 336–37, 375
Memory, 170–74, 176, 188, 202
  computer, 192, 197, 205, 207–9, 215
    automata, 228–29, 230
  decay of, 171–72, 176, 202, 215, 230
Mercury, 36–38, 61–62
  possible life on, 78–79, 332
Mesozoic Era, 130–33, 134
Meteoric evidence:
  of extraterrestrial life, 276–78, 294–300, 308–9
  of origin of planetary systems, 64–65
Meteorological control, 254–55
Milky Way, 8, 9, 15–19, 29, 31, 332, 361–62, 364–66 (see also Earth; Moon; Solar system; Sun)
  communicating civilizations in, 354–56, 369–70
Mississippian Period, 123, 128–29
Moons:
  of inner planets, 34, 36, 41, 76
    exploration, 238, 314, 317–18
    Luna, 37, 39–40, 76, 278, 326–27, 331
  of outer planets, 43, 45, 46–47, 60, 61, 76
Multivator, 282–83, 293

Natural origin theories, 55–66
Nebulae, 9, 32, 62–63, 65, 71
Nebular hypotheses, 58–59
Neptune, 42, 44, 46–47, 61
Neural networks, 211–19, 223
  applications of, 218–19
  automata as, 227–29, 231, 232
  research on, 216–18
Neurons, 211–14 (see also Neural networks)
  artificial, 211, 213–14
  characteristics of, 212–13
Novae, 31–32
  supernovae, 31–32, 54–55, 56
Nucleic acids, 104–5, 154 (see also DNA; RNA)

Output operations, 170, 174–75, 187, 189, 213
  of automata, 226, 229
  computer, 191–93, 197, 206–7, 208, 212, 214
Ordovician Period, 123, 126, 127
Ozma, Project, 325, 341–44, 376

Paleozoic Era, 118, 123-30, 150
Panspermia (see Cosmobiota)
Pennsylvanian Period, 123, 128–29
Permian Period, 123, 129
Phobos, 38, 41, 238, 317–19
Photosynthesis, 79–80, 81, 117, 120
Planetary environments, 74–87, 93 (see also Atmospheric composition)
  biological engineering of, 255–56
  in extrasolar systems, 81–87
  of solar system, 77–81
Planetary systems, 48–73 (see also Solar system)
  extrasolar, 66–73, 81–87, 155, 238, 257, 319–22, 362–70
    communication signals from, 337–44
    environments, 81–87
    existence, 66–73, 257, 319–20
  theories of origin of, 50–66, 93
    accidental, 50–55, 56–57
    natural, 55–66
Planetismal hypothesis, 52–54
Planets, 33–34, 308–9 (see also Asteroids; Planetary environments; Planetary systems; specific planets)
  communications between, early proposals for, 325–29
  inner, 36–41
  outer, 42, 43–47
Pleistocene Epoch, 133–34, 136, 138, 140, 141, 143, 255
Pluto, 8, 34, 42, 43, 46, 47, 61
Pre-Cambrian biology, 119–26
Processing operations, 172–74
Proteins (amino acids), 94, 96–102, 106–7, 154, 285
Proterozoic Era, 122–23, 126

Quasars, 14, 15

Races, intellectual differences between, 184–85
Redundancy, 216
Relative dating, 114–15
Retrieval devices, 203–4, 207–9, 227
RNA (ribonucleic acid), 98–102, 106–7, 187, 373
  components of, 94–96, 98
  messenger, 100–1, 106
Rockets, 276, 281–82, 311

Saturn, 39, 42, 44–46, 60–61
Science fiction, 238, 239, 303–4
Signals and extrasolar intelligence:
  detected signals, as evidence of intelligence, 310–12
  interstellar signals, 333–36
  listening or signaling extrasolar intelligence, 249–50, 329–30
  search for artificial extrasolar signals, 337–45
  signals from extrasolar probes of solar system, 344–45
  signals from spaceships in interstellar space, 344

Silurian Period, 123, 126–27, 150
Simpson, George G., 154–58
Social effects of extrasolar intelligence, 260–72
  effects of artificial intelligence on biological societies, 264–69
  on human society, 269–70
  of interactions of extrasolar intelligence, 270–72
  on social evolution on earth, 262–64
Social implications of intelligent automata, 233–35
Solar system, 8, 34–47 (*see also* Earth; Moons; Sun)
  conditions biologically acceptable to development of life, 77–81
  extrasolar probes of, 76*n*., 344–45
  nonterrestrial solar worlds, 312–19
  search for extraterrestrial life in, 274–86
  theories of origin of, 50–66, 72–73
Space travel, 250–54
  human, 38, 40, 238, 250–53, 276, 281–86, 308–9, 311, 312
    balloons, 281–82, 306–8
    communication and, 332, 346–51
    manned, 286
    rockets, 276, 281–82, 311
    tests for organic survival, 80–81
  interstellar, 244, 250, 271, 344, 346–53
  from outer space, 244, 268, 271, 310, 337, 339
    catastrophic attempted landing on earth, 290–300
  extrasolar intelligence capable of travel, 253–54
  instrumented probes and automata, 253–54
  probes of solar system, 76*n*., 90, 312–19, 344–45
  sightings of alleged, 303–10
Stars, 22–33, 93, 308–9 (*see also* Galactic communication; Galactic systems; Galaxies; Novae)
  controlled reactions of, 256
  extrasolar, 66–72, 257–59, 361–70
    occurrence of solar-type, 66–68
    possible, with planets, 68–70, 84–87, 257, 319–22, 362–70
  signals between, 250–51, 330–44
  young, 70–72
  theories of multiple, 54
  travel between, 244, 250, 271, 344, 346–53

Stars (*cont.*)
  variable, 28–33
Storage (*see* Memory)
Subconscious memory, 171–72
Sun, 8, 16, 17, 22, 26–28, 33–36 (*see also* Solar system)
  development of life and, 78, 79
  solar radiance, 79–80, 115, 123
  ultraviolet solar rays, 117–18
  stars resembling, 66–68
Supernovae, 31–32, 54–55, 56

Telescopes, 191, 329
  radio, 17, 27, 38, 332–34, 335, 341–43, 375–76
Temperature:
  of planets, 37, 38, 39, 41, 46, 62, 65, 78–79, 115, 123, 255–56
  of stars, 25, 26
Thinking (*see also* Intelligence)
  in artificial automata, 224–33
  biological, 177–89
  human, 183–85
  in computers, 190–219
    neural networks, 211–19
  defined, 165–76
Titan, 39, 46, 61, 314
Tracking devices, 218, 308, 311, 312
Transducing:
  input, 168–69, 225
  output, 174–75
Triassic Period, 130, 154
Tungus mystery, 291–300

Unidentified aerial phenomena (UAP) (*see* Flying saucers)
Universe, 6–47 (*see also* Galaxies; Planets; Solar system; Stars)
  nature of features of, 8–33
Uranus, 42, 44, 46–47, 61

Venus, 36–39, 61–62, 86
  communication with, 327, 329, 331–32
  environment of, 78, 79, 255–56
  exploration of, 238, 282, 314
Vidicon microscope, 283, 285
Viruses, 106–7
Vortex theory, 58

Water, 126, 132, 282
  liquid, 79
  ocean, 115–17, 241–42
Weather control, 254–55
"Wolf Trap," 285, 286, 301

DATE DUE